NIETZSCHE'S PHILOSOPHY
OF SCIENCE

SUNY Series, The Margins of Literature
Mihai I. Spariosu, Editor

NIETZSCHE'S PHILOSOPHY OF SCIENCE

Reflecting Science on the Ground of Art and Life

BABETTE E. BABICH

State University
of New York
Press

Published by
State University of New York Press, Albany

© 1994   State University of New York

Production by Susan Geraghty
Marketing by Bernadette LaManna

Printed in the United States of America

For information, address State University of New York Press,
State University Plaza, Albany, N.Y., 12246

Library of Congress Cataloging-in-Publication Data

Babich, Babette E., 1956–
    Nietzsche's philosophy of science: Reflecting Science on
        the ground of art and life / Babette E. Babich.
        p.   cm. — (SUNY series, the margins of literature)
    Includes bibliographical references and index.
    ISBN 0-7914-1865-0 (alk. paper). — ISBN 0-7914-1866-9 (pbk. :
alk. paper)
    1. Science—Philosophy. 2. Nietzsche, Friedrich Wilhelm,
1844–1900. I. Title. II. Series.
Q175.B114   1994
501—dc20   93–17271
                                                          CIP

10 9 8 7 6 5 4 3 2 1

This book is dedicated to the memory of my sister
Colette Barbara Babich
1958–1987

and to all women who, being great of mind and heart,
are denied the life of one in the pain of the other.

CONTENTS

PREFACE AND
ACKNOWLEDGMENTS

This book is the result of influence building upon influence, of ideas resisted, of resistance admitted and relinquished, of encouragement, criticism, and, in all, the inspiration of others.

Patrick A. Heelan, physicist and philosopher, first brought me to the issues in the philosophy of science which currently absorb my intellectual energies. I thank him for many things, for his friendship, as well as for his original and constant support.

I was first inspired by the intense philosophic power and formidable insight of William J. Richardson through a reading of his book on Martin Heidegger's thought. Among other things, I thank him for the (alas: still unmet!) demands of his stylistic exigence and for his personal friendship and support.

I thank David B. Allison for introducing me to the work and interpretive enigma of Friedrich Nietzsche, for his illustrative embodiment of what it is to "have turned out well" in Nietzsche's sense, and for his friendship over many years.

I cannot name all those who have contributed directly or indirectly to this book. However I owe a special tribute to Professors Debra Bergoffen, James Chansky, Richard Cobb-Stevens, Jacques Taminiaux, the late Thomas J. Blakeley, and Robert Sternfeld. Let me also note my admiration for and gratitude to Professor Paul Feyerabend.

I gratefully acknowledge the inspiration and friendly encouragement of Professors Hans-Georg Gadamer, Rüdiger Bubner, Dieter Jähnig, Josef Simon, Wolfgang Müller-Lauter, Ernst Behler, Günter Abel, Reinhart Maurer, Jean Ladrière, Jacques Derrida, Alasdair MacIntyre, and the late Jacob Taubes.

The greater part of this work was conceived and composed while carrying out research abroad in Germany, Belgium, and France. I thank the Fulbright Commission for two fellowships which supported extended visits to Germany, once to do doctoral research in Tübingen and Berlin in 1984–85 and again as a visiting professor at the University of Tübingen in 1991–92. I thank the Belgian American Educational Foundation for a fellowship supporting additional doctoral research at the Université Catholique, Louvain-la-Neuve, Belgium, 1985–86. Finally, I thank Fordham University for a Faculty Research Grant permitting me to visit the Nietzsche archives at the Goethe-Schiller-Archiv in Weimar in early 1991.

Introduction

PROLOGUE:
THE PROBLEM OF THE PHILOSOPHY OF SCIENCE
AND NIETZSCHE'S QUESTION OF GROUND

This is a book on Nietzsche's philosophy seen through the lens of his interpretation of science. By no means do I represent Nietzsche as offering a complete philosophy of science in the contemporary sense. What I do argue addresses the relevance of Nietzsche's thinking for the philosophy of science as such. And this in turn entails a strong claim about what is wrong with the way philosophy of science is currently done within the largely analytic confines of its traditional reception. Thus in a measure that bespeaks the necessity for this book, the deeper problems of the philosophy of science exceed even the questions raised by a review of Nietzsche's reflections on epistemology and scientific method (truth, knowing, and the philosophy of scientific practice and culture) to be discussed in the chapters below.

Although defined by its reference to science, traditional and contemporary philosophy of science lacks a *critically* reflexive orientation to science. I hold that, at a minimum, the philosophic attitude entails critical reflexivity. Lacking such reflexivity, what the philosophy of science lacks is exactly a philosophic disposition. Consequently, it is hardly surprising that for several decades, the philosophy of science has long been declared to be in a state of *crisis,* that is, in transition between "old," positivistic, logical-linguistic, analytic styles and "new," historical, hermeneutic styles. This is a crisis of self-identification. The identity crisis in the philosophy of science does not consist in having *misidentified* the perspective from which an account of scientific history may be written (the interpretive question of textual hermeneutics) or even the nature of scientific practice (the historical question of experiment and its relation to theory). Conceived as it is within the analytic tradition, what the philosophy of science lacks is an *authentic identity* as philosophy. The resolution of this issue falls squarely upon the divide between continental and analytic articulations of philosophy. Because a continental, expressly historical, and hermeneutic approach incorporates philosophic reflection on the nature and doing of science, such an approach offers a corrective articulation of an authentic or genuine philosophy of science.

Where Kant, the inaugurator of modern critique, thought to inquire into the foundations of the possibility of a future metaphysics able to come forth as a science, Nietzsche's critique articulates the foundations for any philosophy of science able to come forth (and able to remain) as philosophy. It is qua philosophy that a critically reflexive philosophy of science is to be instituted, where what has been called the philosophy of science in its traditional guise is in fact not philosophy. But what would count as a critically reflexive philosophy of science (i.e., philosophy as such)? Plainly, the ideal of critical, reflexive discourse is equivocal. For, on the one hand, critique may entail no more than a limited analysis of some subject matter, contained by the limits of its objectivity. But on the other hand, and this is the larger assumption of the present study, it may be possible to conceive critique or philosophic analysis *aesthetically*, instead of in accord with Kant's rationalistic or cognitive convictions. Such a phenomenological critique would exceed the limits of a particularly constituted (or partial) objectivity, and, I argue, only in this radicality may the philosophy of science justify its name.

The point of Nietzsche's perspectivalism entails that neither the reflective vantage nor the subject of the philosophy of science be limited to science. The inquirer must begin at the appropriate point and from a suitable foundation, and this foundation, as Heidegger's phenomenology of inquiry teaches us, involves a consideration of the nature of the inquirer. In the self-critical preface to the *Birth of Tragedy*, Nietzsche declares as foundational the insight that the *problem* of science cannot be recognized on the basis or from the ground of science. From this perspective, Nietzsche identifies the project that would last throughout his life as one of questioning the nature of science in the light of art. If, beyond questioning "science in the light of art," Nietzsche has also to look beyond art to the light of life, this further move emphasises the refractive, reflecting referentiality that is the ultimate significance of Nietzsche's perspectivalism.

If Nietzsche's conviction concerning the problematic ground of science is well aimed—and I think it is—then, by extension, the *problem* of the philosophy of science likewise cannot be recognized on the basis or from the perspective of the philosophy of science. The obliquity of a study of Nietzsche's reflections on science, which might at first glance seem an irreducible handicap for a clear expression of the problem of science, may thus represent an advantage as perhaps being the *only* approach to the philosophy of science. Naturally, to say that an oblique or contextual or perspectival approach may be the *only* way of conceiving the philosophy of science does not commit one to the claim that this approach to approaching the problem of science constitutes a fully fledged philosophy of science. What is claimed in the spirit of critical philosophy after Kant, in the spirit of reflection on reflexive thought, is merely that no other *approach* to the problem of science is possible save one that is not articulated on the ground of science. The value of beginning well is as well known and as frequently emphasised as it is rare. And part of beginning well is knowing where one is to begin—that is the meaning of the philosophical chestnut that getting the answers is not as important as knowing the questions. But if the beginning is important, it is not everything. One could mistake epideictic reflections such as the foregoing considerations for the philosophy of science itself. These epideictic reflections in Nietzsche's philosophy of science represent or illuminate the problem of science *as* a problem. If such reflections are needed in the profession of the philosophy of science, it is because thinking or articulating the problem of science as a problem is itself propaedeutic to a further and ultimately authentic expression of the philosophy of science.

Following Nietzsche's own example in conceiving the problem of science as such, which shifts the basis of the problem of science from science to art and then beyond the question of art to the question of life, the present study of Nietzsche's philosophy of science is itself not to be identified from nor to be articulated upon the basis of science (or limited to the question of science). Rather, the project at hand seeks, in two directions, to pose the question of science in the light of art and to raise the question of art in the light of life. In this way, the philosophy of science explored in this study of Nietzsche's philosophy is expressed as a philosophy of art (artifice and technology, but also culture and creativity) and life.

THE PLAN OF THE TEXT

The first chapter is preliminary, and one can read this book without this background (and grounding) stylistic discussion by beginning with the second chapter. But because Nietzsche's logic of presentation and conception is so very different from expected logical schemes within routine philosophic discourse, a discussion of Nietzsche's style offers the essential context for an understanding of his philosophic project.

The issue of Nietzsche's style must be expressed both as a question concerning the nature of that style and as an achievement of a philosophical and not merely an aesthetic or literary kind. I use the term *concinnity*, a musical metaphor, to emphasise the resonant imperative of Nietzsche's writing. Nietzsche's concinnity is a composer's style or—in what is certainly not a more elegant but simply more suggestive expression—a *conductor's* style of writing. Like a conductor, Nietzsche's style directs the reader's reading of the text.

As this kind of conducting or directing or reader-stylizing style, concinnity also has an architectural metaphorical significance: representing a neat or smooth, well-fitted design. Concinnity requires that the reader, like a singer in a chorus, be part of Nietzsche's echoing musical project. Such a demand imposed upon the reader by Nietzsche's special style of philosophic composition can be answered only by the reader's own interpretive affinity although it remains true that it also elicits or calls for this same affinity. Nietzsche's Zarathustra means and can mean nothing else when he invites his followers/readers to find their own way. The hermeneutic music of concinnity recalls Nietzsche's own description of his writing as a musical composition and his sense of writing for the ears, calling for musically attuned readers. In this way it might be said that a fair reading of Nietzsche's text, at least as Nietzsche himself saw it, would have to be a responsorial or *resonant* reading. And this in turn means that not only Nietzsche's text—as he says of the spirit of his first book—is to sing and not speak—but the reader is called in this musical way to also accord with a phenomenological interpretation or self-deconstructive expression of the working of the text.

Chapter 2, which begins the proper and focussed discussion of Nietzsche's philosophy of science, seeks to pose and to answer the fundamental question of Nietzsche's relevance to the philosophy of science in the wake of the failure of positivism, the received view, the lack of a clear identity, and the project of modernity as such as these issues affect the philosophy of science. Where does *Nietzsche* fit in? If one speaks of *Nietzsche's philosophy of science*, that is perhaps, but not even necessarily, of interest to Nietzsche scholars, to specialists. And to speak of the *relevance* of Nietzsche's thought for the philosophy of science is another thing altogether. In question is the issue of specificity. Perhaps, one might claim, as I have suggested, philosophy of science in its current analytic expression must be broadened and indeed expanded to include other so-called continental styles of philosophy. Yet if a successful defense of this claim can be envisaged, namely that a traditionally analytic discipline such as the philosophy of science *requires* the broadening transfusion of various continental currents, say those of phenomenology or hermeneutics, one nevertheless remains some distance from admitting the relevance of a "perspectivalist aesthetics of truth," much less a "hermeneutic aesthetic of science." For only this last would begin to articulate a properly Nietzschean approach to the philosophy of science.

A *properly* Nietzschean and not a "shock" or stock perspectivalist approach to the philosophy of science exceeds all other approaches. This is true even if we presume the virtues of a continental expression of the philosophy of science in advance, following the efforts of Patrick Heelan, Joseph Kockelmans and Ted Kisiel in this venture, or even in an oblique, still-analytic way, Ian Hacking, who actually refers by name to Nietzsche—although the reference is better directed to Foucault.[1]

With perspectivalism, Nietzsche offers knowledge an infinite domain, but such a perspectivalism offers knowledge seekers no such infinite and no sure method and no truth. To discuss this claim, it is important to develop the difference between perspectivalism and relativism.

I contrast the experimental character of perspectivalism with the (implicitly) absolutist character of relativism. Unlike perspectivalism, most contemporary philosophy of science follows the public rationale of science (taking science according to the specifications of the scientist or else according to the cultural image of science). I suggest that the sophisticated fallibilism exemplified by today's scientists and endorsed by the public conception of "objectivity," features the same turn to absolutism that characterises relativism. Hence, through ever-more-accurate approximations, science and its philosophy claim absolute knowledge or truth via an indirection, a feint concealing the aim that fosters the project at hand. If the method of science does not "yet" yield truth, the point of this disingenuous "yet" affirms that science is on its way to no other goal.

Chapter 3 discusses Nietzsche's epistemology as physio-eco-logical, that is as a product of body and world. For Nietzsche, knowledge must be thought in terms of its organic exigence, that is to say, in effect, its bodily origin. In this ecophysiological concep-tion, the interpretive event of knowledge refers to the constitution of the physiological perspective, including sense perception, along with the embodied structures (physical basis and orientation) of conception and cognition. Environmental or ecological exigencies tailor perceptions: this is the critical foundation of perspectival-ism. If we interpret the world from one (say, because it is easiest to say, our own) perspective, the facts relevant to this projective understanding count as facts only for us within that perspective, and emerge as unique facts only because of this focus. So long as it is effective, our perspectival position de facto overpowers dis-parate perspectives. This exclusive power is evident in the Western scientific and technical world-interpretation, concretized in theo-ries and computations, insofar as they may be known by their results or effects. For Nietzsche, this Western valuation of calcula-tive efficacy is no more than an illusion, that is, merely a dissem-bled interpretation. But other equally or comparable efficient accounts, that is, effectively different interpretive schemes, are possible. Where "truth," for Nietzsche, is an ideal encompassing "all realities"—and not merely what is real for one dominant per-spective—truth does not proceed from and may not be expressed from just one interpretive perspective. The myth of objectivity is the myth of presuming that one's original perspective is universal or unlimited and this exclusive presumption is the distinguishing virtue of illusion, that is, the illusion of truth. For Nietzsche, truth cannot be a singular perspectival construct, and that means that in its nonunivocal fullness, truth includes the ambiguous and the ambivalent dimension of what is Real and is neither abstract nor ideal. What we, in the Western perspective, call (factually or potentially true) "knowledge" expresses, for Nietzsche, no more than our own perspectival life-position, whether it is acknowl-edged as such or not. Accordingly, if this interpretive focus is exclusive (which at its height is precisely what the Western knowl-edge ideal means to be), it is a falsification—or a lie.

Yet, against the traditional interpretation of Heidegger's read-ing on this topic, Nietzsche's philosophic interest vis-à-vis science, art, life, that is, Nietzsche's "Will to Power" is not directed toward earth mastery. To be effective—and what is mastery other than efficacy as such?—such a goal would have to rely upon the antiaesthetic computational models of traditional science and the relevant presumption that there are exhaustive (sufficient), measurable (necessary) truths or stipulable axioms. By dubbing these truths (axioms or facts) "illusions," Nietzsche does not think to contravene their claimed efficacy. His claim is only that everything is interpretation or Will to Power. By understanding interpretation as Will to Power in this way, Nietzsche suggests that interpretive encounters are events structuring the world according to an original interest or power-perspective. The structuring interest of a power-perspective expresses not a passive optic but a selective process; within its specific interpretive domain, a power-perspective is absolutely sovereign. Determining its sovereignty, the selective fiction constituting the efficacy of a power perspective is reciprocally limited by its interpretive function (and force). Negatively, for example, the selective ideal of scientific precision is attained by denying any relevant ambiguity. Ambiguity is the anathematising limit to scientific precision. Be that as it may: the discarded "irrelevant discrepancies" or "tiny errors" between computation and measurement (i.e., the structuring interests of tradition and science) persist as part of the world beyond the scientific domain. We shall see that a concern with these small errors and the limits of scientific sovereignty thus delineated compose what may be regarded as Nietzsche's *hyperrealism*. A further discussion of Heidegger's reading of Will to Power and earth mastery is offered in connection with technology and a reading of the Eternal Return in the final chapter.

In chapter 4, I examine Nietzsche's conception of the scientific object and his assessment of art and science as species of illusion: defining the first as conscious and the second as self-dissimulating *illusion*. The perspective of computational efficiency requires a conservative orientation toward life that in turn must deny change, growth, death, and ambiguity. The domination of nature requires an object, and this object must be (presented as, regarded as) fixed. Nietzsche's conception of an original, persistent dynamic chaos (as what is Real) defines nature (the world) as Will to Power in all eternity—and nothing besides. For Nietzsche, this world-nature can explain the efficacy of subjective or interpretive perception and the possibility of scientific technique in the same effective dimension. The interactive nature of interpretation suggests that as science structures the world of its investigation, so the range of response is likewise structured. This structuring works backwards as well as forwards, and the scientist is absorbed by the congealing thrust of his own pursuit of static truth. The nihilistic or life-preservative (but anti-life) interest of science is self-perpetuating and self-advancing. Hence, the scientific illusion qua illusion is not what is problematic for Nietzsche. What is troublesome is science's insistence that it has (even potentially) absolute truth. Thus it is that, for Nietzsche, science in Western society lacks aesthetic culture (taste) or style.

In chapter 5, I discuss Nietzsche's account of the moral basis or genealogy of science, as the latest expression of the ascetic ideal of Judeo-Christian culture. The ascetic ideal is the means whereby a reactive or weak Will to Power can become superordinately creative, representing its own standard as universally binding. This reactive project is essentially conservative and hence (in a democratic sense) ultimately socially effective. I employ terms taken from the Greek polarization of the common (vulgar) and noble (elite) opposition between coarse, brutal mechanical invention and creative, artistic expression—that is, the classical Greek opposition between βᾱναυσία and τέχνη—to explicate the difference between today's reactive, antiaesthetic science and its technology and what could be brought forth as an aesthetic scientific and perhaps even technological possibility in accord with the perspectival nature of truth. But for the art beyond such artistry, beyond artisanry, beyond Nietzsche, one must attend to Heidegger's song of the earth and before Nietzsche perhaps to Hölderlin, to hear the way poets sing the coming of the gods. I do not address these issues in full, but rather suggest that Nietzsche's question that asks whether we have ears for such singing be heard in this way.

Even if science does not admit its continuity with religion, as the ascetic spirit in another, advanced, and more effective guise, it nevertheless embodies in fact the same purpose: a Will to Truth conceived as the will to the preservation of (and ultimate compensation for) even the most insignificant (human) life. The enduring kernel of Nietzsche's criticism of the *life-preservative* will or impulse is that life affirmed as such cannot be preserved. Life may be affirmed only as lived. Because life entails loss, the preservational instinct opposes life. However successful, however effective science may be, its ultimate end, the preservation of life, is not merely elusive but, strictly expressed, unattainable. Nietzsche's critical point is not the Schopenhauerian or merely literal point that success at preserving life is temporary at best, or that God or some other telos precludes any solution to the riddle of life. Rather the very living of life involves the *expression* of life, which is to say that the cost of living is life.

To live is to embark upon the continuous course of life and death. Living life is the same in the end as dissipating, the same as giving out, as *living* life. The expressive emphasis here is represented by Nietzsche's formula: *amor fati*; and it is important to note that it is not the same as *amor vita*. Like Hölderlin, Nietzsche seeks the music that expresses "the sadness of profound happiness" (*die Traurigkeit des tiefsten Glücks*).[2] Life, the seductress, the insatiable is not to be loved, not because life cannot be loved, but because the love of life does not bless the expression of life. The love of life seeks to possess, to retain, to keep hold of life. Conceived as existence in one of Zarathustra's more pregnant repugnant images, life is a serpent whose golden scaled underbelly slides in the flash of becoming. What is to be loved is not the dream or metaphor of life eternal, but life as it is, life just *as fate*, precisely as stone. Thus the eternal return of life is eternity: at once what has been, what is, what will be.

Up till now, science has opposed the aesthetic ideal of expressive life because this requires the discharge or dissipation of life. A conservative life-focus bifurcates preservative and expressive life and represses expression as far as possible. The term *repression* indicates the significance of a psychoanalytic economy of life-surfaces and life-depths in my reading of Nietzsche. Indeed, the differentiation of the Will to Power may be illuminated in a (particularly or, better, partially Lacanian) psychoanalytic vocabulary. Thus, a reactive force manifests the neediness of desire, the articulation of a lack, while active force manifests the plenitude of desire, or blind articulated affirmation (even as the recognition of the inescapable lack). Where survival is the ultimate value, the dissipating, downgoing moment of life (becoming) is interpreted as inimical rather than as essential to life. Because weakness needs support to be conserved and advanced, the banal, mechanical ideal provides this support as its practice (technique) and its instrument. Hence the conservative perspective works at the expense of the truth of life (and life is inherently ambivalent because death, its presumptive opposite, is not opposite or alien to life but rather a polarizing, self-continuous, manifestation of life).

Beginning with the implications of perspectivalism for the philosophy of science as the affirmation of reality in its multifarious
character and its multiple truths as well as the nefarious deconstructive ideal of truth as a woman, chapter 6 takes up the question of a perspectivalist aesthetics of truth. The scientific value of
the Will to Power works as an effectively conservative hypothesis
affirming the primacy of perspective as an epistemic determinant
of (what can be counted as) reality. Understood as characterizing
the mutually interactive play of world-interpretive (world-cre-
ative) events, the Will to Power illuminates the limited possibilities
of viable interpretations. The character of the world permits infinite interpretations without supplying an applicable domain or a
guarantee for them. To say that the world is Will to Power is not
to claim a subjective or solipsistic Will to Power on any level
because the world is without a predetermined pattern or power.
Hence the existential conflict of interpretations or interpretive
expressions of existence manifests chaos in all eternity. Since no
power can ever win a right for its interpretation apart from its singular and dynamic expression of that interpretation, world-per-
spectives remain mere projective fictions. Likewise, truth, for
Nietzsche, remains illusion.

The value of truth can be no more than its particular charm, a
rational charm that can never be concretized or absolutized apart
from its original, perspectival appeal. It is on this account of an
irrefragibly surface charm, that truth, as Nietzsche says, is a
woman. Her (truth's) calculating appearance is benevolently
affirmed as (deliberately artistic) illusion. But, if Nietzsche writes in
favor of illusion, it is not to champion the life-conservative value of
scientific truth but to work out its aesthetic possibility in the grand
(life-expressive) style. The grand style is a Dionysian affirmation of
life in life's "pessimistic" reality, that is, in its impermanence, ambiguity, and ambivalence. The noble individual of such Dionysian
affirmation is characterized by cruelty but above all by his or her
capacity for suffering, ability to invite suffering into his or her own
life, thus consecrating and celebrating a right to be. In this aesthetic
vision, pain and creation share the same root.

The final chapter considers the possibility of a Dionysian philosophy in terms of the thought of the Eternal Return of the Same.
With the help of a Heideggerian reflection, the fixation of becoming in the life-preservative ethos of contemporary culture (Judeo-
Christian, scientific morality) may be opposed to the aesthetic of

life articulated in the two previous chapters, to illuminate science (will to power) by way of art's confirmation of life. Nietzsche distinguishes between the discontented, wishful, future-focussed, acquisitive ideology of contemporary, technologically and morally sophisticated culture on the one hand, and the affirmative aesthetic character of the self-overcoming, self-creating, self-expressive style of life on the other. The thought of the Eternal Return of the Same provides the possibility of the weight necessary to secure the latter, fearlessly thorough affirmation of life-reality as it is. This affection for or resolution of what is is possible in and through what has been, but without an ascetic, redemptive ambition toward a temporally nihilistic future. Resolute, an aesthetic futural orientation is not frozen in the past, or absorbed in the moment, or transfixed by an eschatological hope. The distinction made between aesthetic confirmation and reactive conservation, I offer a review of Heidegger's understanding of the will to power in the terminology of his reflections on technology in the West. Thereby, I seek to show that the current depreciation of the *poietic essence* of technology may be expressed in (Nietzsche's) terms of the perspectival dominance of decadence.

Opposing the decadent ideal of equality, Nietzsche seeks the Dionysian genius or the heroic, noble human possibility. No more a cosmological principle than a categorical imperative, the thought of the Eternal Return is intended as a selective principle— a philosophic touchstone. As an aesthetic attitude, the Eternal Return is attainable for the human being or the philosopher who lives as an artist of life and its best expression. This reflects the value of life as it is, together with one's lived past and the tradition of one's social being.

But we may well ask, is this solace that of a real comfort or genuine redemption? Or is it merely delusion? I believe that for the most part it must be admitted as no more than a delusionary solace, if Nietzsche would perhaps also insist that the recognition of delusion as such provides the best comfort. The project of aesthetic life, of living well, living the best moment does not reduce to the Christian warning or Roman cliché, *carpe diem!* nor does it express the Epicurean admonition to live for the moment. Rather the aesthetic ethos of life, given Nietzsche's lifelong effort to represent science in the light of art and art in the light of life, is epitomised by Pindar in the special tension emergent in the conflict between divine jelousy and mortal hubris. This human daring is a self-overcoming, self-transformation which sublates the difference between the human and the divine. "Werde wem du bist," in Nietzsche's expression of the Greek γένοι᾽ οἷος ἐσσὶ μαθών (become the one you are). Not because one is truly God, but because the daring that characterizing titanic immortality as much as human mortality encroaches on divinity.

On the one hand, as Nietzsche writes for an elite, his project should not automatically win our sympathy. On the other hand, Nietzsche asserts that his insights must win general approbation. He writes "for everyone," and he even suggests that his highest aspirations, even as Antichrist, representing anti-Christian ideals, still follow the example of the Galilean fisherman. Thus we seek to render the opposition between the artistic creator's advice—Be hard!—as the dynamic of becoming, with its affirmation of expression and extinction together, and the ordinary way of human being, regarded proximally and for the most part. To sing the song of the overman, Nietzsche says it is necessary that human beings as they have been up till now learn to overcome themselves. And modernity has almost done this work for us, as it turns of itself into the postmodern. It is an exoteric or banal misreading of Nietzsche's ideal of the overman or posthuman being which imagines that Nietzsche's claim that ordinary human beings must go under must mean that these same human beings ought to perish—so that some fantastic conception of the overman might then have the room to flourish on the grounds of their ashes. This is an easy confusion, and given its structural metonymic resonance with recent European history, it is perhaps unavoidable, but it remains mistaken. It is not the mass or the average man that must go under. Nietzsche's overman is better regarded as a *post*humanist expression of being human, beyond what has been, up till now, all too human. Nietzsche has, we may recall, no general ethics or universal political theory. Thus he makes no general prescription even where he may be read, as political and philosophical theorists have in fact read him, as making such prescriptions. In its essence, Nietzsche's perspectival project obviates such universal commands. Rather, Nietzsche's claim is that one must die to oneself. The same Platonic, Christian imperative is preserved, if only in inverted form. One, oneself must die to oneself—not in order to lose one's life for the sake of immortality, but rather to lose the fear of losing one's life for the sake of eternity, for mortal eternity. Affirmation is another name for joy, another name for praise.

As human, as recollective, as presciently reflectively being toward what will be in the mode of having been, the human being never will and never can *have done* with what has been. This is the root of melancholy, it is the root of neurosis, the substrate of life now known as art. As the proponents of psychoanalysis teach us, from Freud and Jung to Winnicott and Lacan, the subject of the unconscious is recalcitrantly historical, as the flow of the river that Heidegger names the "house of language."

Beyond the notions of psychoanalysis, we turn to art, to poetry, to music—as Nietzsche understood the lightness of dancing thought. The saving forgetfulness of the future only shuffles the past remembered; the past is not (just as the unconscious may never be) washed away, but hidden, repressed, and real; and that means that the past is ever ready for some potential, dismaying return. As futurally intentional, accidentally recollective beings, an aesthetic favoring of the past, an aesthetic repetition, permits a full or ethical affirmation of the moment (life) apart from any hope in a future return on the past (that is, (mis)understood as a future that makes good or pays one back for what has been). Nietzsche's *amor fati* is a love of mortal, historical, fated life, blessing or affirming, *loving* the life that is because it is. In other words: one must come to win the self one has.

Thus the same benediction discerned in Pindar's dialectical hymn of praise projects the tragic architectonic of life lived in the grand style. For Pindar's γένοι᾽ οἷος ἐσσὶ μαθών (become the one you are) enjoins: be the one you have proven yourself to be.[3] Pindar's injunction is not imposed against human reticence, no matter whether retiring before or never even daring the wrath of the gods whose last curse is the death of all desire. Pindar's word, which Nietzsche makes his own, is *affirmation*. The same song of praise, the same black joy is the tragic word. In the measure poised in the song, still arched by love and sorrow, one is asked to comprehend the philosophic gnomon from all time—that all going and returning agrees with itself, strung like the Heraclitean lyre.[4]

NOTES

1. Hacking, *The Taming of Chance*.
2. See Heidegger's discussion of sadness or melancholy as joy in *Unterwegs zur Sprache*.
3. The phrase is notoriously difficult to translate. The felicity of Nietzsche's version, "Werde wem du bist," using four German words to translate four words of the Greek, even when translated into English "Become the one you are," is obvious in comparison with Sandys's literal and fair translation. What is in Greek remarkably stark is rendered by Sandys with fifteen words: "Be true to thyself, now that thou hast learnt what manner of man thou art" (*The Odes of Pindar*, Loeb Classical Library [1915, 1978]).
4. Cf. Hölderlin, *Lebenslauf*
Hoch auf strebte mein Geist, aber die Liebe zog
Schön ihn nieder; das Leid beugt ihn gewaltiger;
So durchlauf ich des Lebens
Bogen und kehre, woher ich kam.

CHAPTER 1

Nietzsche's Musical Stylistics: Writing a Philosophy of Science

THE HERMENEUTIC CHALLENGE OF NIETZSCHE'S ELITISM: STYLE AND INTERPRETIVE AFFINITY

Nietzsche's writings on truth and Nietzsche's comments on science are routinely dismissed as confused or irrelevant to the substance of Nietzsche's philosophy. Expressed exotically, Nietzsche's understanding of truth is a fluidly protean, quasi-aesthetic, mytho-poietic ideal of nonexclusive truth. Expressed simply, simplistically, we may also say that for Nietzsche there is no truth. The only truth is the lie of truth and the truth seeker is condemned to such a lie, where at its best one has the truth that there is no truth about truth. Such an understanding of truth is beyond the opposition of truth and lie (the logical principle of noncontradiction). Like Nietzsche's idea of the genealogy of morals, which poses the question of morality beyond good and evil, Nietzsche's theory of truth proposes (and an early essay is explicitly titled) an *extramoral* interpretation of truth beyond truth and lie. Recognizing the world-making activity of the expression of power constellations in quantitative/qualitative terms, Nietzsche's account of truth is, in its philosophic origins, archically perspectival. Against those who hold correspondence theories of truth, Nietzsche maintains that truth is only interpretation. This assertion holds even against coherence or Tarskian theories of truth, where these last depend upon ontological riders (if x *is* in fact as it is judged to be). Hence Nietzsche's later perspectival reflections on truth pose the question of the perspectival value or vantage of the truth of nontruth: for Nietzsche once again, there is no truth, there are no facts, only interpretations.

Truth as Nietzsche has it was traditionally misunderstood as separate (separable) from the lie and identifiable without recourse to the merely partial, or to the "false" or merely apparent, or to the ambiguous and illusory. But in place of such an Eleatic and Platonic ideal, Nietzsche conceives truth as the chaotic totality of mutually interpretational power-perspectives, where no one perspective has primacy. Entwined with one another in a contest for supremacy, such a chaos of perspectives excludes the ideal of singular, ultimate truth. Thus, like Heidegger's understanding of aletheic truth, Nietzsche's perspectival truth is nonexclusive or open to the trivial, the discounted, the veiled, or the shadowed. A perspectival or ambivalent expression of hermeneutic truth reflects the epochal expression and free play of event-perspectives in an ever-emergent dynamic of power. Such a perspectivalist expression of truth is both propaeduetic to and the ennabling condition for an aesthetic hermeneutic or philosophy of the theory and practice of science.

But what sense is to be made of such talk? The conceptual difficulty of Nietzsche's claims is not adequately clarified by any author, not because commentators up till now have failed at the task but because this conceptual difficulty is inherent in Nietzsche's expression and cannot be "clarified." This is dramatized in the absurdly analytic project of "cognizing" Nietzsche's theory of truth. Even at its best, such a project inevitably ends up with a representation of Nietzsche's theory of truth as "noncognitivist." But to say this does not mean that Nietzsche's philosophic perspective on truth has no cognitive value. Just the contrary, as we shall see in the chapters to follow. But the paradox, the ambivalence, the contrariety implicit in a perspectivalist theory of truth where there are no facts, only interpretation, where there is no truth characteristically jars or skews such a cognitive value.

The linguistic dissonance is ineluctable. It is not possible to "translate" Nietzsche's talk of truth into analytic style talk about truth. Hence the language employed in the present study of Nietzsche's thinking on science and truth, emphasising his *perspectivalism*, reflects the deliberate challenge to coherence and consistency characterizing Nietzsche's philosophy on truth and language as such (where, for Nietzsche, truth and lie start out as different names for the same thing). The dissonant and contradictory character of Nietzsche's thought, in the best treatment of such a combination to date, led John Wilcox to characterize Nietzsche's thinking on truth and value as "non-cognitivist."[1] If this tactic of naming Nietzsche's views on truth noncognitivist may not be said to have recommended Nietzsche to analytic thinkers on *truth*, it has nonetheless proved particularly fruitful for analytically styled *value* studies.[2] Unlike other reviews of Nietzsche's "cognitive" value, Wilcox's representation does not lead the reader far afield, where, as said, for Nietzsche, truth is a moral value and the *idea* of a truth value or the question of the value of truth follows from the question of morality.

Yet the deliberate ambiguity of the proliferation of allusions, allegorical, metaphorical, metonymic and otherwise, and the necessary elisions that characterize if they do not compose Nietzsche's style remain problematic for the reader interested in the philosophy of science (or, more generally, in epistemology or in aesthetics under the general schematic of philosophy as such), or even a reader with a broad interest in the contours of Nietzsche's thought in particular. What is here problematic is more than a matter of Anglo-American analytic taste in philosophic reading. The problem begins as it ends with the radical difficulty of reading Nietzsche.

I have suggested that there is no transparent way to talk about such notions concerning truth and lie, perspective and illusion except via an explication of the context of Nietzsche's own expression of his thinking on these matters. No one can say what Nietzsche's perspectivalism is about in simple terms without betraying the sense of Nietzsche's perspectivalism, for as Nietzsche indicated in a comment he offered more than once and with more than one published variation: "All truths are simple—is that not a compound lie?" In the present context, this means that the language used to represent Nietzsche's views on truth, science, and power cannot help but clash with the expectations of the philosophic reader with a background in traditional philosophy (of knowledge, language, or science).

To address this stylistic dissonance, in the sections to follow, I explicate the effective value or working of Nietzsche's style on the reader; this I name with the principally musical metaphor of *concinnity*.[3] A concinnous style is not a uniformly effective influence; rather, it depends upon the reader's own affinity for hearing the affects of such a style. As an esoteric conductor's or composer's style of this kind, within the limits of a select range, Nietzsche's style sounds out the "right" readers with, as he puts it, an "ear" for the "music," the tempo, the rhythm of Nietzsche's text. This sounding out finds the appropriate resonance in the reader not now as an affect but as a charge to thought. This idea of challenge will be familiar to readers of Heidegger's phenomenology of thought. Throughout the chapters to follow, I offer specific examples from Nietzsche's writing to highlight Nietzsche's rhetorical, musical style.

Yet with such a claim concerning the musicality of Nietzsche's style of writing, I do not offer an account of Nietzsche's flourishes or rhetorical ornamentation. My claim here is that Nietzsche's style is not at all a matter of rhetorical excess, not given the ordinary elaboration of the meaning of rhetoric. For Nietzsche's writing, like music, may not be separated from the question of style, and that is true in Nietzsche's case more than it is true of any other author just where Nietzsche writes to or for the reader's spiritual ear and not the reader's intellectual eye. All of Nietzsche's (published) texts are composed in this musical way, some, of course, like his *Zarathustra* much more than others.[4] Given this expression, to speak of Nietzsche's concinnity means that to learn to read Nietzsche is more than a matter of learning to see, as for example the poet Stefan George did, that "this soul should have sung and not spoken," using, as so many inspired by Nietzsche's words have, Nietzsche's own words concerning his own first words in his first book. Instead, like George and like other readers philosophical and otherwise with an affinity for Nietzsche's text, Nietzsche's best readers (I might say: Nietzsche's only readers) are brought into the resonant space of an answering harmony. Such a musical expression exceeds the literal expression of the text, sounded over Nietzsche's new seas, tempting those with ears to hear, those he called the philosophers of the "best future."

I differ from an ecstatic reading of the music of Nietzsche's writing where I would name the answering song intrinsic to or needed to understand Nietzsche's musically styled philosophy not an esoteric refinement but the thoughtful correspondence that is the heart of philosophy in Heidegger's understanding of thought and of the meaning of philosophy. To read Nietzsche, to reflect the golden song of Nietzsche's sunrise, the short shadows of his bright midday, and the long shadows of his red and yellow afternoons, the benediction of Nietzsche's sunsets—where even the poorest fisherman rows with golden oars—and the lonely, musically resonant tears of his brown nights, one must be as reader, as quick-eared thinker, as Nietzsche was. One must be a tempter of new horizons and a new dawn, a philosopher. Thus this exigence demands that the reader share the temperament of the thinker who attempts, asking as Nietzsche asks again and again, the question of the value of truth.

PHILOSOPHIC CONCINNITY:
THE SPIRIT OF MUSIC AND NIETZSCHEAN STYLE

Concinnity is the word I have been using to describe Nietzsche's style together with the art of reading appropriate to his style. *Concinnity* is derived from the Latin, *concinnitas*. In its colloquial adjectival significance, the word suggests a consummate, well articulated performance, an elegant or neat accomplishment. *Concino*, the etymologically unrelated but still much more than conceptually associated verb, means "to sing in chorus." And in the sense in which I employ the term, concinnity corresponds to its current technical, musical functionality, that is, the sounding, smooth (ordered, fitted, protentionally, or constitutionally architectonic) harmony of disparate or dissonant or answering themes singing together in chorus or in a round. Thus in the round, in the barcarole, in the chorale, in the symphony, music is liquid architecture. I mention the folk round to emphasise that this fluid arch-structure not only is a characteristic of classical compositions but also emerges in the rag or the jamming of jazz musicians who speak back and forth to one another, each balancing the other's voice across bass strings, piano, and saxophone. In this musical fashion, Nietzsche's style is an example of the rhetorical sublime, an architectonic effected in terms more contemporary to us than to Longinus as a (precociously) postmodern compositional technique. In writing, a concinnous style has two significant registers: in the first place, concinnity refers to what is expressed by the writing as a consequence of the stylistic play and interplay of written texts and, given the importance of Nietzsche's reader-specific, evocative style, beyond the text; in the second place, concinnity refers to the appropriate(d), creative response of the reader, that is to say, what the reader can *work up* or *out of* the text.

Nietzsche's concinnity is a playing of and between his own texts, evoking an echoing reception or choral response by playing among the reader's own background skein woven of anticipations, textual affinities, and reflective/projective recollections. Accordingly, and this is a consequence of capital importance,

Nietzsche's stylistic concinnity or auto-text-deconstructive style has a varying resonance for the general, atonal as well as for the sensitive or attuned reader.[5] Withal, concinnity requires that the reader, like a singer in a chorus, be part of Nietzsche's echoing musical project. Such a demand imposed upon the reader by Nietzsche's special style of philosophic composition can only be answered by the reader's own interpretive affinity if it also elicits or calls for this same affinity. In this way, Nietzsche's Zarathustra, announcing, "This is my way," simultaneously and inevitably invites his followers/readers to find their *own* way.

To use a metaphor from analytic or Anglo-American philosophy, the stylistic achievement of Nietzsche's concinnity or self-deconstructive style effectively makes its elections between readings and readers *like a child*. The analytic rendering of such a choosing, the non-exclusive disjunction, is largely a negative one. Nietzsche repeats only the Hegelian spirit of this negativity. Instead of the aesthetic, ascetic *either/or*, Nietzsche's style invites the mystical ambiguity of the dialectic *sowohl als auch*.[6] In its musically expressive efficacy, Nietzsche's style is pitched to the reader, architecturally as well as harmonically angled in the textual corpus of his own work as a kind of self-subverting (overt content) and reader-subverting (anticipatory expectations) double coding. This last characteristic of postmodern art and literature chooses between *several* styles by retaining them, "quoting" or implicating them all at once.[7]

Nietzsche's self-deconstructing, culture-deconstructive, "double-coded" textual style works as proof against the straight inevitability of the 'grand' narrative. The return of the question from reader to author, speaker to speaker, undercutting authority and thereby undercutting the modern tradition, is a typically Nietzschean stylistic chiasmus. This double-coded style works as a multiregister movement interior to the discourse that not only subverts the reader's self-presumption but also is its own overt, self-subverting reflection.[8]

Nietzsche's importance for an understanding of the post-modern situation is his reflection on the rule of error and the illusion of truth that is not a (weak) skepticism but an affirmative experimentation with illusion (in art) and thus a (strong) confirmational incorporation of error in life (i.e., in the grand style). Thus perspectivalism affirms a multiplicity of perspectives, none of which, including the operating perspective of the philosopher of perspectivalism, has any absolute claim. The emphasis upon the "post-human"—the celebrated *Übermensch*—in Nietzsche succeeds a critical reading of the nature of the human as such, shifting its position to the sliding or precessionally decentered subject of interpretive style.[9] Nietzsche's thought on the use and abuse of history, on the subject of discourse and the discourse of the subject, and, above all, on language, on truth and lie, and so forth, involves a textual inscription/subversion of heterogeneity that can continue to count as postmodern, but its ambivalent regard for that same heterogeneity, manifest in the name of egalitarian democracy, cultural, racial, and even gender pluralism, must continue to confound classification. In a language that qualifies domains and universes of discourse such that it is possible to formulate the propositions, "There is no truth," or "There is at least one truth," there can be no way of pronouncing the end of totalizing discourse apart from such a discourse. The means available are totalizing antitotalizing: hyperbole, parody, the aphorism, the sustained or even catachrestic contradiction. Employing all of these to extraordinary effect, of course, is Nietzsche's protean text.

A postmodern epistemic viewpoint retains the critical, scientific vision of modernity but together with this sophisticated self-reflexive awareness, it avows or more precisely *admits* the impossibility of scientific totalizing or absolutist knowledge. Thus the watchword and touchstone of Nietzsche's anti-totalizing perspectivalism is its resolute *provisionality*. Multivalent, heterogeneous, and above all, as concinnous (multivoiced or choral), the postmodern recoil of incredulous credulity is more than the modernist hyperbolic interrogation, because it also challenges the credibility of its doubt (and, accordingly, does not hesitate to undercut the doubting subject). It is for this reason that the modern scientific habit of modest, tentative declaration or sophisticated qualification (i.e., the "best possible" knowledge) does not qualify as Nietzschean even if it is (may also be) postmodern.

This return, this turning of the question is super-valently Nietzschean. To illustrate this stylistic dynamic, let us turn to an example of Nietzsche's style, from a passage to be more fully considered in a later chapter. Writing on 'The Prejudices of the Philosophers' in his (topologically, and indeed, topically, postmodern) book *Beyond Good and Evil*, Nietzsche challenges the rationalist epistemic presumption of the physical scientist's understanding of nature. The physicist's law of nature, Nietzsche claims, is nothing but a (bad) interpretation. And he concludes with the apparent concession: "Granted this too is only interpretation—and you will be eager enough to raise this objection?" With this, Nietzsche shows his philosophical teeth. The inscrutability of Nietzsche's point is evidenced by the almost complete lack of commentary on the significance of this intensification. For here Nietzsche underlines the spirit of an opposed perspective and incorporates it, harmonizes with it, and, having made it his own, returns it to its projected origin in the putatively self-defensive expression of the physical scientist. Charged with interpretation, with susceptibility to the complexities of hermeneutics and failing to master its complications, that is, accused of the hermeneutic sin of the humanities, the scientist turns into a kind of soft logician and accuses, *tu quoque*. That is the point of Nietzsche's conclusion. In the final chapter, I describe this rhetorical movement as a nonmovement, specifically as the movement that catches one up, halts the speaker, hearer, reader. Thus is Nietzsche's special aposiopesis, and it works as a caesura but is answered by a benediction or coda, "—well," as Nietzsche could reply, "da capo," or as he says here, "so much the better."[10]

THE PROJECT OF COMMUNICATION:
SELF-DECONSTRUCTION AND NIETZSCHEAN
SELECTIVITY

Appropriation is both the redemptive ideal and the danger of the project of a retrieve. For Nietzsche, the ordinary or philological position of interpretive apprehension survives the prospect and fulfillment of a repetition. If, as he says, one is able to understand only what one *brings* to a text in the first place, the ordinary appropriation will be well matched by the ordinary text if by the same token it must conflate the extraordinary text with its antipode. The extraordinary or rare perspective cannot withstand the expositional transfiguration of the ordinary reading.

Because Nietzsche understood the enduring ascendance of the everyday perspective, which he named "nihilism" and we can name the "nexus of modernity/post-modernity," he held that unless conserved by extra-ordinary means, the rarer nature or noble vantage is foreclosed. It is in the melancholic service of this epochal aim that Nietzsche writes. As a result, his text has an ethereal aristocratic orientation, assigned to spur what would be the best reader, whether or not this reader could ever exist.

Opposing traditional communicative efficacy in both challenging context and effective style, Nietzsche's text does not generate the proper understanding in the reader. Instead, in an oblique search for the right reader, Nietzsche disrupts or deconstructs the text available to the general reader. Like a barbed point, a "fish hook," Nietzsche's text penetrates the public reading to spur the "right reader."

Nietzsche conceived his style as a challenge to prospective readers. In this conception, Nietzsche employed an angler's metaphor for his own style: he saw his writing as a kind of cast and his readers as so many fish. For this Antichrist, this Dionysian angler, told himself the same story any fisherman tells himself when nothing takes his line: *"Die Fische fehlten . . . "* (*There were no fish. . . .*)[11] In the pathos of what Nietzsche called "the slow search for those related to me,"[12] the hook, the spin, the swerve of the text is the *esoteric* height of style. The effective selectivity of Nietzsche's style is more than rhetorical intention but claims to be a working aspect of his style.

Yet the concept of a popular selectivity is problematic and today more so than ever. If in the modern landscape, every porter dreams of an admirer, the postmodern vista features a Pulitzer in fact or by right (for by what right, on *whose* authority, is any author to be refused?) for every reporter. The same pluralist conception sponsors every reader's fancy as the singular (multiple) object of every author's (imaginary) selective intention. Nietzsche's *Also Sprach Zarathustra* was written, as its subtitle announces, "for everyone," even if this universal extension is just as quickly qualified via the contradictory conjunction, "and no one." Such a book must have, and its author must expect it to have, its necessary epiphany in the hands of any reader who thinks to take the book at its word. Similarly, a book for "free spirits" or "philosophers of the future" or "men of knowledge" or "artists of the spirit" betrays any vaunted selectivity as the book of casual choice, a counter-commodity for the heightened desires and easy diffidence, the ready disposal of a heterogeneous readership.

The postmodern eclipse of the author's singular relevance changes nothing but further advances the hermeneutic polyphony and ambivalence of reception. A style of reading always takes a text on the terms of the reading. More than an author ever can, it is the reader who styles or constitutes the word of the received text. If Nietzsche's text is to have a selective effect, it is because it acknowledges the reader's self-selective authority and only thus affects, forearms, and so ultimately disarms the reader. Thus, the angling hook of Nietzsche's self-reflective style of authorship is a glancing, teasing anticipation of an egocentric style of reading that can never be foreclosed because it is literally populist. Authorial style against the reader's authority, style against style, Nietzsche cuts the veiled surface of the public text, seeking the reader who would be caught in this way. But the challenge or herald call of Nietzsche's style is an ambivalent passage where mutable maskings, shifting styles advance pointed assaults. The "right reader" of Nietzsche's text is an authorial reader but never an authority—never authorized—a reader-author greeting each mask with the reflection of its inner necessity. Lured by the shifting of such a multifarious text, the "engaged" reader is the reader conceived as thinker: willing to confront and answer the challenge of philosophic thought, the polished challenge of a writing style.

In writing, the rhetorical surface of style projects its underpinnings, suggesting what lies below the turns of the word in the folds of the veil. As the logocentric reader knows well enough, the mask of style is itself a matter of style. The philologist's theoretical aim is the ultimate secret of the unsecreted style—its spur, or meaning—its point. Hence, the philosopher against the philologist, the writer against himself, Nietzsche writes to expose the importunity of the theoretician, the knower bent on truth at any price. By precociously postmodern tropes, Nietzsche's style exposes the excentricity of expositions, where what is suggested in the image of the text is ever undercut in the same turn, at the next turn, or at all turns. The *imaginary Nietzsche* after all exposition, that is, the Nietzsche of consummate stylistic achievement, repeats the project of style to eternity, murmuring against all charges of contradiction, because that is the meaning of his challenge to the principle of noncontradiction: "So much the better."[13] To read philosophy styled in this way (to read Nietzsche) is to engage the daring of thought in a melancholy self-nursing, self-turning mode.

As the consummate author-stylist who invites each reader's interest, Nietzsche foregrounds the motif of barbed trial and masked challenge.[14] It is only interior to its accessibility that Nietzsche's style works as a mask. Dehiscing the esoteric value of the text, Nietzsche's stylised dissemblance addresses the exoteric, the audience that is not the general, last or even higher, run of humanity. Against Nietzsche's apocalyptic assessment of his own writing, "non legor, non legar,"[15] the thoughtful reading counters the querulous challenge, "Am I understood? Have I been understood?" with the claim of understanding. The rare reader's prowess is limned by an accession to the text as a thinker of its thought.

But can such a "rare" reader be found? Is this precious dialectic of conflict and power proclaimed in earnest? If it is meant seriously, is its corollary plausible that would pronounce, as I would seem to be arguing here if only as a subtext of the current text, that none of Nietzsche's current academic readers may be counted as his "right readers?" But these readers are just the ones who must, of all readers, be counted as so forewarned, so admirably forearmed concerning the difficulty and distance of Nietzsche's text that their approach to his text should be the "right" one. Indeed, if any approach is to be right, if Nietzsche scholars cannot "read" Nietzsche, who can? If philologists were not the "right" readers for Nietzshe's philological study of *The Birth of Tragedy,* just who was? Who can have ears for such an author?

Thus the projection of such a skewed hermeneutic nexus of romance and rapture, conflict and accession transcends critique. And the mist of this transcendence collapses in the sun of an ordinary day, that is, in the light of common experience. After all, we are still talking about reading Nietzsche. And throughout the century that has passed since Nietzsche's final philosophical collapse in Turin, reading Nietzsche has been and remains a broadly esoteric (exoteric!) avenue of intellectual enjoyment and achievement. In all sobriety, in all simplicity, where more than a little is needed: reading Nietzsche is a common practice. And understanding Nietzsche is a easily managed feat, demonstrated by any college philosopher, theologian, or poet. Indeed, any beer-hall politician can do it.

Again, if philosophy as such may pride itself on its rigor, Nietzsche is scarcely a representative pinnacle of rigid complexity. As all casual and actual students of philosophy know, perhaps even better than the sometimes turgid students who are the scholars of his work, Nietzsche is fun. Thus it seems that the *elite* or rare power Nietzsche claims indispensable for winning hermeneutic access to his writings is so much routine hype feeding the popularity of his works.

Contrary to his own worst predictions, Nietzsche is more or less read and more or less understood these days—what then? what now? What can be the nature of Nietzsche's style such that, despite his claims for the spareness of its appeal, it nevertheless continues to win such catholic allegiance and, as a corollary to its general fascination, inspires volley after volley of conservative assault?

NIETZSCHE'S STYLE: A MECHANICAL MODEL

More than an epi-textual flourish for publication, Nietzsche's style of philosophy emerges through the deliberate composition of his texts. This deliberate presentation is less that of calculating artifice than it is of artistry. But because Nietzsche offers his own cautions to the reader,[16] it would seem that it is the force of the author's self-interpretive urgings rather than critical literary insight that pronounces Nietzsche's style the consummation of stylistic mastery.

Nietzsche ensures that his readers will be apprised of his style by overtly announcing the range of his virtuosity, just as he underlines the necessity of an adequate response to that style where *he explicitly warns against easy presumption. Nietzsche's text is thus a kind of* gauntlet, thrown forth in a double sense: cast down as a challenge, it is also meant as a course to be run. In this twofold rite of passage, by assuming the traditional challenge of pride and desire, by running the course of bravery, the reader is bound. There is nothing surprising in such a double effect: between two metaphors, the gauntlet is cast, either to be taken up or to be run. In literary transactions, the reader's interpretive complicity, which I have already named as prerequisite to any hermeneutic account, and the author's demand for this 'good will'—in running or taking up the gauntlet—is essential to an expressive consummation. The rhetorical legacy emerges as the sensitive reader's patrimony. In the good will attending the rhetorical turn, style is the selective transmitting medium of Nietzsche's philosophic intention.

Analogically expressed, the selectivity of Nietzsche's style may be compared to the industrial chemical technique of vacuum filtration. Like gel chromatography, vacuum filtration is a selective technique. Like Nietzsche agonistic style, a vacuum is used to intensify the effect of a mechanical sieve. If the grade of the filter can be compared to the difficulty of Nietzsche's style, a highly permeable filter would be one that transmits everything (all readers) at the lowest level, while at the ultimate stages or levels of difficulty, very little is transmitted by the increasingly selective and near impermeable (in industry: micropore) filters. On the terms of this analogy, the popular reception of Nietzsche's text may be seen as the operation of the first, broad transmission. This corresponds to Nietzsche's image as being easily or generally accessible among philosophic authors. But at the higher levels, representing the difficulty of discriminating and advancing to the esoteric operation of the style of writing, the same transmission is possible only for a few readers, corresponding to the development of more than an average, educated style of reading.[17]

Nietzsche's postmodern, parodic style, is the rhetorical movement of his text against the wrong, insensitive, and impotent reader. An active filter, the text draws and then evades the possibilities of reactive understanding. Seducing the reader with "ears to hear"—that is, the reader who can *think*—by means of the mutable allure of a shifting text, Nietzsche simultaneously diverts or deconstructs the public character of the text. The return of the question from reader to author, undercutting authority and the tradition, and, again, the reader, is a Nietzschean chiasmus on the level of style that was earlier reviewed as a postmodern, multi-register, concinnous movement interior to the discourse that is not only self-reflexive but self-subverting.

In the experimental register of the aphoristic and throughout his textual ventures, Nietzsche writes for the reader *able* to understand. Ultimately, it is by means of a technical, misdirection in the apprehendable text, an alogical, metaphorical, or metonymical composition, that the style of Nietzsche's irony excludes the general reader. In this way, readers with differing capacities for understanding are distinguished by their interpretive response to the text.

Nietzsche's coordination of style and exigent reading may be seen in the following brief illustration. We may recall that, in the concluding section of the preface to his *Genealogy of Morals*, Nietzsche declaims the style of reading needed to match his style of writing: "one thing is necessary above all if one is to practice reading as an *art* in this way . . . *rumination*."[18] That is to say, the style of reading suited to the style of Nietzsche's writing/thinking is not discursive but rather a 'digestive,' appropriative, or incorporative style. Appositely, too, a ruminative style of reading is needed because, long received as a texture of oppositions (e.g., the perfect case and classic locus is his self-touted aphoristic), Nietzsche's proclaimed style is an announcement that should never be skimmed over or, worse, swallowed down but only sounded out. By shifting the metaphor of intestinal cyclings to the labyrinthine *ear*, we may review Nietzsche's style as unfamiliar music: seeking the harmony that gradually reconstitutes dissonant strangeness. The "music" of a text is not just the rhythm that moves the reader along, it is also the harmony that brings the reader back to a text, that draws the bowing of reflection back upon itself. Read once and, within the reflective memory of the first reading, read again, the first essay of the *Genealogy* illustrates the postmodern signature of advance and demurral, deliberate inscription and covert subversion that is Nietzsche's style.

In the very first section of the same essay, Nietzsche betrays the direction of his writing. Its first address is to the general reader. That is to say, given the explicit context of the currency, time of publication and appeal of the text, Nietzsche's average reader shares the casual psychologizing interest of the educated reader looking for something new about morality, something revealing the nature and cause of good and evil. Nietzsche writes, then, first of all with reference to the texts preceeding his own text in the same genre. These were the texts of his readership's prior acquaintance, authored by or influenced by the then- and still-current interpretive scientism, a brave modern sophistication Nietzsche ascribes to those (English psychologists!) with "a self-deceiving instinct for belittling man," out of "the mistrustfulness of disappointed ideals," or from a "subterranean hostility and rancor toward Christianity (and Plato)" or maybe derived from a "lascivious taste for the grotesque." Perhaps, Nietzsche suggests, these impulses are also present in the readers of such books, the past-present reader, his readers.

Strikingly or laughably risque, such psychological (today's own sociological, culture-critical) readings are at the same time the stalest, most mechanical readings of humanity: taking moral instincts according to their animal, physiological, or habitual origins. Today's popular accounts of propagation, gene-serving generosity are equally unflattering. *Ad hominem*, Nietzsche moves then to argue that the motives of such an analysis of morality require examination. Why? Then as now, the fixation on mechanism requires an accounting. Hence the first section of the *Genealogy* turns the reader toward the reticular alleys of low suspicion.

In the wake of this modern romantic psychologizing, invoking and reproving the prurient fascinations of the Victorian scholar/reader, we have learned to inquire. Postmodern, we are suspicious by nature. Just whose imaginary genealogy does Nietzsche trace here? The "English psychologists" named as such or, rather, the reader reading this work with a taste for such secret origins? Because unresolvable ambiguity is essential for the play of Nietzsche's rhetoric, the questions, who is the subject? who the author? who the reader? do not matter. On the terms of a hermeneutic reading projected according to the metaphor of rumination or resonant audition, the coils of the text have already doubled, redoubled, and collapsed. Choking the linked momentum of his rhapsodic psychologizing, Nietzsche deflects any conclusive attribution or original analysis: he is, he tells us, "*told* they are simply, old, cold and tedious frogs" Now, Nietzsche draws back, boldy launching the text into a blind passage. In this turn, it can be said that he gives himself away to the sensitive, retentive reader (only later returning in succeeding sections to address the philologist and the man of science). Thus writing for *his* "readers," he writes of his noblest, vainest, impossible hope for these men of (English) science, among whom we must count his current readers, then and now. For the reader interested in the genealogy of moral reflection, Nietzsche offers his best wishes,

> that these investigators and microscopists of the soul may be fundamentally brave, proud, and magnanimous animals, who know how to keep their hearts as well as their sufferings in bounds and have trained themselves to sacrifice all desirability to truth, *every* truth, even plain, harsh, ugly, repellent, unchristian, immoral truth.[19]

The double valencing of Nietzsche's style (broad appeal/narrow focus) is of interest in tracing the textual movement of this first passage. The broad appeal of Nietzsche's style is cast out like the reel of a fishing rod, but even on a good day the hook never even makes contact with, the lure never even entices all the fishes circumscribed by its arching passage. Thus if the mocking tone of the introductory litany of self-deception forbade the reader's identification with the (English) investigators of the psyche, this last account softens this very and indeed already implicit (German) alliance. The play of the first rhetorical turn is superficially ironic. The reader is deflected from an identification with his prior (and that means immediate) reading interest. Because this first move works against both readers and original authorities, its irony lacks the arch of an even appeal, and the trope is not casually understood or romantic irony but rather a didactic or Socratic irony. Thus, just as the reader could begin to catch the direction of Nietzsche's appeal in the flatness of its duplicity, we must be ready to change metaphors, where Nietzsche deflects the direct impact, hooks, or puts "English" upon the spirit of his challenge, proposing the last passage as the generously willed ambivalence of a high melancholy reserve.

The common experience of familiarity with Nietzsche's texts is one of a certain mutability. When the reader returns to a familiar text by Nietzsche, whether one reviews a textual locus, context, phrase, or term, the reader finds Nietzsche's text renewedly new. I have elsewhere described this phenomenon as an effect of Nietzsche's self-deconstruction. This protean quality is characteristic of poetry, music, and the other arts. It is also characteristic of philosophy and any written text, in the wake of deconstructive readings, if not in the same way and to quite the same extent.

By speaking of Nietzsche's text as musical, it has been argued that Nietzsche's text is so composed that, like a musical piece, it is always different. To read Nietzsche, as to play or to perform a musical piece, is to interpret Nietzsche: and each time that interpretation as an interpretation must differ. The achievement, the interpretation of the text testifies as much to Nietzsche's philosophic artistry as to the reader's.

To offer an observation that I would venture as part of a phenomenology of Nietzsche's texts (and in this way confirmable by any reader), which is also a claim repeated in the essays on Nietzsche's style, whether naming Nietzsche a poet (and so excluding him from philosophic currency) or celebrating Nietzsche as consummate stylist, Nietzsche's texts are emminently mutable. Such mutability is characteristic of a philosophic text when the context of interpretation can be altered, either through a subsequent claim that forces a revision of preceding statements or because of a specific interpretive focus. Nietzsche's text has features of both philosophic and musical, poetic style and is thus susceptible to different readings (contexts) and interpretations. But this is also to say again—and Nietzsche does say it—that Nietzsche is not for everyone. And this limitation is not a matter of elitism or blind esotericism. Instead this limitation derives from the question of direction, it is a matter of a text given to those who can read differently.

Some read philosophy and poetry and listen to music from a distant, foreign perspective, as the reflection of another. These are vicarious readings. But just as the genuine reader of poetry is neither the scholar nor the critic nor even the general reader but only the poet, so the reader of philosophy must be (ideally) a philosopher. The poet is the one for whom the poet sings the whole music of poetry's past in the first place. The lyric poet is thus the proper hearer, the one addressed. In our era, poetry is no longer heard, indeed, some would say that music itself is no longer heard as music. But in its epic, primordial emergence, the poem is a song sung not only by the poet but also by all its hearers. It is this that is meant when Nietzsche, appropriating the claim of recent writers, of modern poets and critics like Shelley or like Schiller and Lessing, would imagine that in the beginning every man must have been a poet. This potency represents what Nietzsche regards as the lordly right of *giving* names. When the poet first sang, everyone hearing also sang, not vicariously, inspired not by the poet's charm but by the song itself. In the same way, it has been said that one must be a musician oneself, with music in one's heart and head just to hear music properly. This is not to say as Adorno would claim, with unimpeachable right in some other context, that one must be a scholar capable of reading or envisioning the score in order to be able to "hear" an opera. For the matter of "expert listening" does not yet address the question of hearing. A response to this question is given only with the tears Adorno invoked as the matter of negative knowledge. This is the knowledge granted, Adorno suggests, when one listens, say, to Schubert's music. The question of proper musical hearing can achieve only what Adorno calls a "negative knowing" that is, like Nietzsche's, an anti-epistemology exhausting the whole meaning of a "knowledge which would not be power."[20] One must be a musician in this sense to hear music. Only then can one speak as Adorno does of a reference to whole world "in weeping and singing," which thus inaugurates one's entrance "into alienated reality."[21]

In the present context, this means that one must be a philosopher to read Nietzsche. There is no other way; one must be able, as Heidegger would insist, to think in order to read Nietzsche. And thinking, as it may be said from Heidegger's many reflections on thinking, is not a matter of calculation or of being able to understand cognitive claims (these last are never too demanding). Instead what is at stake is the task (what Nietzsche calls the "adventure") of thinking without guides, without axioms, before an open horizon. Thinking is always thinking against foundations; it is always the step back, another beginning.

If science, if calculative philosophy is a matter of translating, as Nietzsche has it, the unknown into the known, the ennabling condition of philosophical thinking for Nietzsche as for Heidegger will be the capacity to discern the unknown in the known and to ask after it. This is the heart of Nietzsche's philosophical questioning, which Deleuze for one has discerned in Nietzsche's skill in finding the question of value, of worth, of moral right. This question is not to be articulated in the manner of Foucault, that is, not as such or as if it were the highest (the most worthy or truest) question but, rather, posed to the question of truth. What is the value of truth? What is truth good for? What power is granted by the kind of knowledge we call "truth"—what do we gain or win by it? Thus the meaning of science, and scientific truth, is indeed for Nietzsche to be found to begin with Bacon (rather than Galileo or Newton), but the value of knowledge as power over nature is precisely what is to be put in question. And we are still engaged in this task—as if for the first time. And Nietzsche can well ask his readers then as now, who of us has ears for such a question?

In all, the topological complexities of Nietzsche's turns cannot be traced in their entirety. But they can be sounded out, in a preliminary fashion, given, of course, as Nietzsche prefaces the assertion of the possibility of this first accessibility, "that one has first read [Nietzsche's] earlier writings and has not spared one's efforts in doing so."[22] The echo of the sounding tone, the course of rumination is not linear: self-interfering, self-reinforcing, the style of recollective attention complicates its own pattern. Because Nietzsche is writing for philosophers, driven by a "*fundamental will* of knowledge," he questions the value of values, or admits truths beyond truth. "For such truths do exist."

NOTES

1. Wilcox, *Truth and Value*.

2. Cf. K. Westphal, "Was Nietzsche a Cognitivist?" In this connection, it is significant that the strongest chapters in M. Clark's recent *Nietzsche on Truth and Philosophy* are not the central chapters addressing truth but the latter chapters on the ascetic ideal, will to power, and the Eternal Return conceived with reference to value.

3. See my "On Nietzsche's Concinnity." Cf. my "Nietzsche's Self-Deconstruction."

4. This designation of *Zarathustra* as a text that should be included within the rubric or the discipline of music is Nietzsche's own as much as it accords with my own analysis.

5. Hence Nietzsche's musical stylistics is a deconstructive strategy in advance of Derridean deconstruction.

6. This, some would say, specifically Swabian spirit was not Nietzsche's own but was one he could have admired—even without the taste he asserted he did not have himself for Hegelian dialectics.

7. This is the irony of the postmodern. This coded coding, this having it both ways or, less frivolously, with a disposition closer to diffidence than to the tragic, this life program that knows better but acts, goes along anyway, embodies the only style of life-election remaining for our age: postmodern, post-Saturnine, past all lived melancholy that Umberto Eco sees as the resonant "age of lost innocence." The only way to approach the sober innocence and ideological idealism of the past is by the gently dehiscent way of irony, which invocation both sustains poetic reference and underscores our tacit recognition of its shimmering illusion. For Eco, what we know now is that rather than the modern occlusion of the past in the anticipatory service of an eschaton, "the past . . . must be revisited." This inevitable revisitation—there is no other way to the past other than the way of return—opposes the monotony of Habermas's protests, and never aspires to the delusions and the nostalgic vision of romantic neoclassicism, because it is clear that the past is to "be revisited with irony, not with innocence" (Umberto Eco, *The Postscript to the Name of the Rose*, trans. W. Weaver [New York: Harcourt Brace Jovanovich, 1984], p. 67).

8. In its literary, philosophic, cultural expression, the postmodern is the modern in its fullest extension. That is to say, and this expression must be emphasized, the postmodern is the *enduring* failure of the modern. It is the rupture of the project of rationality—which yet retains its surface function, featured like pieces of a fractured mirror or like the reticulation of a cracked piece of acrylic or auto—glass, which yet preserves the superficial contours of the original surface in destruction.

9. For a postmodern position, the anti or post-humanism of Nietzsche's perspectivalism reveals its unimpeachable ambivalence.

10. JGB 22, KSA V, p. 37; BGE, p. 34.

11. EH, KSA VI, p. 350; E, p. 112. As such a masking deflection, cutting through all readers, Nietzsche's style is meant for no one in this publicality—Nietzsche's style is a selective device: a recurved barb, an angling hook.

12. Ibid. "der langsame Umblick nach Verwandten." Ibid.

13. In the tradition of postmodern restraint, it should be said that there is no interpretive progress-ideal that can confirm the *reality* of this stylistic achievement.

14. This barbed trial and masked challenge is similar (in affect not reality) to what shocked French Jesuits named the Northeastern American Indian practice of the "gauntlet." This ritualistic torture confirmed a captured enemy as an equal. Among the Five Nations of the Iroquois, running the gauntlet—two rows of armed warriors constituting the assaulting framework of a narrow passageway, raining clubs or fire brands upon the prisoner as he passed—although often a prelude to death was also a ritual of cultural respect and honor.

15. "I am not read, I *will* not be read" (EH, KSA V, p. 259).

16. For one example, Nietzsche's delayed preface to his first book, *Die Geburt der Tragödie: "Versuch einer Selbstkritik."*

17. So reviewed, Nietzsche's style is an exemplary vehicle of textual self-deconstruction, articulating a selective hermeneutic. The notion of a selective hermeneutic opens a critical avenue toward distinguishing the possibilities for understanding within Nietzsche's style. The notion of a self-deconstruction does not follow Derrida's trajectory but is deflected by the rhetorical strategy of Paul de Man. The self-reflective, self-critical position of the deconstructive self-exposition rhetorically assumed against the reader, describes a super-version, that is, an overcoming and extension, of Derrida's own critical deconstructive project.

18. GM viii, KSA V, p. 256; G, p. 23.

19. "Dass diese Forscher und Mikroskopiker der Seele im Grunde tapfere grossmüthige und Stolze Thiere seien, welche ihr Herz wie ihren Schmerz im Zaum zu halten wissen und sich dazu erzogen haben, der Wahrheit alle Wünschbarkeit zu opfern, *jeder* Wahrheit, sogar der Schlichten, herben, hässlichen, widrigen, unchristlichen, unmoralischen, Wahrheit" (GM I:1, KSA V, p. 258), G, p. 25.

20. Theodore W. Adorno, *Philosophy of Modern Music,* Trans. A. Mitchell and W. Blomster (New York: Seabury Press, 1973) p. 129.

21. Ibid.

22. "dass man zuerst meine früheren Schriften gelesen und einige Mühe dabei nicht gespart hat" (GM viii, KSA V, p. 255), G, p. 22.

CHAPTER 2

Science as Interpretation: The Light of Philology

THE QUESTION OF A NIETZSCHE-STYLED PHILOSOPHY OF SCIENCE

An adequate representation of Nietzsche's perspective on science presumes a broader conception of the meaning and culture of science than is current in today's expression of the philosophy of science. Reflecting the difference between the German word *Wissenschaft* and its English counterpart, *science*, Nietzsche's concern with science was not limited to the influence of the physical or so-called hard sciences but includes, in addition to the social or "soft" sciences, disciplines not considered proper "sciences" at all but regarded in Anglophone academic divisions as belonging to the "humanities" or even to the "arts." Such disciplines for Nietzsche include philology, theology, poetics and literary theory, history, and philosophy, including metaphysics and epistemology, ethics and aesthetics. For Nietzsche these are to be counted as sciences along with natural philosophies such as the today typical and proper sciences of physics and astronomy, biology and chemistry.

What should be the meaning of the philosophy of science? What is its authentic task? If one takes the philosophy of science to be a reflexive enterprise, the philosophy of science is primarily not an analytic but a regulative phenomenological project: the effort of the philosophy of science questions the project of science as such. What is more, the questioning of the philosophy of science reflects upon the philosophy of science and the philosophy of science reflects upon science, where, for its part, science reflects the working structure of reality. Thus through a series of removes constituting hermeneutic reflection, the question of the philosophy of science examines the discipline that examines those disciplines charged to examine all reality. A hermeneutic effort to understand the philosophy of science is an effort to understand the interpretive role philosophy should play in questioning or coming to understand science. If philosophy is a search for the meaning not only of what is (ontic register) but of the ground of being (ontological register) as well, the philosophy of science is addressed both to the beings that scientific inquiry makes known (ontic level) and the Being of scientific inquiry (ontological level). Yet even here, even at the beginning of such a reflection, it is obvious that ordinary philosophy of science in its today predominantly analytic expression ignores the ontological in favor of the ontic.

Nietzsche's fundamental insight has the capacity to transform the character of the philosophy of science. Nietzsche's claim is that the way of science cannot bring us to an understanding of science: "the problem of science cannot be recognized on the ground of science."[1] In consequence, the judgment Nietzsche offers us of his own first "scientific effort" in his late-written preface to *The Birth of Tragedy* must be regarded first and foremost as a scientific critique of a scientific venture. Nietzsche's first scholarly, philological project confronted the problem of science as a problem—and let us not fail to note that confronting the problem of science *as* a problem is already a radical conceptual approach— and attempts the resolution of this problem on the artistic level, as the only resolution possible. But, to be effectively understood, the reference to the artist requires a further turn to the perspective of life. In this way, the aesthetic-life orientation guiding Nietzsche's entire philosophic effort—"*to look at science in the perspective* [Optik] *of the artist, but at art in that of life,*"[2] offers us the foundation for a perspectivalist and, thereby, for an authentic or proper philosophy of science. This possibility of a perspectival revisioning of science, that is, of the philosophy of science, may be articulated by way of what Martin Heidegger has named *retrieve*.

Expressed in the spirit of Heideggerian retrieve, a Nietzsche-styled philosophy of science does not follow Heidegger's interpretation of Nietzsche.[3] Neither does it entail the effort (impossible here) of thinking Heidegger's own project for thought in its authenticity, but simply accords with Heidegger's *way* of doing philosophy. Reflecting on a thinker's thought requires a radical attention to what is thought in its originating project. Such a disciplined devotion rethinks, recovers possibilities already outlined but not given expression in the original thought. The rethinking of a thinker's thought is never expected to be exhaustive, nor is it presumed to reveal the essential core of the thought: the project of retrieve is a modest effort. One does not claim to surpass the thinker's own achievement—even where one must go beyond or overcome the original thought. The spirit of Heideggerian retrieve or repetition suggests a recuperative, experimental endeavor; retrieve is an experiment of thought articulating the basis of a critical philosophy of science.

TOWARD A NIETZSCHEAN CRITIQUE OF SCIENCE

Minimally, a philosophy of science concerns the institution of science in two ways: science as a method of inquiry or investigation and science as a worldview or world-construction. Science is not only the investigation but also a representation of nature: the discovery and expression of the facts. Yet even this minimal account of science runs into difficulties for there are no facts to be had. For Nietzsche: "Against positivism, which sticks to the phenomenon: 'There are only *facts*'—I would say: No, facts are precisely what there is not, only interpretations."[4] Similarly he declares with respect to physics, the most "scientific," so to speak, of the sciences—indeed, the paradigmatic science today: "physics too is only an interpretation and arrangement of the world (according to our own requirements, if I may say so!) and *not* an explanation of the world."[5] As Nietzsche sees it, all knowledge-claims including those of physics itself, are matters of interpretation, not fact; exegesis not explanation.

But one can well object—in the company of many literalist or over-ingenuous or simply more impatient readers—that Nietzsche's claim that "facts are precisely what there is not" is nothing more than Nietzsche's own interpretation. Isn't Nietzsche's statement, as such an interpretation, consequently "false" no matter what? Claiming that there is no truth, Nietzsche presumes the criterion he seeks to undermine. *It is true* that for Nietzsche there is no truth.

The standard Nietzschean counterfactual encountered in the literature of serious scholarly repudiations and in casual academic dismissals of Nietzsche's philosophic significance may be identified in the following three steps or moves. First, for Nietzsche there are no facts, "only interpretations." This is equivalent to the claim that "everything is interpretation." Second, but it is a "fact" (it is true) that for Nietzsche "Facts are *precisely* what there is not." Nietzsche's first claim is thus not "only interpretation." Third, hence it is not true that *everything* is only interpretation because for Nietzsche there is at least one truth, namely that there are no facts, "only interpretations." A philosophical discussion of what is a fact and what is not, is sufficiently close to a discussion of what is true and what is not for if what is asserted as true is not material, that is, is not a so-called matter of fact, then it is at least asserted to be sufficiently like a fact to correspond to it in its brute inescapability. Nietzsche's position on truth, Nietzsche's position on the "facts," reduces to what some (but not Heidegger) might call Nietzsche's "metaphysics," that is, Nietzsche's position on what there is. And what there is, for Nietzsche, is precisely no facts: there is no truth. Thus Nietzsche contradicts himself, and thus we have the logical repudiation, as it were, of the Nietzschean counterfactual. All this is very easy. But despite the appeal of simplicity, be it a canon advanced by taste or laziness, the foregoing logical reservation should be treated with reserve. For it is a consequence of Nietzsche's terminological fluidity or (musical) style that such quick efforts to formalize Nietzsche's claims often run aground.[6]

Logical exercises like the above suggest that if what Nietzsche says is an interpretation (if we mean what most people mean when they talk of interpretation), what Nietzsche ventures *as* an interpretation is as such possibly, potentially false. And an interpretation can be an interpretation as such (i.e., potentially false) only with respect to some other given or presumed absolute account—even if that account reads: "Everything is interpretation." This point is perfectly reasonable, and Nietzsche himself concedes that it is. It is this concession that is problematic. Thus the further and ultimately the more interesting point remains to affirm that everything is interpretation and not fact *and* to admit this not so much as a "claim" as an interpretation works in effect as a frontal assault on the standard notion of rationality. Unlike the standard Cretan self-contradiction, Nietzsche's self-contradiction is not merely explicit. Nietzsche does not say, "This too is an interpretation," as the Cretan says, "Everything I say is a lie." Nietzsche is already on the other side of the rhetorical gambit that is the logical process of legal evaluation. In this way, Nietzsche's epideictic appeal is also a forensic expression. *That* epideictic, forensic projection is the point of Nietzsche's self-contradiction.

It is "perfectly reasonable" to say that Nietzsche contradicts himself here, but it is exactly this contradiction that Nietzsche affirms. What is the consequence of *that* deliberate precision? If Nietzsche deliberately contradicts himself, the result undermines "perfect" reason itself. Beyond this, it is relevant to note that to reveal a perspectivalist statement as perspectivalist, assuming a "perfectly reasonable" perspective by assuming Nietzsche's perspective in its full, logical extension, cannot reveal perspectivalism as self-contradictory in the ordinary way of self-contradictories just because a perspectivalist position as such is assumed from the outset. When one plays by the enemy's rule (Nietzsche's perspectivalism) one cannot of a sudden change the rules (to a nonperspectival order of articulate expression) in order to claim a triumph. If the logical problem with Nietzsche's position can be classified and set aside as a logical contradiction or an instance of circularity, such a classification might be of interest to logicians (as indeed a warrant for noninterest) but nothing more.

In any case, there is a rather perverse disingenuousness in the effort to set out to criticize Nietzsche's logic when that effort assumes that this logic must fit within or work in accord with the schemas of standard propositional logic despite Nietzsche's axiomatic critique of the axioms of the same schema. Beginning from such a logical conviction concerning Nietzsche's critique of logic, that is, by denying the relevance of Nietzsche's epistemological critique as such, one is able to end—as Maudemarie Clark so often concludes in her book—by describing what Nietzsche *could not have meant* on these same standard terms. Thus Clark notes, "If I seem to be saying that what appears as radical in Nietzsche's position on truth is actually mistaken or confused . . . this is in fact what I believe."[7]

The specific point of the perspectivalism Nietzsche suggests in opposition to positivist perspectives (which, so says Nietzsche, affirm that there is "nothing but the facts"), is the claim that self-sufficient facts are not to be found in reality. I have suggested that for stylistic reasons this claim is not affected by a charge of self-contradiction. To review the reasons why, we may begin by asking why Nietzsche would publish a position that he himself, by the expedient of suggesting it as perspectival (i.e., as itself a biased interpretation), does not support as true but rather celebrates—all too the better! he says—as nothing but *interpretation*? We have seen that the standard response to this question cites Nietzsche's failure to understand the logic of thought, going so far as to assert that Nietzsche is confused and, therefore, that he contradicts himself. Nietzsche does indeed contradict himself, but I hold that what must be taken into account in this self-contradiction is the working of his stylistic concinnity as a style posed for the reader's response, a position that is completed only by the reader's response. This radical open-endedness obviates the charge of confusion or error. It is a grave mistake (committing what might be called the only "cardinal sin" from Nietzsche's perspective: the sin of excess gravity) to hold as Clark does that Nietzsche's "position on truth is mistaken." One simply does not have the whole of Nietzsche's position when one reads Nietzsche.

It is an essential feature of Nietzsche's style that he is writing not for the general reader but instead, as he puts it himself, for certain rare readers. These readers are to be identified as those, as he says, who are related to him, possessing the ability to think along with him, those with a musical affinity for his thought. This would seem to mean that Nietzsche is writing only for those who would have an understanding of what he writes almost in advance, and this is indeed what Nietzsche claims.

Nietzsche means to challenge these readers, writing to their affinity with him, eliciting their reflective response and their attention by means of a style described, from this resonant sounding-out of other like souls; this I have expressed with a musical metaphor, as *concinnity*. It is this style that resonates in the self-contradiction that threatens in the claim that there are no facts, only interpretations. Recognizing the threat of self-contradiction, as he does in advance, Nietzsche ends by rejoining this same charge of self-contradiction with a rhetorical shrug. His philosophic position is held to be self-contradictory because it states as a fundamental, universal claim what can be only (even as such a universally valid claim) a merely individual opinion or what can be proffered only on the metalevel. And Nietzsche acknowledges just this limitation and gallantly saves "perfect reason" the effort of challenge by charging himself with self-contradiction: "Granted this too is only interpretation." As an impossibly unflinching reply, comes the undaunted, deliberately daring or mocking "—Um so besser" (Well, so much the better).[8]

Let it be emphasised that the categorical statement "everything is interpretation" does not claim that all interpretations are the same (or interchangeable). Nietzsche's point in the retort noted above is less an effort to reveal the prevalence of interpretation than an attempt to reveal the origin of physics (in interpretation). Just as there are less and more obvious interpretations, there are better and worse ones; that is, life-affirming and merely life-preserving interpretations. If Nietzsche claims that there is no "correct" interpretation, he does not assert that there are no faulty or false interpretations. Just the opposite.[9]

In the last chapter, discussing the effect of Nietzsche's style against the strictures of ordinary logical discourse, we saw that the objection offered by the logical precision "this too is only an interpretation" turns upon the assumption of the incoherence of self-contradiction. This is a contingent and not an essential incoherence. The dissonance in Nietzsche's thought conversely is an *essential* and therefore coherent incoherence. If self-contradiction is anathema within traditional logic, it is not inconsistent with a philosophy of interpretation, or perspectivism. In his blithely self-conscious, deliberately confrontational way, Nietzsche proposes a coherent self-contradiction, that is, a dissonant harmony, or harmony-in-and-through-contradiction. This claim will be familiar to readers acquainted with German Nietzsche interpretation. For Karl Löwith and Karl Jaspers and, more recently and nuancedly, Wolfgang Müller-Lauter, Nietzsche's philosophy is intrinsically oppositional or "contradictory."[10] This dissonance resounds in Nietzsche's own text(s) as a harmony of self-reflexive oppositions contrasted with other more traditional understandings: all the while selecting readers attuned to possibilities for thought beyond Nietzsche. In Nietzsche's text, as in the experimental music of modern times, dissonance is rendered harmonious—at least to certain ears, at least following the slow habituation of these ears in the right attunement—over time.

Nietzsche's stylistic concinnity is articulated by a thematic persistence in dissonance through texts and against the reader's expectations. In the explicitly, consistently coherent self-contradictory context above, Nietzsche's position on science is stated in such a way as to show that the ideal vision of natural laws, even in physics, is rooted in the nihilistic sociopolitical tendency of the modern soul to reduce everything to a common or average level. Thus Nietzsche presents our understanding of nature's conformity to law as a representation of our own democratic interests and wishes functioning on the level of nature: "'nature's confor-

mity to law' of which you physicists speak so proudly, as
though—exists only thanks to your interpretation and bad
'philology'—it is not a fact, not a 'text,' but rather only a naive
humanitarian adjustment and distortion of meaning with which
you go more than halfway to meet the democratic instincts of the
modern soul! 'Everywhere equality before the law—nature is in
this matter no different from us and no better off than we.'"[11] For
Nietzsche, the universal necessity of natural law is not an essential
expression of the nature of things but our own projection—an
interpretation based on our democratic political views: "But, as
aforesaid, that is interpretation, not text; and someone could
come along who, with an opposite intention and art of interpretation, knew how to read out of the same nature and with regard to
the same phenomena the tyrannically ruthless and inexorable
enforcement of power-demands."[12] Physics is (necessarily)
expressed from the interpretive circumstance of our own cultural—hence, political—perspective. The political perspective
seeks a sovereign expression opposing other perspectives and—
employing the terms of this metaphor—as so sovereign, necessarily overlooks the influence other cultural possibilities might have
on physics.[13] What is more important, given the subject matter of
physics, is that we fail to realise that the interpretive dimension
describes the dynamic of the Will to Power—which in turn, for
Nietzsche, describes the world.[14] The phenomenon of world—
nature, cosmos, all that is, the universe—does not correspond to
the sovereign conatus of a political expression.[15] Only an inexorable play of forces can fully account for the world: "every center of force—and not only human beings—constitutes the entire
world from itself outward that is to say that according to its own
force it measures, handles, arranges. . . . You have forgotten to
take this perspective-*setting* force into account in 'true being.'"[16]
"*Reality*," explains Nietzsche, "consists precisely in this particular action and reaction of each individual against the whole."[17] It
is essential to stress the reciprocal character of this interaction or
we will fail to see the radical direction of Nietzsche's argument
against the anthromorphising tendency of modern, as of ancient,
science.

As noted above, Nietzsche claims that physics as we know it is
"an interpretation and arrangement of the world (according to
our own requirements . . .)." His focus upon the interactional

character of the world in contrast to our interpretation of it does not represent to the world as it is in-itself but interprets the world-process as being itself interpretive. This means that every striving for, or expression of, or Will to Power is affected not only by other centers of force (on the model, say, of a traditionally mechanical worldview) but by the entire universe.[18] The world, for Nietzsche, "is essentially a world of relations. It has in this condition, a *different appearance* regarded from any point outwards. Its being is essentially different at each point, imposing upon each point and opposed by each point—and in each case the sum total is altogether *incongruent.*"[19] Our efforts to express this world succeed because we, too, participate in this fundamentally interpretive cosmic dynamic.[20] However, this success is by (necessary) chance for it could have been otherwise.[21] With this in mind, we return to Nietzsche's dramatization of the conflict of interpretations: the Will to Power. Instead of reading nature's conformity to law from the phenomenon of the world, it is possible to read nature as an expression of the Will to Power:

> an interpreter who could bring before your eyes the universality and unconditionality of all "will to power" in such a way that almost any word and even the word "tyranny" would finally seem unsuitable or as a weakening and moderating metaphor— as too human—and who nonetheless ended by asserting of this world the same as you assert of it, namely that it has a "necessary" and "calculable" course, however *not* because laws prevail in it but because laws are absolutely *lacking*, and every power draws its ultimate consequences every moment.[22]

In this alternative interpretation, we see that Nietzsche's speculative interest is directed to challenging our ordinary empirical expectations and what we conceive to be the possibilities for thought. This challenge is effected by Nietzsche's stylistic concinnity, contrasting a primary conceptual expectation with the expectations arising from different perspectives. Because no perspective is privileged, there is no knowledge as such in any one of them. This assessment includes Nietzsche's own position because it is that position in its full extension. Using an expressive concinnity to contrast expectations of nature's conformity to law with the source of such expectations in our own democratic wishes, Nietzsche says necessity results from the Will to Power, not the obedience of nature. So he advises, "Let us beware of saying that there are laws in nature. There are only necessities: there is nobody who commands, nobody who obeys, nobody who trespasses."[23] Necessity, then, results from an absence of laws rather than a prevalence of laws. By this Nietzsche means to denounce the rational idea of a rule-governed cosmos. What happens happens, and every event determines its ultimate consequences, but there are no reasons, no regulations, no rationale. There are no rules, only necessities.

As our previous reading of the concinnous logic of Nietzsche's challenge to scientific reason demonstrates, the idea of a law of nature is contrasted with that of natural necessity in an exhibition of contrasts that includes Nietzsche's own position. Using the rhetorical artifices of concinnity, Nietzsche both articulates and replies to an internal contradiction: "Granted this too is only interpretation—and you will be eager enough to raise this objection?—well, so much the better.—"[24] In this way, both confrontationally and dialogically, Nietzsche describes his own position vis-à-vis traditional positions, that is, his own philosophy of science (as well as his philosophy of nature),[25] vis-à-vis traditional approaches to and understandings of science. Nietzsche's concinnity is fundamentally propaedeutic: what Heidegger would call the "ad-venture" of the question in Nietzsche's musical style that challenges the reader to echo, to counter or magnify the claim advanced in Nietzsche's writing is itself essentially a preliminary questioning of logic and science. It prepares the way to put the morality of truth in question. We see again that what is said by Nietzsche's text (traditionally, its content) is inseparable from the mode of its expression and (in a chiasmatic precision) the style of expression articulates the perspectival openness that is for him—as it can be for us—a new possibility for thought, philosophically beyond the tradition.

It is Nietzsche's controlling perspective that everything is an interpretive or moral perspective—or Will to Power—including his own interpretive understanding. And he is irreproachably, almost compulsively consistent in this. If Nietzsche is quick to say that there are, strictly speaking, "no facts, only interpretations," he is even quicker in stating that traditional logic is nothing more than a physiological/ecological[26] exigence: "to regulate [rectify, justify] a world for ourselves, in which our existence is made possible."[27]

The originating impulse of our "language of signs" or "sign system," as Nietzsche calls traditional logical conventions,[28] is nothing more, he says, than its life-preserving and life-enhancing efficacy. In this context, the concepts of 'identity' and 'substantiality' have logical primacy because they are indispensable for the survival of the organism.[29] Logical thought is a method not for revealing but for determining what should count as, what is called, "truth": "Life is no argument. The conditions of life might include error."[30] Thus, for example, the principle of noncontradiction is a rule for determining the value of logical thought by delimiting what counts as valid logical thought from impermissable formulations. For Nietzsche, such an "inability to contradict [oneself] demonstrates an incapacity rather than a 'truth'."[31] The principle of non-contradiction is a bio-pragmatic limit (governing the conditions of organic life), rather than an absolute.

The apparent self-contradiction in Nietzsche's perspectivalism is self-referentially consistent, hence it does not, strictly taken, have a self-contradictory effect.[32] This consistency highlights the experimental character of perspectivalism. The challenging claim that there is no truth *but* much rather "—only interpretation" expresses Nietzsche's experimental process of thought. The experimentation, the temptation of thought frames, while daring the extension of, the thought against and in a tradition.

Nietzsche opposes the comforting construction of stasis created by projecting parochial belief into universal law (i.e., by construing perspectival truths as scientific absolutes). Because philosophy is (or must we say, should be?) an experiment of thought, of thinking in-process, the challenging position offered through Nietzsche's perspectivalism liberates philosophy from the rule of grammar as an absolute, frees its project from a faith in time worn ideals (truths) and from practical dictates ordered merely to a painless comfort-oriented life (values). So Heidegger explains the character of the "new" thinkers who follow Nietzsche's challenge: "the creators, the new philosophers at the forefront, must according to Nietzsche be experimenters; they must tread paths and break trails in the knowledge that they do not have *the* truth."[33] There is danger here because the drive characterizing such seekers after knowledge, those who seek knowing that that they cannot know (hence tragic seekers: Nietzsche's philosophers of the future), is inherently ambivalent.[34] This drive has a widespread expression in nihilist culture. The difference between nihilist culture and the tragedy of its self-overcoming is to be found in the daring of the latter. Seekers of tragic knowledge dare the danger that is knowing and not-knowing. So precisely apposite in this context, Gadamer describes Nietzsche as "a genius of extremes, a radical experimenter with thought. He himself characterized the figure of the coming new philosopher as that of the attempter [*Versucher*], who brings not truth but risk."[35] Because Nietzsche's project shares the drive characteristic of philosophy in its fierce desire to know, his project, however ambiguously posed, remains in that tradition.

NIETZSCHE'S PERSPECTIVALISM: THE SPECTRE OF RELATIVISM AND THE SPIRIT OF DIFFERENCE

By using the word perspectivalism, my purpose is to distinguish Nietzsche's philosophical reading of the ontological, aesthetico-epistemological implications of "position," "point of view," "perspective" from the crypto-relativism of the mere endorsement or repetition of perspectives as such (*perspectivism*).[36] I employ the *adjectival* noun form of *perspectivalism* in an attempt to avoid the relativistic confusion inherent in the word *perspectivism*. As I construe the term, perspectivalism (a perspectival philosophy) is not an instance of perspectivism (perspective philosophy) but rather a reflective collection of perspectivisms, that is, a philosophy built up on the idea that the world is replete with different viewpoints and different from every perspective. In another word, where perspectivism expresses the perspectival condition, perspectivism both reflects and reflects upon the perspectival condition. Thus perspectivalism does not, as perspectivism seems to, connote the view that all knowledge is no more than interpretation, the representation of a particular perspective. Clark defines "perspectivism" as the *claim* that "all knowledge is perspectivism."[37] However, perspectivalism comprises reflection *on* the consequences *of* this claim concerning knowledge (which for its part is not a knowledge claim) and not merely the expression of the claim itself. Thus articulated, perspectivalism suggests the properly philosophic, higher-order viewpoint or thinking of the question of perspective that adverts to the primacy of perspective and its implications. In the critical spirit of Kant, a perspectivalist philosophy reflects the epistemic fact of perspective and traces its origins and seeks to outline its critical consequences.

Yet apart from the semiotics (or politics) of word analysis, the issue of perspectivalism as a philosophic position must be distinguished from that of relativism. This is essential in the context of the present book, that is, conceived with reference to the philosophy of science, Nietzsche's perspectivalism in his grounding/ ungrounding claim that there are "no facts, only interpretations" might appear not only as a kind of relativism but also as an irrationalism, simply because the ordinary and proper concern of science is the facts. Nietzsche's declaration, undermining the domain of science, is, as we have seen, typically and self-contradictorily antiscientific. In particular, Nietzsche's perspectivalist position is methodologically problematic. Thus denying the "facts" in order to valorize interpretation effectively advances an ultimate interpretation: the claim or "fact" that there is only interpretation. This is a classically circular procedure: a *circulus vitiosus veritatis.*

We have already noted that against this reduction to the level of commonsense interpretation, Nietzsche's query of the ordinary scientific concern with the facts is more than a circular contradictory aspect of the Nietzschean program uncritically tolerated by logical nihilists: it forms the keystone of his philosophic reflections on science. Thus although the self-contradiction implicit in the totalizing claim that everything is interpretation is obvious, it is just as patent—and critically so as I have pointed out—that this same implication did not escape Nietzsche. Nietzsche himself pointedly, stylizedly adverted to this inherent self-contradiction and, in an anticipation of a programmatic analysis of the genealogy of propositions and grammatical presumption Derrida would subsequently turn to the capital of a movement in literary criticism, namely, deconstruction, prototypically employed the reflexive structure of contradiction to advance the thetic point of the irreducibility of interpretational vantage. In such a nascently postmodern parodic context, the rhetorical claim of perspectivalism is not assertoric but *confrontational.*[38]

Thus what finally recommends a defense of Nietzschean perspectivalism as a non-absolute position is its nonrelativistic dimensionality. Multidimensional rather than relationally two-dimensional and effectively monological, perspectivalism is essentially opposed to (typical) relativism. It may be further suggested that a seeming self-contradiction can be self-referentially consistent only if it is conceptually provisional or aesthetically experimental, or if it is articulated in a higher-order interpretive framework. Here the issue ranges from the ontic to the ontological.

Beyond the hermeneutic turn, Nietzsche's perspectival position proposes an interpretational aesthetic ideal and challenges the self-evidence of any axiom (or axiomatic horizon) affirming an ultimate or absolute ideal of truth. Insofar as perspectivalism is non-(self-)conservative, Nietzsche offers knowledge an infinite domain, but—and this is Nietzsche's stylistic precision—he offers knowledge seekers no such infinite and no sure method and no truth.

For Nietzsche, in its most radical expression, the valuation of truth as an exclusive ideal is no more than a moral prejudice; its prescription through method, only a grammatical stricture.[39] In other words, Nietzsche takes the scientific ideal of truth, in opposition to its own self-definition, if it holds itself to be unique,[40] to be merely one interpretation of life and reality, among many possible interpretations. This is a view of the scientific ideal of truth that seems patently relativist. Rather than a progress, that is, rational sequence, of scientific theories (e.g., in the positivist works of Carnap, or in the writings of Sellars, Hempel, Popper, and the early Schapere), or the progressive evolution of research "programs" (e.g., in the work of Lakatos, Kuhn, Feyerabend, Toulmin, and McMullin) ordered toward more and more comprehensive truth, or toward what we take to be the truth, Nietzsche views scientific theory and research activity as a fundamental expression of human needs and interests. Such an expression is not valid in itself or for itself but is instead for Nietzsche wholly arbitrary. This arbitrariness should not be taken to imply that science lacks any necessity, only that it must lack a metaphysical mandate.

Viewed perspectively, science is a world-interpretation. But the perspectivalist position cannot be identified as a relativism, because, in its valuation of the ideal of apodicticity, relativism depends upon the ideal of reason. The ideal of this dependency is expressly taken to be ideal, that is, ideal in the sense, like the Platonic Good, of the unattainable. Qua unattainable, relativism can share the ultimate objective ideal of science. Both conventionalist and realist manifestations of relativism correspond to the critical and historical demands of contemporary interests in the philosophy of science. For these reasons, the following conceptual considerations may be expected to differentiate Nietzsche's perspectivalism (not relativist and not realist, that is, an approach unformed by any methodic ideal)[41] from other approaches to the philosophy of science.

To distinguish perspectivalism and relativism, Alwin Mittasch, the German historian of (specifically chemical) science and its cultural influence, invokes the classical critical tradition within German philosophy. Within this context, Mittasch is able to argue that Nietzsche's standpoint be regarded not "as 'relativism' but rather as 'perspectiv(al)ism' . . . as a perspectiv(al)ism thoroughly in line with Leibniz-Kant."[42] To see the relevant orientational difference, we will take a brief look at the ordinary understanding of relativism.

The relativistic position takes all philosophical viewpoints (and Nietzsche's perspectivalism would not be excepted) as ultimately equivalent, that is, *equal* one to another as well as to its own position.[43] This conception equates all perspectives and presupposes an absolute, conceptually essential, if admittedly or at least declaredly unattainable, standpoint. The unattainable absolute is the imaginary Archimedean standpoint from which all perspectives are relatively equivalent. Axiomatic for any relativism, this mediating absolute permits the relativist to claim the same right or value for all philosophic positions. The comprehensive character of the relativistic viewpoint manifests its dependence on this ulterior, structuring absolute: the apodictic ideal demands a complete extension to all contexts. For a relativist, in other words, *everything* (absolute idealism, empirical realism, or Nietzschean perspectivalism) is ultimately—when criticized from the relativist's absolute standpoint—relative.

Rather than the perspectival claim that there is no truth, which given the perspectivalist standard always entails that some non-truths (interpretations) are better than others (art, illusion, deception, and delusion), the implicit claim of relativism is that there is a truth above all positions to which no particular position has any privileged claim. Like its divine correspondent and model, the image of a *veritas abscondita* is a condemnation. The relativist's position is the tempered outcome of a *truth-agnostic* but still truth-directed despair: the truth *about* truth is unattainable. For Nietzsche, the resolution of this despair will be the "cock-crow of positivism." Humanism is brother to this kind of positivistic relativism.

TRUTH, PRAGMATISM, AND RELATIVISM: REALISM AND THE REAL

Although Nietzsche's perspectivalism shares certain characteriztics of relativism's more reflective manifestations, we have seen that perspectivalism is not what is often called "cognitive relativism," as this might be expressed in a science-critical context. Conventionalist (Pragmatist) and realist views within the philosophy of science are, however, infused in various ways with a relativism of this kind. Below I seek to make some of these distinctions and clarify the difference between Nietzsche and pragmatic (originally Dewey, but also James) approaches to truth, as well as the difference between Nietzschean perspectivalism and relativism, all in distinction to what philosophers of science call (ontological) "realism."

Pragmatist approaches to truth embody the commonsense pragmatism, or respectful disinterest, of a technician. This means that claims to truth are deferred. The truth of an account of the world, for the pragmatist, is simply not the immediate business or responsibility of the scientist. Realist approaches to truth are more ambitious. Truth is ultimately attained through scientific accounts of the world. In spite of this promissory character—orientations meant to define the object of scientific investigation independently of those interests that lead merely to communal or pragmatic consensus—realists do not deny the practical limitations of the scientist's approach to truth. It is not my intention to stress the differences between pragmatic and realist views of the scientific truth. I am concerned instead to draw attention to the commonality of their attitudes toward (consummate) truth. Fundamental here, the point is that the notion of truth as such, as a unitary and ultimate, or absolute ideal, is not in doubt for pragmatist or conventionalist or realist philosophers of science. The pragmatist or formalist perspective assumes truth to be independent of—if not irrelevant to—a scientific perspective. This incidentally independent character assumes the objective value of truth.[44] The realist orientation approximates (or approaches asymptotically in principle) objectively defined truth, and hence its achievements are correspondingly provisional. But in spite of all programmatic differences, no matter whether the ideal of truth is to be attained by empirical-theoretical or by praxical-semantical means, the ideal remains that of objective truth.

Despite an express commitment to objective truth, the many varieties of realism often employ some version of relativism (or conventionalism) to account for theory change in the history of science, along with shifting bases of "fact"—different rosters of defensible phenomena. The incorporation of relativist perspectives within realism does not contradict the realist program. Rather, it underscores the philosophical tenability of relativism. A limited relativism is a serious, rational response to contextual problems, such as, for example, those differences in opinion to be found among individuals or emergent within the historical development of science, requiring a rational way to "tailor" or fit divergent accounts to reality.

Of course, there are "realistic" limits to the expression of relativism in philosophically sophisticated positions. But relativism is very far from being *tout court* cognitively unsophisticated, and relativists are consequently often more sophisticated than their opponents concede. Feyerabend offers a very powerful defense of his version of relativism in his recent *Farewell to Reason*. But the distinctions Feyerabend makes in his first chapter ("Notes on Relativism"), listing eleven points of view and nicely following analytic protocol by offering definitions and elaborations of R1 through R11, will hardly win him any points with those unsympathetic to such treatments.[45] Indeed, it does not save him from being lumped in with the "highly relativist 'subjective' views of Kuhn" by Frederick Suppe in his recent book.[46]

With respect to the role of relativism in epistemology and thus in the philosophy of science, consider the blandly misleading representation of relativist truth offered by William Newton-Smith: "what is true depends in part or entirely on something like the social perspective of the agent who entertains the hypothesis or on the theory of the agent."[47] Newton-Smith equivocates: the term *true* is assigned a stronger sense than is needed for the expression of a relativist position. Relativists know very well what it would mean for something to be true in an absolute sense, because they depend upon that insight for their own justification. What they dispute—and it is this disputation that defines relativism—is how human, that is, historically and circumstantially limited judgment, is ultimately to be evaluated as an adequate expression of what really is the case. We have noted that both pragmatic and realist visions of the philosophy of science are sensitive to this question. Newton-Smith deliberately exaggerates when he says that, for the relativist, "from theory to theory, what is true changes and not merely what is taken to be true."[48] Newton-Smith is speaking not merely of conceptual relativism but of perceptual relativism.[49] Newton-Smith's failure to recognise differences between relativist positions or between relativist positions, on the one hand, and historical and critical perspectives, on the other hand, allows him to claim, for example, with Suppe, that "both Kuhn and Feyerabend articulate relativist positions."[50] Yet both Kuhn and Feyerabend, together with many conventionalists (and relativistic realists) address epistemological questions of a critical structural kind. In effect, Newton-Smith ignores relativism's claim to an absolute ground, taking its putative project to cancel its philosophic value.

It is significant that few positions or claims, apart from Nietzsche's "truth-claims" or perhaps those made for the "scientific" value of astrology, have taken as severe and repeated a philosophical beating as relativism. Yet despite recurrent efforts by a number of authors, despite qualifying (ameliorating) precisions by Feyerabend and others, the threat of relativism is not about to go away. For relativism has the genuine philosophic warrant that it has because its most compelling inspiration—and here we may as well count in Nietzsche's perspectivalism as a relativism—derives from a sober consideration of the limits of knowledge as such. Feyerabend suggests that the important distinction to be made is between concepts and human relations.[51] In this spirit, R. L. Gregory observes that relativism proceeds from a critical rational viewpoint, taking "critical" here in the Kantian sense. Gregory admits to accepting:

> an extreme form of relativity of belief. This follows from the proposition that all observations depend on assumptions and that no assumptions can be independently justified. This situation extends even to formal proofs. The logician Kurt Gödel . . . showed that the consistency of a formal system cannot be proved from within the system. Proof of consistency requires justification from meta systems. But, this leads to an infinite regress for formal proof. . . . So even consistency cannot finally be proved for any logical or mathematical system.[52]

In a later chapter, I shall have cause to suggest that it is not merely ironic but inevitable that the ally Gregory chooses in support of relativism is the logic of formal systems.

If relativism may be said to correspond with a kind of critical rationalism, where does this place grand old scientific realism? In both its formalist and realist variations, the philosophy of science has a general commitment to objective truth. The limited relativism adopted by any realism, for example, "to save the appearances," varies with reference to the variety of realism in question. Without violating the realist claim to offer true representations of reality, specific infusions of pragmatism, conventionalism, and istrumentalism, along with hermeneutics in general, serve as (critical, rational) ways to account for historical and cultural change, that is, the process-oriented character of science and its social image. It is significant that some variety of historical relativism is pretty well accepted without reservation by writers in the philosophy of science today. Critical or otherwise qualified realisms are thus able to claim an approximation to the "facts" while—reasonably—accomodating the philosophical fallibilism required today in both scientific practice and philosophy.[53]

From the perspective of a Nietzsche-styled philosophy of science, the problem as I have asserted is not relativism as such but the desire manifest in relativist positions for the absolute. The problem, in other words, is that today's philosophy of science, no matter whether primarily pragmatist (empiricist) or realist, asserts an unknowing knowing in its dedication to the ever advancing ideal of progress. Without making a claim to a finished truth, pragmatists and realists alike "have" what is just as good as knowledge, for the former (i.e., relative knowledge), or knowledge plain and simple (the "facts"), for the latter. The conviction unifying relativism and its variant manifestations as pragmatism (or as formalism) in the philosophy of science is that there is a Truth to which these relative approaches have as much right as realism—and any other position—may presume. Nietzsche's critical approach denies this Truth and every "right" to it.

The difference between positivistic pragmatism (or "constructive empiricism," as one recent writer describes his position)[54] and realism (of varying "sophistication" in the above sense) emerges first of all with reference to questions of truth and belief. Relativists, after all, may themselves be defined as lapsed realists who have been burnt or stung by scientific changes. For comfort, they focus on their formalist heritage: in logic and the possibility of reconstruction,[55] or perhaps translation and/or interpretation,[56] and stipulation.[57] Loyal realists are less shocked by the conceptual challenge of theory change and the dynamic of scientific advance (considered in its negative, retrospective dimension).[58] They are the "proper" heirs of the British empiricists and reflect their stubborness. With such an attitude they acknowledge the magic change in scientific styles that either alters the spots on entities formerly considered scientifically real or whisks them away entirely, leaving some other thing or even nothing at all in its place. (Newton-Smith's charge is "what is true" and not merely what is "taken to be true.") Realistic realists—as distinguished from relativists—accept such theoretical hocus-pocus because something (at least) remains on the scientific stage. And thus they find the metaphysical comfort that inspires most realists: it shall be that, whatever that is, that is real! But one should not imagine that the ideal of the absolute thereby vanishes.

The need to adapt to cultural (emphatic) and factical (phenomenal) change is important because it explains the necessity of hybrid philosophies of science. A pure realism is impossible; to adapt to changing climates of scientific theory and fact as it does and must, realism must incorporate certain traits typically associated with relativism.

In a "dominant"-realist hybrid philosophy of science, if one can (in some way) physically apprehend an object—if one can still "kick" it, as it were—then it is *real* in some, that is, in whatever, sense. In *Representing and Intervening*, Ian Hacking offers a contemporary expression of this realistic empiricism as a kind of optimistic pragmatism: if you can do something with or to an object, it is real. In place of direct kicking, Hacking substitutes "spraying," with a bow to the contemporary instrumental mediation of technologized research. Here, Hacking describes the procedure used by experimental physicists for varying the charge on a supercooled niobium ball in a contemporary version of Millikan's oil-drop experiment. By their own account, the research physicists *spray* a supercooled niobium ball "with positrons to increase the charge or with electrons to decrease the charge." This expression persuades Hacking that the experimental project has to do with real entities all round. Now it may be that Hacking simply likes the image of spraying with positrons or else electrons, rendering the latter two subatomic particles less theoretical but approximately real (realistic) entities, but in any case he thinks he understands what might be meant by a "super-cooled niobium ball"—and he declares, "If you can spray them they are real."[59] For Hacking, it is unimportant that the relevant electrons and positrons operating in this experiment as a kind of dynamic spray paint might someday be set aside as inadequate representations of reality, or as illusions, that is to say, as unreal. Then the talk might be of waves oscillating between dimensions, or quark composites, or dynamic field quanta with no fixed but only a probablistic nature, or some other thing employing different imagery. This possibility as such is unimportant to Hacking and other (dominant) realists because what matters to them is *efficiency*. The efficient question with regard to niobium balls, hedgehogs, and beach balls asks only if you can spray, pat, or kick them. This emphasis (corresponding in a fundamental way to practical scientific interests, as a general commonsense view) is less important for other ("dominant" formalist) philosophers of science for whom logical structure does not accommodate such flexibility in experimental and theoretic variations.

Perspectivism, regarded from realist or pragmatist/conventionalist positions, can only be understood as yet another brand of relativism. Thus defined, the empiricist realist critique of Nietzsche's perspectivism is obvious: the affirmation of a multiplicity of interpretations in place of the "facts" ignores the persistence of those facts (i.e., their resistance to kicking, or suitability for spraying, and so on, no matter what they are called) and so affirms (potentially) uncontrolled and even irrational possibilities. For its part, relativism does not bother with a critique: it does not recognise any distinction between its own position and that of perspectivism. It assumes that perspectivism is undertaken from a position like its own with the same interest in absolute values. When claimed as a relativism, perspectivism is thought to be no more than the negation or subversion of a dominant position leaving the hierarchical structure of valid and less-valid accounts of things intact, altering only their relative positions.

But we have seen that this may not serve as a definition much less as a dismissal of perspectivism. In the self-referential assertion that everything is interpretation, an absolute claim or frame of reference is irretrievably excluded. Nietzsche's perspectivism is not derived from an absolute, by originating in its negation, as a kind of Hegelian *Aufhebung*, nor does it posit an absolute, as relativism posits the truth (in the fashion of what Nietzsche calls *"Ressentiment"*). The twin dialectical movements of *Aufhebung* and *Ressentiment* preserve the idea of an absolute. This same (mediating) absolute is relativism's ultimate warrant. Nietzsche's philosophy is conceived as an "attempt," an experiment, daring to question what is least (hence most) questionable in questioning what has hitherto remained unquestioned.

In this respect, Nietzsche projects a philosophy that is truly open, that makes a virtue of this openness, and that although scorning all reserves is yet ready to justify itself in an open encounter with life as art. In his refusal of an absolute, Nietzsche affirms the infinite possibility of knowledge: "Rather has the world become 'infinite' for us all over again, inasmuch as we cannot reject the possibility that *it may include infinite interpretations.*"[60] It must be acknowledged that the perspectivalist project is as frightening in this as it is invigorating. Yet there can be no "safer," no more conservative alternative for Nietzsche. The daring orientation of perspectivalism is indispensable, since philosophy itself is at stake, since it is the eros of understanding that moves us: "At long last the horizon appears free to us again, even if it should not be bright; at long last our ships may venture out again, venture out to face any danger; all the daring of the lover of knowledge is permitted again; the sea, *our* sea lies open again; perhaps there has never yet been such an 'open sea.'"[61] There is a clear difference between a perspectivalism that scorns preservative measures taken for their own sake, daring all perspectives and every destruction including its own, and the relativism that proposes only its own equal right to the same right as opposed values and refuses any possibility of its own destruction or diminution (relativism trades an explicit absolute sanction for an absolute right to its claims). Nietzsche's perspectivalism, daring its own destruction from its inception, does not claim any equal right for its position nor, indeed, any right at all. This is its fundamental difference from relativist positions.

The convicted belief in the commonality of right, and so too in an egalitarian opposition to any hierarchical ordering of right, characterizes the position of relativism. For Nietzsche, this represents the "demokratischen Instinkten der modernen Seele," epitomized by the cry of the "terminal human being": "No herdsman and one herd. Everyone wants the same thing, everyone is the same"[62] and so revealing "a nice piece of mental reservation in which vulgar hostility toward everything privileged and autocratic, as well as a second and more subtle atheism, lie once more disguised."[63] For Nietzsche, relativism is the appropriate corollary to this democratic instinct. Nietzsche's perspectivalism, however, is the product of a fundamentally elite, esoteric interest. This hierarchic difference in rank further prohibits the identification of relativism and perspectivalism.

Nevertheless the elitism here is a matter, a principle of thought and not an ethos of science. Instead the ethos of scientism manifests an extreme autocracy inhibiting critique directed to the social value of science or, as Nietzsche has it, the value of science for life. As an idealization of science, the philosophy of science permits no criticism of science in either theoretical or practical expression. Accordingly, any critique of science that is ventured inevitably draws the denigrating typification of an "antiscience." This is the fate of a critique of science offered under whatever philosophical rubric, and indeed even when offered in practical terms, reflecting the social-progressive interests of political organizations such as the Greens or other liberal-cum-ecological movements, or the empowerment projects of public-interest groups such as Science for the People.

Where science is taken to be ideal knowledge, the philosophy of science, for its part, becomes or approximates a "science of science," which, it is thought, provides the basis for a "theory of theory." I have sought to argue that philosophy even as the philosophy of science should be conceived more broadly, that is, historically, hermeneutically, phenomenologically, and withal perspectively not as an expression of a theory or maxim or law of life but rather as the discipline/festival of thought—the adventure of questioning.

THE MEANING OF CRITIQUE: NIETZSCHEAN POSSIBILITIES FOR PHILOSOPHY

As an introduction to the account of Nietzsche's ecophysiological epistemology in the following chapter, it should be observed that Nietzsche takes the collective events or happenings of the world as so many events of interpretation or style. Beyond the perspectival aesthetic representation there is nothing. Thus Nietzsche declares the importance of seeing, "that *what things are called* is incomparably more important than what they are. The reputation, name, appearance, the usual measure and weight of a thing, what it counts for—originally almost always wrong and arbitrary, thrown over things like a dress and altogether foreign to their nature and even to their skin."[64] Naming as such is creative: there is nothing that can be known (named or interpreted) beyond what is interpreted or what can be named. So Nietzsche explains that his epistemic insight into our arbitrary aesthetic relation to things as an interpretive event is not a revelation after which one can see things as they really are. Nietzsche's point is perspectival and not metaphysical: "How foolish it would be to suppose that one only needs to point out this origin and this misty shroud of delusion in order to *destroy* the world that counts for real, so-called reality."[65] Nietzsche concludes with what we can see to be a perspectival emphasis: "We can destroy only as creators.—But let us not forget this either: it is enough to create new names and estimations and probabilities in order to create new 'things' in the long run."[66]

For Nietzsche, "the lordly right of giving names"[67] is the cornerstone of intellectual interpretation, understood as the manifestation of perspectival Will to Power. This is the case whether the Will to Power expresses a reactive, relationally descriptive representation,[68] or a free, active creation. In this way Nietzsche understands the human propensity to take the pattern of subjective (ego) action analogically as paradigmatic for events. In passive experience, from our own experience of willing (desire, or impulse), we uncover an active subject willing, desiring, performing, and, in all, doing. The presumptive relationship between doer and deed is then extended to a description of nature at large.[69] Accordingly, we can conjecture things that either act or are acted upon (and, for Newton, are free not to change, i.e., to continue in motion or remain at rest). Again, (on the basis of our own experience), we posit the identity of things in themselves over time (the "same" thing); furthermore, we assume the identity in character of similar distinct things (the "same" things). So, Nietzsche continues, what originally was an arbitrary error, "grows from generation unto generation, merely because people believe in it, until it gradually grows to be part of the thing and turns into its very body. What at first was appearance becomes in the end, almost invariably, the essence and *works* as such.[70] In our substantiating and identifying conceptual processes (*Gleichsetzen*), the world of science emerges. But this is arbitrary, and it is so in both objectively incidental and subjectively unconscious senses of the word. We cannot do otherwise, so Nietzsche says, than conceive the world from our own perspectival angle.

The critical insight into the analogistic root of scientific knowledge and human subjective experience is not Nietzsche's own discovery. He has this insight from Schopenhauer, and it reflects an older (and noncritical) tradition in the history of philosophical thought of a parallel between the microuniverse (meaning humanity) and the macrouniverse (the cosmos as such). Exceeding both Schopenhauer and the tradition, Nietzsche questions the anthropomorphism of anthropological knowledge. If it is an illicit process to move from our experience of ourselves to understand our experienced encounters with the world, for Nietzsche, the very familiarity that we enjoy with ourselves is also to be put in question. Nietzsche challenges the self-evidence, the egoic certainty of the subject. Even the experience of willing—classically the most direct ego-expression of subjectivity—tells us nothing about that subjectivity (as object) for Nietzsche. We cannot look round our own corner, because we cannot descry the deep structure of a surface phenomenon by remarking upon and from the level of that surface alone; even our best perspective cannot completely encompass multiple foreign perspectives. Nietzsche differs from more catholic accounts of interpretation, because the notion of a thoroughly public perspective together with its apodictic ideal can only be a decadent fiction. The limit of perspective cannot be overcome. This is the fundamental insight of Nietzsche's radical epistemic critique.

The perspectival limit of critique is itself problematic. Indeed, given this perspectival limit to the possibility of philosophic reflexivity, can one ask the kinds of questions Nietzsche dares or attempts to bring to expression? Even if such questions are possible, how hope for an objective answer, given the irreducibility of our human circumstance as perspectival being in the world? If we take our purpose to be philosophical, even radically philosophical and so reflexive, how reconcile the recognition of the unavoidability of the surface limit and the impossibility of full reflection—of reflexivity itself!—with the height[71] needed for the philosophic vision Nietzsche demands as a sign of recognition from those with a right to read his work? Do we know how to read, that is, to ponder and to celebrate, thinking above and even beyond Nietzsche's original thought? By considering the meaning of critique in further detail, a preliminary answer to this fundamental projective question may be found.

What is the influence of perspectivalism on the possibility of critique? It is certain that analysis for its own sake (criticism, along with other aspects of professional philosophy), is not intended by Nietzsche when he speaks of the possibility of doing philosophy. He writes, "critics are the philosophers' instruments and for that reason very far from being philosophers themselves!"[72] By this Nietzsche means to outline the narrow dimension of critique articulated in a scientific (here, noncreative) perspective. We have noted the essential limits of scientific critique above. Ordinarily, Nietzsche takes critique to be nihilistic—thus, he denies the title of critic to his "philosophers of the future," describing their willingness to challenge and dare the most difficult questions: "nonetheless they still do not want to be called critics on that account."[73] However, it will not do to forget here that critique can be (as thinking occasionally is) "something easy, divine, and a closest relation of high spirits and the dance!"[74] The same name is ambivalently used for two different activities carried out by two different types. These two possibilities for critique must be kept distinct. It is this distinction, then, that inspires Nietzsche's demand: "I insist that philosophical laborers and men of science in general should once and for all cease to be confused with philosophers."[75] In this way, Nietzsche can contrast the practical scientific worker (even within professional philosophy) with the (genuine) philosopher. For Nietzsche, one is reactive, the other active. Only the latter can be creative. Rather than a scientific critique of science, my effort in subsequent chapters will be to work out the active possibility of a philosophic critique. Since I have shown that, by and large, contemporary philosophy of science offers only a scientific critique, subsequent chapters challenge the ruling role of science in culture and in philosophy. For Nietzsche, this ruling role can be revealed as follows: "now that science has most successfully resisted theology, whose 'handmaid' it was for too long, it is now, with great high spirits and a plentiful lack of understanding, taking it upon itself to lay down laws for philosophy and for once to play the 'master'—what am I saying? to play the *philosopher* itself."[76] Opposing the reactive domination of science, a Nietzschean philosophy of science would restore the reflexive critique proper to philosophy, emphasizing the distinction between philosophy and science without assigning the role of secondary or lesser importance to philosophy. Their roles are not "separate but equal." The hierarchy is clear: in a Nietzschean aesthetics of truth, the only truth is that which admits— affirms—its lie and thereby proposes a life-enhancing or artistic project. Only philosophy —or a science conducted from the playful vantage of this perspective—can be creative on the terms of this artist's aesthetic. Below, I address the question of rendering Nietzsche's reflection on science from the perspective of either science or Nietzsche studies.

NIETZSCHE AND SCIENCE: THE QUESTION OF VALIDITY

It is the firm conviction of the mature Nietzsche that science is the most recent expression of the nihilism of Western culture.[77] Yet this seems extreme. If Nietzsche perceives science as nihilistic, does he not then fail to see its positive, culture-determining or informational potential? W. Joos expresses this as an objection: "In light of the great cultural and civilizing achievements that we owe to rationality, to choose to represent knowledge as [Nietzsche does as:] 'a beautiful means for decline' is obviously unusual." It is difficult for today's rational human being, thoroughly captivated as he is by positivistic scientificity, to take such an epistemological defeatism seriously."[78] How seriously are we permitted to take Nietzsche's nihilistic vision of science today? I have already indicated that a good many of Nietzsche's readers along with (one assumes) all of his nonreaders would claim that we ought not or cannot take Nietzsche seriously on this matter. Among these dissenting opinions, two basic positions may be distinguished: first, the scholarly tradition of Nietzsche research; second, the scientific tradition that bears the brunt of Nietzsche's criticism.

The first subjective tradition might be represented as saying, for example, and positively or optimistically enough, that Nietzsche just wasn't interested in the logic, epistemology, and ontology of the natural sciences but limited himself to humanistic questions or to the social sciences. Proofs for such claims within this tradition might refer to the evidence of library records and both serious and casual references made in letters, along book margins, and on note cards. For these scholars, representing the general conviction concerning this matter, Nietzsche's vision of the sciences was ultimately and irremediably distorted by the predilections of the scientific writers of his day such as Spir, Boscovich, Roux, Mayer, Helmholtz, and Darwin, among others. Nietzsche's views are not his own, then, but the poorly digested representations of such theoretical accounts of science that were at the time indistinguishable from popular accounts. These authors, especially taken pell-mell, are often dismissed, as they represent currents that do not coincide with modern interests.[79] In a similar way, Rüdiger Safranski claims that Nietzsche was no more than a "child" of the pro-scientific enthusiasms of his times with, as Safranski reads it, disastrous and drastic historical consequences. For Safranski does not hesitate to underline that Nietzsche's view of science with reference to racism and biologism was appropriated by Nazi philosophy.[80] If a Nietzschean philosophy of science is to be expressed from this perspective, even if not Nazi-recidivist, it would not be a particularly useful innovation. Thus with respect to other less inflammatory associations, (the philosophies of Spir, Boscovich, and others) one can either say that a Nietzsche-styled (traditional) philosophy of science can wait for Nietzsche's day to come, arguing that his thought has a special affinity for contemporary scientific trends,[81] or insist instead on the significance for our time of scientific writers who have been eclipsed for years.[82]

The second (objective) ground for negating Nietzsche's value with respect to current science is the judgment of the (general) scientific or modern tradition. This view is broader than (and, at least potentially includes) the first and holds simply that as a philosopher Nietzsche was not scientifically inclined. Whatever Nietzsche's intentions in writing may have been, they do not include a serious philosophical consideration of science and scientific culture. Nietzsche scholars, for their part—and this is why this second, objective view can comprise the first, subjective viewpoint as well—are often pleased to admit that Nietzsche was irresponsible in his use of scientific theories (e.g., his doctrine of the Eternal Return and the idea of the Will to Power),[83] yielding the explanation of physical reality to the scientific tradition. Both viewpoints offer the conclusion that Nietzsche has no more than a rhetorical or metaphorical interest in science, even where he writes explicitly on the topic. Where the philosophy of science itself embodies a scientific ideal, the dismissal of Nietzsche's efforts as "nonscientific" is plainly problematic.

For analytic philosophers working in the philosophy of science (or other areas in philosophy) it would appear obvious that there could be no linguistic or conceptual basis for uniting Nietzsche's philosophy and the philosophy of science. Nietzsche's thinking is regarded as too emotional or poetic to express a serious philosophy of science. And looking at Nietzsche's brusquely antagonistic conception of truth, it seems plain that traditional philosophers of science would be dismayed by the idea of a Nietzsche-styled philosophy of science:

> What is truth? a movable host of metaphors, metonyms, anthropomorphisms: in short, a sum of human relations which have been poetically and rhetorically intensified, transferred, and embellished, and which after long usage seem to a people to be fixed, canonical, and binding. Truths are illusions which we have forgotten *are* illusions, they are metaphors that have become worn out and have lost their sensuous force, coins which have lost their image and are now considered as metal, and no longer as coins.[84]

For Nietzsche, as we shall see in detail below, the progress of knowledge is a train of metaphors and tropes that do not serve even an heuristic function but are themselves the substitutions for truth. This vision of truth is patently unacceptable to a traditionally rigorous (i.e., in the Anglo-American analytic tradition, logical empiricist) scientific worldview. From this perspective, Nietzsche is a convicted "anti-scientific" thinker. Regarding truth as metaphor, he opposes the contemporary progressive as well as the classic absolute ideal of truth. Nietzsche's antipathy to scientific world-constructions would thus aggravate the negative judgment of his presumed incompetence in his understanding of scientific methods and goals.

Likewise, in historical context, Willamowitz's attacks on Nietzsche's scientific (philological) competence inaugurated both Willamowitz's own public career and echoed the damning judgment offered by an older colleage still in Bonn that anyone who dared to write as Nietzsche did "*sei wissenschaftlich Tot,*" that is, such an author would have to be considered "scientifically dead." To this day, we may note, the characterization of Nietzsche's thinking as fundamentally antiscientific and therefore, in the widest sense too and not only with reference to philology, to be likewise named scientifically stillborn—is foremost among the simplifications of his thought. If indeed, as I argue here, Nietzsche was not uncomprehendingly unscientific and therefore not mindlessly but much rather philosophically critical of science, it nonetheless seems that his fears of being taken for "what he was not" have had and continue to have substance.

The two orientations mentioned above ground the negative position on the relation of Nietzsche's philosophy to natural science. Both derive their rational claims from the "democratic" appeal to commonality, which we have seen to be characteristic of relativism. It is my proposal, in the light of the "elitist" nature of Nietzsche's stylistic project, to resist this appeal and so to raise the question of the possibility of a Nietzsche-styled philosophy of science. But this proposes an attempt to think such a philosophy of science. It is well to ask how this is possible?

The negative tack offers us, as it usually does, a preliminary purchase on the question of such a "possibility." But an attempt to think a Nietzsche-styled philosophy of science is, at the very least, not made by the limited effort of trying to guess Nietzsche's ability to comprehend the science texts he had access to based on his fellow students' or even his Pforta teachers' estimations, or to estimate his own scientific talent on the same limited basis.[85] Nor do we wish to make premature concession to the hegemony of accepted philosophy of science by limiting the field of reference of a Nietzschean philosophy to the social or historical sciences. And, finally, to adopt in advance the view of science itself as the source of a salvific technology would undermine the attempt to understand Nietzsche's philosophy of science (and indeed, as we will have cause to show below, any philosophy of science at all).[86] To think science philosophically, in the light of Nietzsche's thought, one must adopt a critical stance toward accepted philosophies of (natural) science as well as toward (natural) science itself.

As I shall argue further in the chapters to follow, the critical project of articulating a Nietzsche-styled philosophy of science corresponds to the neediness of current philosophy of science, and it is thus more than a project reflecting Nietzsche's philosophical/personal development. Moreover I reject the description of Nietzsche's philosophic work as something that underwent three or more stages of metamorphosis, with the so-called positivistic period occupying the central position.[87] Yet although I reject the ultimate utility of this schema of ontogenetic analysis, I would not deny but must affirm that Nietzsche had a great and ever-increasing fondness for, sensitivity to, and affinity with science and its process of discovery and description. And I grant that, just to this extent—and it remains to show the irony of such an extension—Nietzsche may be aligned with the positivists. Yet one must be careful to note the precise nature of the project one serves when one asserts such an affiliation: unlike (the ultimately more perceptive) Safranski or Baeumler before him, I cannot link Nietzsche with "biologism" and "naturalism" *as such.*

Hence I am in full accord with Müller-Lauter's milder assertion of the critical relevance of the scientific problems of Nietzsche's day,[88] and I would echo Abel's judgement where he observes that "it is undeniable that Nietzsche's phsyiological-chemical manner of thinking not merely in vocabulary but in many substantive questions indicates a clear connection to biology as well as the physical sciences of his day."[89] In addition, corroborating the opinion of Abel Rey, the French philosopher of physical science, Mittasch had also enthusiastically explained Nietzsche's more than casual or dilletant's association with science by affirming the aforementioned affinity with and anticipatory proclivity for science evident in Nietzsche's philosophic character. For Mittasch, "Nietzsche eagerly sought to assimilate unfamiliar scientific knowledge and thought [*Naturwissen und Naturdenken*] . . . to boot he posessed the gift of genius: constructing from the few colours at his disposal a painting in accord with his own plan."[90] For Rey, in almost the same terms but with a different focus regarding science and art, Nietzsche "in 1881, with an intuition of genius, one of the greatest philosophical geniuses of human history, would consecrate ten years of his life to a study of natural science in order to found his theory of return on atomic theory."[91] Rey's expression of Nietzsche's "intuition" of genius is bolder than Mittasch's judgment of genius at the limit (in limits). Such an anticipatory genius is what Nietzsche describes as the characteristic achievement of the ancient Greeks, a people whose inventiveness was primordially original, even where it was demonstratively derivative, because their singular talent was the ability to take up the spear from where others had left it to throw it further. The metaphor is one that I take for my part to describe the practice of philosophy.

NOTES

1. "Das Problem der Wissenschaft kann nicht auf dem Boden der Wissenschaft erkannt werden" (GT 2, KSA I, p. 13), BT, p. 18.

2. "Die Wissenschaft unter der Optik des Künstlers zu sehn, die Kunst aber unter der des Lebens" (KSA I, p. 14), BT, p. 19.

3. If Heidegger gives the rule to my reading, I will join at the head of those readers who affirm that Heidegger reads Nietzsche—and pretends nothing less—for his own purposes not for others. See my recent article "Heidegger's Interpretation of Nietzsche and Technology: Cadence, Concinnity, Playing Brass" in *Man and World*.

4. "Gegen den Positivismus, welcher bei dem Phänomen stehen bleibt 'es giebt nur Thatsachen,' würde ich sagen: nein gerade Thatsachen giebt es nicht, nur Interpretationen" (KSA XII, p. 315), WP 481.

5. "Dass Physik auch nur eine Welt-Auslegung und -Zurechtlegung (nach uns! mit Verlaub gesagt) und nicht eine Welt-Erklärung ist" (JGB 14, KSA V, p. 28), BGE, p. 26.

6. In addition to Wilcox and Grimm, two notable recent examples rehearsing the same family of Nietzschean counterfactuals may be cited here including White, *Within Nietzsche's Labyrinth*, and Clark, *Nietzsche on Truth and Philosophy*, among, of course, others.

7. Clark, *Nietzsche on Truth and Philosophy*, pp. 22–23.

8. JGB 22.

9. Noting that, "derselbe Text erlaubt unzählige Auslegungen: es giebt keine 'richtige' Auslegung" (KSA XII, p. 39), Nietzsche confirms the falsifying spirit of ordinary interpretations, including perception and cognition, which ignore this essential ambiguity: "O sancta simplicitas! In welcher seltsamen Vereinfachung und Fälschung lebt der Mensch! . . . wie haben wir Alles um uns hell und frei und leicht und einfach gemacht! wie wussten wir unsern Sinnen einen Freipass für alles Oberflächliche, unserm Denken eine göttliche Begierde nach muthwilligen Sprünge und Fehlschlüssen zu geben!" (JGB 24, KSA V, p. 41).

10. Müller-Lauter, *Nietzsche*.

11. "Aber jene "Gesetzmässigkeit der Natur," von der ihr Physiker so stolz redet, wie als ob—besteht nur Dank eurer Ausdeutung und schlechten 'Philologie,'—sie ist kein Thatbestand, kein "Text," vielmehr nur eine naiv-humanitäre Zurechtmachung und Sinnverdrehung, mit der ihr den demokratischen Instinkten der modernen Seele sattsam entgegenkommt! 'Überall Gleichheit vor dem Gesetz—die Natur hat es darin nicht anders und nicht besser als wir'" (JGB 22, KSA V, p. 37), BGE, p. 34.

12. "Aber, wie gesagt, das ist Interpretation, nicht Text; und es könnte Jemand kommen, der, mit der entgegengesetzten Absicht und Interpretationskunst, aus der gleichen Natur und im Hinblick auf die gleichen Erscheinungen gerade die tyrannisch-rücksichtslose und unerbittliche Durchsetzung von Machtansprüchen herauszulesen verstünde" (KSA V, p. 37).

13. That Western science alone is titled a science confirms the narrowness of our perspective. Whatever different standards or values inspired the world-interpretations of other cultures, Chinese, Indian,

Mayan, and so on, we do not regard them as scientific except where convergences are found. The general lack of such convergence (see Needham, *The Great Titration* for an example) supports Nietzsche's point. However, in affirming the limitation of the human perspective: "Die Wissenschaft der Natur ist 'Menschen-Kenntniß' in Bezug auf die allgemeinsten Fähigkeiten des Menschen" (KSA XI, p. 115), Nietzsche embraces the significant possibility of other-than-human knowledge perspectives.

14. It is not just subjectivity that structures perspectives but the corresponding objects as well. Against the stipulation of the constants necessary for science (not only experimental constants here, such as the speed of light or Avogadro's number, but atoms, chemical compounds, amoebae and leaves), Nietzsche maintains "Wir haben das Unveränderliche *eingeschleppt*, immer noch aus der Metaphysik, meine Herren Physiker" (KSA XIII, p. 374). Cf. G-D, KSA VI, p. 9.

15. The apocryphal Canute's tidal vanity is no more an illusion than the ideality of the scheme of natural law. The idea that the king's command must be obeyed as law is the extension of royal power beyond its real range. That language can lead one to err, that the laugh is our ready response to such an error, these are significant details.

16. "Jedes Kraftcentrum—und nicht nur der Mensch—von sich aus die ganze übrige Welt construirt d.h. an seiner Kraft mißt, betastet, gestaltet ... Sie haben vergessen, diese Perspektiven-*setzende* Kraft in das 'wahre Sein' einzurechnen" (KSA XIII, p. 373). Cf. WP 567.

17. "Die *Realität* besteht exakt in dieser Partikulär-Aktion und Reaktion jedes Einzelnen gegen das Ganze" (KSA XIII, p. 371).

18. On the basis of this generalized interpretation/interaction structure, a number of writers in the German language tradition of Nietzsche-scholarship find parallels between Nietzsche and Leibniz. A review and an assessment of these parallels would require a more extensive treatment than is possible in a note to the work at hand. Authors on this topic include G. Abel, W. Gebhard, and F. Kaulbach.

19. "Ist essentiell Relations-Welt: sie hat, unter Umständen, von jedem Punkt aus ihr *verschiedenes Gesicht*: ihr Sein ist essentiell an jedem Punkt anders: sie drückt auf jeden Punkt, es widersteht ihr jeder Punkt—und diese Summirungen sind in jedem Falle gänzlich *incongruent*" (KSA XIII, p. 271).

20. See chapter 6 below.

21. GS 46.

22. "Ein Interpret, der die Ausnahmslosigkeit und Unbedingtheit in allem "Willen zur Macht" dermaassen euch vor Augen stellte, dass fast jedes Wort und selbst das Wort 'Tyrannei' schliesslich unbrauchbar oder schon als schwächende und mildernde Metapher—als zu menschlich—erschiene; und der dennoch damit endete, das Gleiche von dieser Welt zu

behaupten, was ihr behauptet, nämlich, dass sie einen 'nothwendigen' und 'berechenbaren' Verlauf habe, aber nicht, weil Gesetze in ihr herrschen, sondern weil absolut die Gesetze fehlen, und jede Macht in jedem Augenblicke ihre letzte Consequenz zieht" (JGB 22, KSA V, p. 37), BGE, p. 34.

23. "Hüten wir uns, zu sagen, dass es Gesetze in der Natur gebe. Es giebt nur Nothwendigkeiten: da ist Keiner, der befielt, Keiner, der gehörcht. Keiner, der übertritt" (FW 109, KSA III, p. 468), GS, p. 168.

24. "Gesetzt, dass auch dies nur Interpretation ist—und ihr werdet eifrig genug sein, dies einzuwenden?—nun, um so besser" (JGB 22, KSA V, p. 37), BGE 22.

25. The object of natural philosophy is nature, that of the philosophy of science, science. The effort of natural philosophy to comprehend nature is expressed in theoretic elaborations of observations or experiences within mythic, theological, metaphysical, or logical structures. In the latter case, natural philosophy may be considered "scientific." Thus one may speak of science as "natural philosophy." The philosophy of science is then the philosophy of natural philosophy. They are not identical fields, but the one includes the other: the philosophy of science questions the nature or ground of science, while natural philosophy shares the project of science and can itself be a science, although it tends to (and largely is given such an association in the literature) the transcendent rather than the empirical in its various elaborations from the pre-Socratic cosmologies, to Schelling's and Goethe's *Naturphilosophie*, to its recognizably modern manifestations in Whitehead and Lonergan.

26. FW 121, KSA III, p. 477.

27. "Uns eine Welt zurechtzumachen, bei der unsre Existenz ermöglicht wird" (KSA XII, p. 418).

28. KSA XI, pp. 209, 505; G-D "Zeichen-Convention", KSA VI, p. 76.

29. Cf. FW sections 110, 111.

30. "Das Leben ist kein Argument; unter den Bedingungen des Lebens könnte der Irrthum sein" (FW 121, KSA III, p. 478), GS, p. 177.

31. "Das Nicht-Widersprechen-können beweist ein Unvermögen, nicht eine 'Wahrheit'" (KSA XIII, p. 334).

32. Again we see that unlike the Cretan who reveals, "All Cretans are liars," Nietzsche is not affected by the fact that his own assertion is also an interpretation, just because his affirmation is not implicit, but *explicit.*

33. Heidegger, *Nietzsche I*, "Die Schaffenden, voran die neuen Philosophen, müssen nach Nietzsche Versuchende sein; sie müssen Wege gehen und Bahnen aufbrechen, mit dem Wissen, daß sie nicht *die* Wahrheit haben" (Heidegger, *Nietzsche I*, p. 37; Heidegger, *Nietzsche Volume 1*, Trans. Krell, p. 28).

34. "Philosophen des gefährlichen Vielleicht in jedem Verstand" (JGB 2, KSA V, p. 17).

35. "Ein Genie der Extreme, ein radikaler Experimentierer mit Gedanken. Er hat selbst die Figur der neuen kommenden Philosophen als die der Versucher charakterisirt, die nicht Wahrheit bringen sondern Wagniss" (Gadamer, "Drama Zarathustras," p. 2; cf. same title in *The Great Year of Zarathustra*).

36. Hence relativism is the morally current doctrine that fixes the value of viewpoints as equivalent to one another because there is no proof against the superiority of any one but only relative superiority vis-à-vis all views, meaning in the end that all should be entertained as sharing the same quasi-absolute value. As a philosophic conviction, this view might well be named "perspectivism."

37. Clark, *Nietzsche on Truth and Philosophy*.

38. See my "On Nietzsche's Concinnity," "Nietzsche's Self-Deconstruction," and "Nietzsche and the Condition of Postmodern Thought" essays.

39. In Paul de Man's account, truth is only a tropic structure. Epistemology and other truth-valuing accounts (such as the philosophy of science and science itself) are then subordinate to a "tropology."

40. Note that because Nietzsche's critique of scientific truth involves its definition as correspondence and its projection as method, this critique applies equally well to post-Kuhnian evolutionary theories of the progress of science.

41. Nietzsche's perspectivalism does not advocate an anarchy in the sense intended by Feyerabend. The latter's precise aim is to preserve the practical and efficient ideal of method by the device of abandoning its monistic image.

42. "Als 'Relativismus,' sondern als 'Perspektivismus' . . . als ein Perspektivismus, der durchaus in der Linie Leibniz-Kant liegt." Mittasch continues to note that from this same critical perspective, humanization or anthromophization is inevitable: "Alle Naturerkenntnis ist in ihrer Art und in bezug auf ihre Schrank durch die seelisch-geistige Organisation des Erkennenden Menschen bedingt und kann darum gewisse Züge des 'Anmenschlichung' nicht loswerden" (Mittasch, *Nietzsche als Naturphilosoph*, p. 47).

43. Whereas, by way of contrast with relativism, Nietzsche's perspectivalism remains intransigently individualist.

44. Tarski does not compromise with subjectivism simply by defining truth semantically and underestimating the difficulties of providing a satisfactory theory of reference.

45. Feyerabend, *Farewell to Reason*, pp. 19–89.

46. Suppe, *Semantic Conception*, p. 301.

47. Newton-Smith, *Rationality of Science*, p. 34.

48. Ibid., p. 35.

49. For a discussion of the differences between conceptual and perceptual relativism, see the introductory essay to Hollis and Lukes, *Rationality and Relativism*.

50. Newton-Smith, *Rationality of Science*, p. 35.

51. This is Feyerabend's disclaimer in *Farewell to Reason*, p. 83. As indicated in the text above, Feyerabend's forthcoming expressions should advance even more detailed qualifications.

52. Gregory, *Mind in Science*, p. 556.

53. R. Bernstein observes that "if we focus on the history of our understanding of science during the past hundred years, from Pierce to Popper, or on the development of epistemology during this period, we discover that thinkers who disagree on almost everything agree that there are no non-trivial knowledge claims that are immune from criticism . . . Absolutism is no longer a live option" (*Beyond Objectivism and Relativism*, p. 12). So much for defeasibility. But it is worth drawing the further consequence of the analysis of relativism given above, that the current broad-minded fashion of intellectual accommodation is intrinsically disingenuous. It is prudent to keep the precision and sophistication of today's analytic philosophy of science well in mind. For it is this very sophistication that makes its continued hegemony subtle and difficult to pinpoint. Thus, for example, if the analytic tradition takes over (or is induced to assume) continental perspectives or problems, a professional pluralism might be made redundant.

54. van Fraassen, *Scientific Image*, Introduction.

55. Cf. Lakatos, *Methodology of Scientific Research Programmes*.

56. Cf. here: Quine, *Word and Object*, and Davidson, "On the Very Idea of a Conceptual Scheme," respectively.

57. Putnam. See *Mind, Language, and Reality*.

58. Newton Smith writes, "all physical theories in the past have had their heyday and have been eventually rejected as false . . . There is inductive support for a pessimistic induction . . . any theory will be discovered to be false within, say 200 years of being propounded" (Newton-Smith, *Rationality of Science*, p. 14). In the literature of the philosophy of science, this projected "half-life" is known as the "Duhem Problem."

59. I. Hacking, *Representing and Intervening*, p. 23.

60. "Die Welt ist uns vielmehr noch einmal 'unendlich' geworden: insofern wir die Möglichkeit nicht abweisen können, dass sie unendliche Interpretationen in sich schliesst" (FW 374, KSA III, p. 627), GS, p. 336.

61. "Endlich erscheint uns der Horizont wieder frei, gesetzt selbst, dass er nicht hell ist, endlich dürfen unsre Schiffe wieder auslaufen, auf jede Gefahr hin auslaufen, jedes Wagniss der Erkennenden ist wieder erlaubt, das Meer, unser Meer liegt wieder offen da" (FW 343, KSA III, p. 574), GS p. 280.

62. "Kein Hirt und Eine Heerde! Jeder will das Gleiche, Jeder ist gleich" (*Zarathustra*: Vorrede 5 [KSA IV, p. 20]); TSZ, p. 46.

63. "Ein artiger Hintergedanke, . . . die pöbelmännische Feindschaft gegen alles Bevorrechtete und Selbstherrliche" (JGB 22, KSA V, p. 37), BGE, p.34.

64. "Dass unsäglich mehr daran liegt, *wie die Dinge heissen*, als was sie sind. Der Ruf, Name und Anschein, die Geltung, das übliche Maass und Gewicht eines Dinges—im Ursprunge zuallermeist ein Irrthum und ein Willkürlichkeit, den Dingen übergeworfen wie ein Kleid und seinem Wesen und selbst seiner Haut ganz fremd" (FW 38, KSA III, p. 422), GS, pp. 121–22.

65. "Was wäre das für ein Narr, der da meinte, es genüge, auf diesen Ursprung und diese Nebelhülle des Wahnes hinzuweisen, um die als wesenhaft geltende Welt, die sogennante '*Wirklichkeit*,' zu vernichten!" (KSA III, p. 422).

66. "Nur als Schaffende können wir vernichten!—Aber vergessen wir auch diess nicht: es genügt, neue Namen und Schätzungen und Wahrscheinlichkeiten zu schaffen, um auf die Länge hin neue 'Dinge zu schaffen.'" (ibid.).

67. "Das Herrenrecht, Namen zu geben" (GM I:2, KSA V, p. 260), G, p. 26.

68. The creative ascendance of *Ressentiment* through an inversion of values is effected through the appropriation of descriptive characteristics and the redefinition of other individuals. See below: chapter 5.

69. It is interesting that Nietzsche is traditionally accused of such a psychologism in his notion of the Will to Power in its cosmological expression. In addition to the analysis of the error of assigning a psychologistic valence to Nietzsche's concept given here, it should be considered that such a contention can only be presented from a position different from Nietzsche's that distinguishes between the two realms of "man and the universe." Nietzsche does not make this distinction; he holds that the distinction itself derives from an anthropocentric (subjective) psychologism to begin with.

70. "Ist durch den Glauben daran und sein Fortwachsen von Geschlecht zu Geschlecht dem Dinge allmählich gleichsam an- und eingewachsen und zu seinem Leibe selber geworden: der Schein von Anbeginn wird zuletzt fast immer zum Wesen und wirkt als Wesen!" (FW 58, KSA III, p. 422).

71. In the section on the "Free Spirit" in *Beyond Good and Evil*, Nietzsche explains the esoteric insight as the view from above. But this perspective is not commonly available.

72. "Kritiker sind Werkzeuge des Philosophen und eben darum, als Werkzeuge, noch lange nicht selbst Philosophen!" (JGB 210, KSA V, p. 144), BGE, p. 122.

73. "Trotzdem wollen sie deshalb noch nicht Kritiker heissen" (JGB 210, KSA V, p. 143), BGE, p. 122.

74. "Etwas Leichtes, Göttliches und dem Tanze, dem Übermüthe, Nächste-Verwandtes!" (JGB 213, KSA V, p. 148), BGE, p. 126.

75. "Ich bestehe darauf, dass man endlich aufhöre, die philosophischen Arbeiter und überhaupt die wissenschaftlichen Menschen mit den Philosophen zu verwechseln" (JGB 211, KSA V, p. 144), BGE, pp. 122–23.

76. "Nachdem sich die Wissenschaft mit glücklichstem Erfolge der Theologie erwehrt hat, deren 'Magd' sie zu lange war, ist sie nun in vollem Übermuthe und Unverstande darauf hin aus, der Philosophie Gesetze zu machen und ihreseits einmal den 'Herrn'—was sage ich! den Philosophen zu spielen" (JGB 209, KSA V, pp. 129–30), BGE, p. 110.

77. GM III:25. Nihilism is here analyzed as the equivalent of the "ascetic ideal."

78. "Angesichts der grossen kulturellen und zivilatorischen Errungenschaften, die wir der Rationalität verdanken, nimmt es sich seltsam heraus, die Erkenntnis als 'ein schönes Mittel zum Untergang' darstellen zu wollen. Dem rationalen Menschen der modernen Zeit, der ganz in einer positivistischen Wissenschaftlichkeit aufgeht, fällt es schwer, einen solchen Erkenntnisdefaitismus ernst zu nehmen" (Joos, *Die desperate Erkenntnis*, p. 1; cf. Nietzsche, KSA VII, p. 476).

79. Beyond Mittasch, *Nietzsche als Naturphilosoph*, reflecting the relationship between Nietzsche's thinking and the thermodynamics of J. R. Mayer, see Bauer, "Zur Genealogie von Nietzsches Kraftbegriffe," Barbera and Campioni, "Wissenschaft und Philosophie der Macht bei Nietzsche und Renan," Henke, "Nietzsches Darwinismuskritik aus der Sicht gegenwärtige Evolutionsforschung"; Müller-Lauter, "Der Organismus als inneren Kampf," and so on. For his part, Müller-Lauter follows this connection sympathetically. Yet his answer works, as the others do, on the detractor's own ground: defending Nietzsche's scientific sensitivity in spite of his (exaggerated) lack of ground-level or primary training. This focus retains the problem in a restricting transformation of its expression. Success with such an alliance then depends on having got a champion with a respectable scientific pedigree to defend Nietzsche's insights. At one extreme among English readers, Stack's *Lange and Nietzsche* presents Nietzsche as little more than Lange's better publicised epigone. Moles of course corroborates Stack's interpretation, and Clark echoes this same perspective to a lesser degree. For my own reading, the difficulty of justifying this attribution is the question (not yet posed because Lange is not yet read except in connection with Nietzsche) of Lange's (superior) competence/relevance for the problems of contemporary reflection on science.

80. Safranski, *Wieviel Wahrheit braucht der Mensch?* It is notable

that Safranski's views fail to appreciate the complexities of Nietzsche's understanding of science. But this distinction cannot be properly made until later in the present study.

81. See Löw, "Die Aktualität von Nietzsches Wissenschaftskritik," Mittasch, *Friedrich Nietzsche als Naturphilosoph;* and Rey, *Le Retour eternel et la philosophie de la physique.* In addition to R. Grimm, *Nietzsche's Theory of Knowledge,* see also Stack, Moles, Bittner, and Abel on Nietzsche's epistemology. From the side of science, as it were, I. Prigogine and I. Stengers cite Nietzsche approvingly and with some genuine understanding in *La Nouvelle Alliance.*

82. Thus, from yet another perspective one can attempt to demonstrate that Nietzsche, in fact, saw through "errors" in certain "scientific" writers, or the dying currents in the natural sciences of his day, or that he distinguished his philosophic position from his support of certain (today unacceptable) viewpoints. Bauer (reference above) illustrates how even a critical dissociation or demonstration of negative influence serves a reactive function.

83. Cf. Magnus, "Nietzsche's Existential Imperative" and Danto, *Nietzsche as Philosopher;* see too Abel, *Nietzsche,* which takes a position straddling both perspectives with respect to Nietzsche's relevance/irrelevance to "real" science.

84. "Was ist Wahrheit? Ein bewegliches Heer vom Metaphern, Metonymien, Anthropomorphismen. Kurz eine Summe von menschlichen Relationen, die poetische und rhetorisch gesteigert, ubertragen, geschmückt wurden, und die nach langem Gebrauch einem Volke fest, canonisch und verbindlich dünken: die Wahrheiten sind Illusionen, von denen man vergessen hat, dass sie welche sind, Metaphern, die abgenutzt und sinnlich kraftlos geworden sind, Münzen, die ihr Bild verloren haben und nun als Metall, nicht mehr als Münzen in Betracht kommen" (UWL, KSA I, p. 880), TL, p. 84.

85. Such a basis for interpretive assessment bears a striking similarity in the prurience of its motives to the eternally recurrent discussion of Nietzsche's sexual preference/health. And in fact, Nietzsche's interest in science was not a juvenile one. As a student it seems, Nietzsche concentrated on classical philology, a focus that so impressed the classically oriented interests of the school that even with the natural side-effects of such a concentration (inadequacy in other subjects receiving less attention from the student) they passed him despite an unbalanced record and so latterly impressed Ritschl at Leipzig that Ritschl's enthusiasm in writing on Nietzsche's behalf won him a job. Perhaps this success came too easily; certainly it did not assure Nietzsche's certification in or understanding of the philological guild. In hindsight, we can say that a professorship at Basel, achieved solely on the basis of an effusively written letter of recommendation was "too early." If so, the same hindsight would

also argue that in the light of Nietzsche's (relatively short) life, this turn of events was fortuitous. Fortuitous or not it was also fateful: Nietzsche knew better than most the consequences of Hölderlin's warning, cited for us by Martin Heidegger at the conclusion of *Introduction to Metaphysics*: "For the mindful God abhors untimely growth"; "Denn es hasset/Der sinnende Gott/Unzeitiges Wachstum." "Hölderlin, *Aus dem MotivKreis der 'Titanen'* 1V, 218, cited in Heidegger, *Einführung in die Metaphysik*, p. 157. The effects of this too-early success were momentous. Nietzsche's bitter disappointment at the judgment and perceptiveness of his colleagues and "superiors" (in both cases found lacking even where laudatory) would only grow. In any case, let it be reserved emphatically: *Nietzsche's scientific turn occupied his university and productive work-life, not his school days.*

86. The view of science as the source of a salvific technology would be an approbative view. From such a perspective, we are to regard science as positive and attempt to understand and advance its efforts rather than criticise its limits, or, still worse, set limits to it. Science (we are told), after all, "delivers the goods." This "cargo cult," or, science-as-panacea attitude characterizes even those individuals or organizations who fear the antilife or antinature consequences of science and uncontrolled technology. To meet the danger of science one prepares more research programs, more technical innovations. The solution to "science" is provided by still more science. Where an adamant philosophic line is adopted (cf. Mauer, "The Origins of Modern Technology in Millenarianism" in Durbin and Rapp, *Philosophy and Technology* [which see, pp. 253–65]) it finds a facile expression in a Manichean (negative) affirmation of this same presumption of scientific power. Scientific technology is thought either to contain the seeds of its own redemption within itself or to be capable of unleashing ultimate damnation.

87. The notion of Nietzsche's "positivistic period" (usually cited as the second stage in this metamorphic process) is as absurd as the vision of a heroic (or perhaps pathetic) finale. Such a view usually implies a telic hermeneutic that regards previous efforts in science and philosophy, including the development within those efforts, as stages advancing toward the hermeneut's (in this case, the Nietzsche interpreter's) own prescient position. The problem of Nietzsche's developmental periods will be discussed further in later chapters.

88. Müller-Lauter, "Die Organismus als innere Kampf," p. 190.

89. "Zwar ist nicht zu leugnen, daß Nietzsches physiologisch-chemischen Denkungsart nicht nur im Vokabulär, sondern auch in mancher Sachfrage eine deutliche Verbindung zur Biologie sowie zu den Naturwissenschaften seiner Zeit aufweist" (Abel, *Nietzsche*, p. 5).

90. "Nietzsche geradezu aufnahmehungrig für fremdes Naturwissen und Naturdenken war . . . dabei besaß er die Gabe des Genies, aus

einzelnen dargebotenen Farbtupfen eine Gemälde eigener Planes zu gestalten" (Mittasch, p. 44).

91. "Et Nietzsche, qui en 1881, par une intuition de génie, d'un des plus grands génies philosophiques de notre histoire humaine, voulait consacrer dix ans de sa vie à étudier les sciences de la nature pour fonder son idée du retour sur la théorie atomique." p. 309. (*Le Retour éternel de la philosophie de la physique.*) Rey concludes with a reference to the inspiration of J. Muybridge which would prove so suggestive for the futurist (and not only futurist) movement in art: "Nietzsche n'aurait aujourd'hui qu'a, enregistrer les conséquences de la théorie cinétique."

CHAPTER 3

On the Ecophysiological Ground of Knowledge: Nietzsche's Epistemology

THE QUESTION OF NIETZSCHE'S EPISTEMOLOGY: CRITIQUE AND GROUND

For Nietzsche, one can know that absolute truth is an illusion and one can know that the ideal of absolute truth is an impossible ideal and the expression of a nihilistic culture. This much can be suggested, communicated, reflected upon, and, to this extent, said to be "known." It is only problematic to ask what it is that is known in knowing this about truth or knowledge? Nietzsche's perspectivalist position commits him to the position that, apart from interpretation, there is no truth. Knowing that our knowledge is limited does not transform our knowing position into truth: "Life is the condition of knowing. Error, the condition of life and erring *indeed* in the deepest way. Knowing about error does not eliminate it! Nothing bitter there!"[1]

When Nietzsche asks in a later note, "What distinguishes true and false beliefs?"[2] he suggests that there is no heuristic device for moving from the realm of doxa to the heights of truth. Thus he denies not only Plato but the metaphysical impulse evident in Kant's own critical project. So he writes, "The πρῶτον ψεῦδος how is the fact of knowledge possible? Is knowledge a fact at all? what is knowledge? If we do not know what knowledge is, we cannot possibly answer the question whether there is knowledge. Very well! But if I do not already 'know' whether there is knowledge, whether there can be knowledge, I cannot reasonably put the question, 'What is knowledge?'"[3] Nietzsche questions our capacity for understanding, not in the sophisticated, skeptical sense addressing the competence of (conditions for) our understanding as objective knowledge (or of the knowledge claim itself), but, in our ability to say what knowledge is. This question addresses the intelligible possibility of knowledge, not its conditions. Nietzsche's critique follows Kant's. If this relationship is often overlooked in reviews of Nietzche's theory of knowledge, it is because Nietzsche does not repeat but rather exceeds Kant's critical philosophy.

Without knowledge of what knowledge is, the possibility of absolute truth becomes a metaphysical fantasy—a "fable" in Nietzsche's words. Without knowledge of knowledge one may not hold fast to the actuality of any partial truths or hope for the possibility of such truths (whether in part or in total). This last is because from Nietzsche's philosophic position the idea of a conscious lie (the other side of truth, or the singular instanciation from the multiplicity of perspective) taken in the "extra-moral" sense requires a suspension of disbelief.

Nietzsche does not advocate the "lie" in the place of "truth." Instead, what he advocates is that the usual attitude toward the lie, toward illusion, the attitude of suspicion, anger, or disbelief, be suspended. The result is not belief. Nietzsche's epistemology looks only to the truth of illusion, recoiling as he does from the illusion of truth. He is inspired by art in this: "If we had not welcomed the arts and invented this kind of cult of the untrue, then the realization of general untruth and mendaciousness that now comes to us through science—the realization that delusion and error are conditions of human knowledge and sensation—would be utterly unbearable."[4] Nietzsche's aim is always life-enhancing. But to this end, Nietzsche's epistemic question is counter-metaphysical. Why is the presumption of knowledge necessary for belief? Why is the *belief* in (true) knowledge necessary—why not a *belief* in lies?[5] Why is belief within the framework of the ideal of knowledge not much more catholic, not much more pluralist? We might well ask, as Nietzsche does, why belief takes itself so seriously—why it in fact *believes* in itself? This series of questions is significant because it leads beyond dogmatic belief (illusion of truth) to the aesthetic possibility of willed or conscious illusion (truth of illusion).

Nietzsche affirms not that human beliefs are "lies" opposed to an unique possibility of truth but rather that reality can only be known from our organic perspective. Thus, he affirms nothing but the inescapable fact of perspective. The truth beyond perspectives is the ambiguity (*Abgrund*) of existence. Unmediated knowledge of this truth would be (from an organic perspective) fatally disruptive. We will examine this connection in its life-affirmative aesthetic emphasis in a later chapter. Yet, when Nietzsche speaks against the illusion of truth, he is not battling the deconstructionist's imaginary dragon of classical metaphysics (this is a different project). Instead, he is challenging the belief in the ultimate power of knowledge (and thus the confidence in its truth) that characterizes the contemporary technological, scientific era.

Nietzsche claims that there is no knowledge that is true because human beings are fundamentally unequipped for knowledge.[6] If Nietzsche's philosophic standpoint thus opposes the tenability of the epistemic enterprise altogether, what can be meant by speaking of Nietzsche's epistemology?[7] One of the few commentators who address the issue, Jochen Kirchhoff, prefaces his own venture not so much with an ironic demurral as with the qualifying deflation of any attribution of epistemological intent on Nietzsche's part: "Not only did Nietzsche fail to offer a systematic, fully articulated, and self-consistent theory of knowledge, but his philosophical self-understanding radically contradicted such a project."[8] And in *Nietzsche on Truth and Philosophy*, Maudemarie Clark finds that the problem as such defies preliminary conceptualization. Yet Clark continues to argue that so far from obviating her own project, this vagueness, "namely that Nietzsche's claims about truth seem hopelessly confused and contradictory,"[9] is instructive because it points to the ground of a resolution. In Clark's view, Nietzsche rejects only "the existence of ontological truth—correspondence to the thing-in-itself—but not truth itself."[10] Staked out for the benefit of an analytic tradition run desperately dry on theories of truth, conceptual schemes, networks, and paradigms, Clark's claim is that Nietzsche has no "theory of knowledge." Against Clark, I argue that an assessment of Nietzsche's theory of knowledge can only be made in the context of Nietzsche's ecophysiological reflections on the nature of the human world-interpretive dynamic.[11]

The rhetorical efficiency of Clark's argument is worth noting. Armed with a kind of truth, now reduced to the level of an analytic nicety, Clark can make a cognitive claim for Nietzsche's own philosophy. Of course such a strategy does deny the ultimate coherence of Nietzsche's various representations of truth. Hence Clark's extreme contention repeated once again is that—and she takes pains to underline that she *does* mean it exactly the way it sounds—"what appears as radical in Nietzsche's position on truth is actually mistaken or confused."[12]

Thus, just as Clark suggests, Nietzsche never denies the virtuality or even actual, effective currency of knowledge "claims." Neither does Nietzsche refrain from offering an account of the formative genealogical interest of knowledge claims. Instead, and Clark overlooks the significance of this extended distinction, Nietzsche criticizes the notion of truth (correctness) altogether, and he opposes it precisely as a *claim*, just as he opposes the general concept of right thereby entailed. In this way, Nietzsche raises the larger (which Clark above dubbed the more *radical*, "mistaken or confused") epistemic question of truth as such. In *this radical* context, Nietzsche's more commonly acknowledged effort to consider morality as a problem comprises in its scope the possibility of knowledge (or science) as a problem.

Beyond the above comments, I will not offer further criticism of Clark's project. Instead, I wish to show that Nietzsche's proper concern with truth is a concern with the ground of knowledge. And for all Nietzsche's radicality, it is important to emphasise that Nietzsche is writing within the canonic German philosophical tradition when he writes on the question of truth. The question in that tradition is the question of ground. And this question is Nietzsche's question. But what makes Nietzsche unique, as we have seen, is his recognition that a *problem* (whether that be the question of morality, of science, or, as in this case, of truth) to be regarded *as* a problem first must be recognized as such and that such a recognition is not possible from the basis or ground of the problem itself. This insight into the traditional problem of ground constitutes Nietzsche's Copernican revolution.

For this reason, Nietzsche's epistemic position is both fundamental for his own philosophic project and generally significant. We are returned to what Nietzsche would call the "crucial nexus" or foundational Schopenhauerian kernel: the unbearable, uncrackable *nut* of the problem of the problem, that is, the problem of ground. Nietzsche contends that knowledge—even as a limited perspective—is impossible. Is this position philosophy? is it an epistemic perspective of any kind? is it not in fact much rather "actually mistaken and confused"—as Clark has it and as so many other readers, for their part sternly unsympathetic to Nietzsche's perspective, are likewise securely convinced?

Maintaining the impossibility of knowledge as knowledge, Nietzsche contradicts himself: he *holds*, he *knows* that there is no knowledge. We must beware of thinking that this contradiction is without sense, but this is possible only if we are prepared to bear the force of its implications, its consequential weight, its resonance. In order to understand Nietzsche's claim that there is no (true) knowledge, one must be prepared to follow the idea in its complete and so in its discordant and full extension. The attention to the thinker and thinking in what is said and what is unsaid, which hermeneutic reserve is prescribed by Heidegger,[13] is the interpretive effort that seeks not to expose oversights—this is both too easy and too limited a task—but to recover the possibilities of the original thought for the further project of thinking. This requires that we understand the task of thinking as it is lived and actual among thinkers. The thinking of philosophy, even philosophic interpretation or hermeneutics, is more than just an exegetical exercise. Hence, when a thinker employs contradictions and when the same thinker declares the judgment of contradiction itself an organic, nonessential limitation of thought, the destructive attitude that adverts to the logical rule of noncontradiction is not merely trivial but a failure of basic understanding. Following a hermeneutic reflection in Heidegger's sense, rethinking the thinker's original thought demands more than a hypercritical challenge to the thinker's putative good will and logical sportsmanship; it must be prepared to admit the critical claim of Nietzsche's thought.[14]

Yet it must be noted that as radical as the discussion of Nietzsche's *epistemology* in the present and following chapters is, a scholarly review of Nietzsche's theory of knowledge and truth is not without important precedent. More clearly than Walter Kaufmann in this regard, R. J. Hollingdale, Nietzsche's translator and commentator, introduces Nietzsche as "a philosopher . . . whose instinctive contribution to European thought was to recognize and face the consequences of a radical change in Western man's apprehension of the attitude toward 'truth.'"[15] For Hollingdale, even if Nietzsche "never arrived at a formal 'theory of knowledge' his thinking on this subject is unmistakeable and can be detected at every stage of his development."[16]

In Rüdiger Grimm's *Nietzsche's Theory of Knowledge*, which today remains the most focused and developed English-language commentary on Nietzsche's theory of knowledge, Nietzsche's epistemic position is explained in the expressive terms of the Will to Power. Advancing beyond the traditional qualitative reading of the Will to Power, Grimm presents a *quantitative* exposition that

has the particular material virtue of representing Nietzsche otherwise than as metaphysical theorist. It is, however, troublesome
that despite its considerable didactic advantages, Grimm's expression of the Will to Power as the sum of what might be called
"punctuated form" (for Grimm, "quanta") presents a substanti-
fying image that fails to emphasise the ecophysiological
active/reactive dynamic in Nietzsche's worldview. The so-called
Will to Power demands a quantitative representation that is
thought in qualitative terms. For Nietzsche, the quantity of power
is its interpretative efficacy or evaluative *force*. As the quantitative
dynamic of the Will to Power is reciprocally interpretive or qualitative, the world of nature (*physis*) is amenable to (and indeed in
Nietzsche's radical version of the world as will to power-and-
nothing-besides, the world is itself an expression of) multiple
interpretations. While Grimm recognizes the dynamic nature of
the Will to Power as the fundament of all being, he fails to think
beyond this to a nuanced affirmation of the becoming character
of the world and the interpretational manifestation of power-perspectives in this process. To be true to Nietzsche's thought, a *ecophysiological* focus on the interpretive becoming (*physis*: understood as self-generating, self-manifesting) of the world *and* the
body (the last understood as the expenditure of organic force
expressing a particular perspective) is needed. Here, in a parallel
that follows Spinoza as much as Leibniz, *being as such* (understood as the concrescence of manifold moments in the achievement and preservation of each individual perspective as an *ecological*, world-building force) *is exactly epiphenomenal to the
interrelational dynamic of every world-interpretive event.*

In spite of the necessity of a refinement of this kind, already
routine in German and French Nietzsche scholarship following
Wolfgang Müller-Läuter and Gilles Deleuze, Grimm's basic reading is sound. Moreover he offers the special interpretive advantage of tracing the origin of Nietzsche's epistemology to its basis
in Nietzsche's considerable academic and theoretical training in
philology. And philology, in the broad German sense, represents
Nietzsche's "scientific" background. As Grimm observes, if "we
substitute 'world' for 'text,' the epistemological relevance . . .
becomes quite obvious."[17] Thus Grimm's outline of Nietzsche's
philological program can be constructively mined as an extended
metaphor for Nietzsche's epistemological interests.

In *Truth and Value*, J. Wilcox offers an analytic or Anglo-

American, sympathetic, noncontinental approach to Nietzsche's epistemology. In this sphere, Wilcox's work provides an important access to the conception of Nietzsche as a thinker on knowledge—if not precisely on the exigent (where Nietzsche has an offensive name for this kind of exigence) terms of the Anglo-American, analytic philosophic definition of any kind of proper epistemologist. In a move I have already characterized as ingenious, Wilcox saves Nietzsche (as non-cognitivist) for the sphere of value but it is also true that as Wilcox reserves (cognitive) cogency to the sphere of truth, he thereby excludes Nietzsche from that domain.

Apart from the irreducible and problematic limitations of stylistic congeniality, there is a long tradition of analytic-style writers on the topic of Nietzsche's epistemology. Hence there are influences of other authors such as Kaufmann, Arthur Danto, and, further afield, Bernd Magnus to be found in Wilcox, Grimm, Clark, and most recent authors in English-language Nietzsche scholarship. Not surprisingly, nonanalytic German and French scholarship on the topic has a considerably broader range and manifests a number of different directions.[18] I shall below make special reference to Jean Granier's important work on the problem of truth in Nietzsche's philosophy.[19]

THE KNOWER AND THE KNOWN

The connection between the metaphysical ideal and the scientific ideal of a mathematically expressible absolute truth in terms of a Socratic-Platonic logos is evident. This epistemic connection dictates a unitary and positive correspondence (the perfect or suitable adequation or ὁμοίωσις) between the project of knowing and its true object. In the metaphysical tradition, this "truth" is what justifies or indeed makes an articulation (judgment) of what is "Real" possible. Such a capacity for judgment is based upon a correspondence between the human knower and its object. It is this preestablished fit that founds the efficacy and validity of knowledge. As we have seen, Nietzsche challenges this neat arrangement, together with the theistic connection so easily made between such a preordained order and a divine instigator/guarantor; there is no wholly adequate (in the sense of ὁμοίωσις distinguished above) referential connection for Nietzsche beyond perspectival, subjective "adequacy." The Cartesian turn, grounding the certitude of knowledge in the subject, merely represents the ideal connection in the subjective fantasy of self-knowledge. Thus where Nietzsche asks: "Are designations congruent with things? Is language the adequate expression of all realities?"[20] he remains true to his philosophic mandate—challenging the meaning of any posited correspondence or relation between knower and known. In this, as we have seen, he suggests the possibility of an arbitrary designation, a selective reference, an aesthetic (if, let it be added, nonartistic as well, when self-deceptive or dissimulative) perspective accounting for the world we know.[21]

For Nietzsche, our world is described without metaphysical remainder by the term *appearance,* that is, if one term of a classic opposition may be taken on its own. The plausibility of Nietzsche's perspectivalist possibility is ecophysiological, that is, determined by the *physiological* or physical constitution of the interpreting perspective and its relative (*ecological*) position in the world (or cosmos). This means that he is concerned to work out the critical consequences of error or illusion as the ground of any truth. Thus construed, the path Nietzsche takes is not difficult to follow. His project is physiologically genealogical: he expresses the genesis of logic according to its ecophysiological exigence or importance for life (ecological economy). What grounds the establishment of our own faith in the correct fit of our knowledge and the Real that is represented as known reality is no more than our own ecophysiological needs and presumptions:

> Fundamentally, what the investigator of such truths is seeking is only the metamorphosis of the world into man . . . His method is to treat man as the measure of all things, but in this he proceeds from the error of believing that he has these things before him as pure objects. He thus forgets that the original perceptual metaphors are metaphors and takes them for the things themselves.[22]

This interpretive, organic perspective is prerational. Our need to function in a world (our ecophysiological circumstance) requires that it be falsified (that is, made over, represented, comprehended—choose your metaphor!) or rationalized: "We have arranged for ourselves a world in which we can live."[23] Such representations, as human projections, are necessarily anthropomorphic achievements.

The ideal of objective truth beyond the adequacy of interests represented according to an originating perspective presupposes a fantastic selective affinity between the human knower and its object. The ideal projection of such an affinity is a product of human interest: in itself this affinity is simply not given. Empirically and ideally oriented epistemic reflections provide no argument to such a connection but proceed rather from such an assumption to the validity and significance of what is thus known. The connection between knower and known is axiomatic in the sense that it is merely assumed on faith. So Nietzsche continues to describe our arrangement of the world for the sake of sustaining a life in it: "by positing bodies, lines, planes, causes and effects, motion and rest, form and content; without these articles of faith nobody now could endure life."[24] The axiomatic connection between knower and known is either guaranteed through the presumption of a metaphysical or divine order or, in a less romantic, more nihilistic yet as presumptuous a fashion, it is simply referred to and presumed upon the basis of its efficiency, or usefulness for life.

This critical focus is not unique to Nietzsche. Following Hume, Kant's Copernican turn in philosophy and through science, for example, challenges the presumption of an adequate correspondence between knower and known, while taking the fact of such a functional correspondence for granted. But for Nietzsche this connection is fictional: "Or, more clearly, crudely and basically: synthetic judgments a priori should not 'be possible' at all: we have no right to them, in our mouths they are nothing but false judgments."[25] It can be observed that Heisenberg's extension of the Kantian limitation of the conditions for the possibility of scientific knowledge corroborates Nietzsche's perspective. Today's scientific position affirms its own limits: we cannot know the object of scientific study without interference. Likewise, we cannot know our process of knowing. Yet this is not to suggest Nietzsche's philosophy of science as presaging or paralleling a quantum mechanical vision, as a corrective paradigm or model for the philosophy of science. This interpretation should be rejected, although both Grimm and Kirchhoff (among others, such as Günter Abel most subtly as well as more recently at greater length and rather a good deal more ingenuously Alistair Moles) hint at such a possibility. It is significant that these last parenthetically named commentators also conceive or articulate the value of Nietzsche's epistemology only through the lens of the teaching of Eternal

Recurrence. Thus Kirchoff and Grimm as well as Abel and Moles regard the Eternal Return as being the only point of departure for a consideration of the "scientific" value of Nietzsche's critique of epistemology.[26] This enthusiasm mistakes (at the very least) the scientific intention of quantum mechanical theory for its more popular religious or more recent new-age expressions. Where quantum mechanics works as a "mechanic" producing exact knowledge of one component in a complex of momentum/position, then approximating the probable value of the nonprecised component, it is firmly on the trail of invariants and efficient calculations. Talk of indeterminacy, as talk of fractals and "chaos" theory, in this context retains the sober interest of (predictive, power-oriented) determination.[27]

Nietzsche challenges (1) the possibility of subjective knowledge, (2) the possibility of objective knowledge; and, further (rendering the subject/object distinction pointless), 3), the possibility of reflexive knowledge as such. This tripartite challenge to the possibility of knowledge at all illuminates the (rhetorical) suggestion Bernd Magnus offers when he describes Nietzsche's perspectivalism as a "rhetorical device" writing, "A theory of knowledge is not something Nietzsche has, it is what he parodies."[28] In its broadest, most generous interpretation, this suggests that Nietzsche's critical epistemic interest is *therapeutic* or prophylactic. Playing his own position against traditional asseverations, Nietzsche criticizes the arbitrary convention, "Was sich beweisen läßt ist wahr,"[29] as tautologous and thus absurd if taken to be a foundational truth. Accordingly, he offers a helpful roster of the practical functionalization that counts as knowledge for us, in sum "that there are enduring things; that there are equal things; that there are things, substances, bodies; that a thing is what it appears to be; that what is good for me is also good in itself."[30] A critical, reflective, or, indeed, scientific perspective is a latecomer in this ecophysiological interpretive sphere. But that does not mean that the "pure desire for truth" operates in another domain than that constituted by the organic/inorganic perspectival interests of the Will to Power. The difference between organic perspectives and pure epistemic (scientific) interest is merely a matter of finesse. In this sense, Nietzsche speaks of "truth" apart from presumptions and illusions and, ultimately, a belief in an absolute: as "critical" truth, as "the weakest form of knowledge." Such a weak form of knowledge must be evaluated in terms of the critical honesty

(*Redlichkeit*) Nietzsche reserves for his own loyalty to the metamorphic complexity of events and ambiguous (*metonymic*) differences between things.[31]

However closer science may bring us to this reflective perspective, it never engages this perspective except as a limit. Nietzsche's vision of truth as the revelation of the "frightful and questionable character of existence," the life-challenging chaos of nature, remains what the French psychoanalyst Jacques Lacan would call an "impossible" perspective just because it is a perspective abutting the register of the Real. Ordinary notions of true and false function as logical norms within the terms of scientific rationality. As interpretational perspectives within the ecophysiological sphere, they are erroneous precepts. As Nietzsche explains, "Such erroneous articles of faith, which were continually inherited, until they became almost part of the basic endowment of the species . . . even in the realm of knowledge, these propositions became the norms according to which 'true' and 'untrue' were determined."[32] For Nietzsche, the analytic motto (as it might be expressed from Frege to Wittgenstein) "what is true is what can be proven," which Nietzsche expresses as "What can be proven is true," is "an arbitrary determination of the concept 'true' *cannot be proven*. It is a simple 'this *should* count as true, should be called 'true' . . . *Accordingly*, 'what can be proven, is true' presumes *truths as given in advance.*'[33] Logical truths reflect their ecophysiological foundation: their demonstrations are those practically operative for common sense "because what can be proven appeals to the most universal in all minds (to logic): wherefore it is naturally nothing more than a measure of utility for the interest of the many."[34]

THE PROBLEM OF KNOWLEDGE
IN ITS ECOPHYSIOLOGICAL GROUND

In the preceeding chapter's discussion of Nietzsche's perspectivalism it was seen that Nietzsche's criticism of the possibility of epistemology does not prohibit one from considering this criticism as an epistemic perspective on its own. For Nietzsche, declaiming "Against the prejudice of science" could pronounce that "The biggest fable of all is that of knowledge."[35] Nietzsche thereby means to remind us that all talk of knowledge, from Plato through Kant, presumes knowledge in advance (as a kind of μάθησις). In particular, and this is what is important for Nietzsche, what is presumed is that we *already know* what counts as knowledge. We already possess a standard for knowledge. That is, we are capable of recognizing knowledge. This presumption means that we can assume both internal and external epistemic positions: internally knowing ourselves as knowers (on the knowing subjective side) and externally knowing what thus counts as knowledge of things (knowing knowledge objectively). But Nietzsche challenges:

> One would have to *know* what being *is*, in order to *decide* whether this or that is real (e.g., the "facts of consciousness"); in the same way, what *certainty* is, what *knowledge* is, and the like.—But since we do *not* know this, a critique of the faculty of knowledge is senseless: how should a tool be able to criticize itself when it can use only *itself* for critique? It cannot even define itself![36]

The presumption of a transcendental epistemic locus contradicts what, for Nietzsche, is the essentially *perspectival* nature of knowledge. Thus, although the requirements for transcendent knowledge are in opposition to the ecophysiological condition of human knowing, human knowledge is historically expressed as such a subjective transcendence. The critical impossibility of knowledge claims, along with their functional incorrigibility, presents a condition of paradox that is humanly unavoidable. Indeed, for Nietzsche, the basis of the belief in conceptual being and perceptual reality, that is, intellectual certainty and empirical knowledge, is to be found with its roots in organic nature. It is important to define this "nature" as an "animal consciousness." This animal consciousness presents an insurmountable horizon, or *Gefängniss* as Nietzsche early on expressed these constraints in *Daybreak*:

> My eyes, however strong or weak they may be, can see only a certain distance, and it is within the space encountered by this distance that I live and move, the line of this horizon constitutes my immediate fate, in great things and small, from which I cannot escape. Around every being there is described a similar concentric circle that has a midpoint and is peculiar to him. Our ears enclose us within a comparable circle, and so does our sense of touch.[37]

For Nietzsche, the nature of our senses is such that it is impossible to reconcile their superficiality with "an honest and pure desire for truth . . . [the] eye merely glides around the surface of things and sees 'forms'; their perception leads nowhere to the truth, but is satisfied with receiving stimuli."[38] Thus,

> The habits of our senses have woven us into lies and deception of sensation: these are the basis of all our judgments and "knowledge"—there is absolutely no escape, no backway or bypath into the real world! We sit within our net, we spiders, and whatever we catch in it, we can catch nothing at all except that which allows itself to be caught in precisely *our* net.[39]

Nietzsche denies the title of "knowledge" in the traditional sense to organic perceptions because they are interpretations, that is, expositions from the vantage—for the advantage—of the perceiving being. Truth as such, that is, truth considered in its traditional sense as the neutral, innocent, "adequate expression of all realities"[40] is not at stake. The only effective truth desired by the human being (and the only truth for an animal) must reflect "the pleasant, life-preserving consequences of truth. He is indifferent to pure, inconsequential knowledge; toward those truths which are possibly damaging and destructive, he is in fact hostile."[41] Not truth but life-preserving errors, consequential truths, serve as the original basis of (scientific) knowledge.

Yet just as Nietzsche concedes that our truths are consequently, effectively "adequate" to our interests, he does affirm a certain, ironic *adequatio*, if this affirmation also sacrifices the properly scientific ideal of (Platonic) ὁμοίωσις and of μάθησις as the ideal of complete knowledge. A formal correspondence is impossible: the ideal of truth required would have to be as multifarious as reality. The understanding of truth as conformity, where what is known conforms to its object, requires that things resemble one another. Nietzsche locates this similarity in our own perspective, rather than in any objective character: "We should speak of similar rather than 'equal' qualities. And 'similar' for us. Nothing happens twice: the oxygen atom is without an equivalent. The truth is that the assumption that there are countless equal cases *suffices* for *us*."[42] Although our truths are matched to our interests, they are not adequate (there is no ὁμοίωσις in truth) to "all realities"—they are *our* truths. For Nietzsche, "similarity is not a degree of equality: but something completely different from equivalence [*Gleichen*]"[43] The distinction Nietzsche makes here is precise one.

Because Nietzsche's concept of 'truth' is not a bridge to a beyond, it embraces without annulling the small inconsistencies, differences, disconnections, and so the ambiguities that make up and go beyond the appearances of a phenomenal world. By claiming that there is no (ultimate) knowledge and no (absolute or ideal) truth, Nietzsche does not oppose the results of scientific observation. Instead he criticizes the presuppositions that limit such observations. Above all, he challenges the presumption that rules regularity, that is, our scientific presumption of law. This presumption takes a limited "adequacy" as equivalent to the "adequate expression of all realities." One assumes constancy and identity by suppressing—forgetting but not explaining— "minor" differences. For Nietzsche, "Overlooking the individual and the real gives us the concept, as it gives us the form, whereas nature is acquainted with no forms and no concepts, hence with no species, but only with an x that is inaccessible and undefinable for us."[44] Yet, for his part, and this particular distinction is important, Nietzsche does not posit an unknowable thing in itself as reality, as some have concluded.

Nietzsche is adamant, coming again and again to the assertion against Socratic prudence, that to assert that one does not know is to claim too much.[45] Although he takes his epistemological starting point from Kant, Nietzsche does not share Kant's confidence (as Schopenhauer did) in the thing in itself, apart from the knowable, phenomenal world. Yet this move does not place the thing in itself in the phenomenal world. Because Nietzsche challenges both the traditional metaphysical ideal and the skeptical-cum-positivist counter-ideal, if we persist in viewing his philosophy as the mere articulation of fundamentally metaphysical concepts, we will end by losing our way in tracing his thought. The concept of a thing in itself (whether knowable or unknowable) lacks meaning for Nietzsche. Neither the true world nor the apparent world has any viability as an alternative: both are codeterminate oppositions.

Because all human knowledge is perspectival, human beings are part of the perspectival nature of the "relational" world.[46] Apart from this relational chain and web, there is nothing.[47] The point of Nietzsche's argument not merely repudiates the knowledge of the "true" in human knowing but challenges our (sophisticated and therapeutic) assertions of unknowing, as equally presumptuous delusions of an "unknowing knowing": "For even our contrast between individual and species is something anthropomorphic and does not originate in the essence of things; although we should not presume to say that this contrast does not correspond to the essence of things. For that would be a dogmatic assertion, and, as such, just as unprovable as its opposite."[48] In a later note complementing the important and brief "History of an Illusion" in *Twilight of the Idols,* Nietzsche writes that the distinction between the (true) nature of things (the real world) and the phenomenal world is based upon an absurd presumption:

> In order to make such a distinction, one must conceive our intellect as equipped with a contradictory character: on the one hand, oriented to perspectival seeing, as it must be in order that beings such as ourselves can preserve themselves, and simultaneously on the other hand constituted by a capacity to see precisely this perspectival seeing as perspectival, to grasp the appearance as appearance.[49]

The question of fundamental error is not the relevant emphasis here: we recall that for Nietzsche, error is necessary for life. An ecophysiological perspective does not pronounce natural judgments to be either best or truest (in a pragmatic sense) judgments. Instead, an ecological vision of knowledge requires an account of knowledge in its relation to error:

> Those, for example, who did not know how to find often enough what is "equal" as regards both nourishment and hostile animals—those, in other words, who subsumed things too slowly and cautiously—were favoured with a lesser probability of survival than those who guessed immediately upon encountering similar instances that they must be equal. . . . The beings that did not see so precisely had an advantage over those that saw everything "in flux."[50]

The inspiration for Nietzsche's ultimate preference for "exact perception" (*genau sehenden*) is doubtless due to his philological background.[51] But exact perception (seeing everything in flux, as in a chaos of constant becoming—*again* Nietzsche's view of nature as protean chaos must disallow the Platonic ideal of ὁμοίωσις in perception!) is hardly a general characteristic of our organic perspective, nor indeed does exactness describe human conception. In an earlier context, Nietzsche had written "our thinking is superficial and content with the surface; indeed it is unaware of it."[52] Any effort to criticize the superficiality of our knowing process is confounded by this structure:[53] "When we try to examine the mirror in itself we discover in the end nothing but things in it. If we want to grasp these things we get hold once again of nothing but the mirror. —This, in the most general terms, is the history of knowledge."[54] The perspectival limitation is thus intractably ecophysiological, it is not a result of scientific or even sociologically determinist "knowledge interests." The reality reported by our senses structures what we know and the very same filtering event that characterizes perception streamlines conception. We are now prepared to begin to consider Nietzsche's own expression of his perspectival phenomenalism:

> This is the essence of phenomenalism and perspectivalism as I understand them: Owing to the nature of *animal consciousness*, the world of which we can become conscious is only a surface and sign-world, a world that is made common and meaner; whatever becomes conscious *becomes* by the same token shallow, thin, relatively stupid, general, sign, herd signal; all becoming conscious involves a great and thorough corruption, falsification, reduction to superficialities . . . You will guess that it is not the opposition of subject and object that concerns me here: This distinction I leave to the epistemologists who have become entangled in the snares of grammar (the metaphysics of the people). It is even less the opposition of "thing-in-itself" and appearance; for we do not "know" nearly enough to be entitled to any such distinction. We simply lack any organ for knowledge, for the "truth."[55]

Nietzsche's interest lends itself neither to relativistic skepticism nor to a transcendental metaphysics. So he introduces the *Nachlaß* note partially cited above (paralleling his emphases in *Beyond Good and Evil*[56] and *Twilight of the Idols*):

> It really wouldn't disturb me if today someone with the modesty of philosophical skepticism or with religious conviction declares: "The essence of things is unknown to me" or another, rasher sort, one who hasn't yet learned enough critique and mistrust: "To a great extent, the essence of things is unknown to me." I maintain, in opposition to both, that they presume to know, under the circumstances, they fancy themselves as knowing, much too much by far.[57]

The problem, he continues to say, is a matter of a contradictory double vision—regarded from the vantage of a sophisticated and let us not forget to add critical and even Kant-inspired) perspective:

That is to say: outfitted with a belief in "reality," as if it were the only possible belief, and with the additional insight concerning this belief as no more than a perspectival stricture regarded with reference to a true reality. However belief conceived on the basis of such an insight no longer remains belief but is rather extinguished as such. In short, we may not think our intellect as thus contradictory: that is as a belief that at the same time is knowledge concerning belief as belief.[58]

What this means, of course, conceptually rendered in accordance with the simple exigences of analytic English syntax for the taste of a positivistic and journalistic style of reading, is that Nietzsche's own perspectivalism does not itself lay claim to an ultimate representation of reality. What is more, this means that Nietzsche would have little patience for the current (analytic) distinction between knowledge and so-called knowledge *claims*. The whole notion of a *claim* to knowledge (i.e., belief vs. "knowledge concerning belief as belief" [*Glauben als Glauben*] disintegrates in this context.[59]

It is in the spirit of this same critical limitation concerning belief and knowledge "claims" then that Nietzshche declares,

It will do to consider science as an attempt to humanize things as faithfully as possible; as we describe things and their one-after-another, we learn how to describe ourselves more and more precisely. Cause and effect: such a duality probably never exists; in truth we are confronted by a continuum . . . The suddenness with which many effects stand out misleads us; actually these are sudden only for us.[60]

The discriminative or evaluative project of scientific or philosophic knowledge (whether its claims are positive or negative) depends upon foundational knowledge for its very possibility. Nietzsche supposes that we necessarily lack yet assume such foundational knowledge. This assumption is a presumption in the most prosaic and general sense. To have knowledge at all, one must transgress the limits of perception in abstractive generalization, while perception itself, for Nietzsche, is nothing but a filtering or simplificational activity. The knowledge that is taken to be true is delusionary, at least twice removed by metaphoric distance from its "object." Nietzsche claims that, given the essential falsification of perspective, that is, the selective character of all perception, it is difficult to imagine, much less to admit "that the insect or the bird perceives a completely different world than man does, and that the question which perception of the world is more correct is completely senseless for this would have to have decided in advance in accordance with the criterion of the *right perception*, which means in accordance with a criterion *which is not available*."[61] The failing standard is a god-like perspective (i.e., for Nietzsche, in the same theological parallel, an omniscient, transcendent knowledge of true and false). In its stead, we have merely our ecophysiological interests: our experience as perspectival beings. Judgment is possible—or, better, unavoidable—on this basis, but such judgment is ineluctably interpretive. What we lack is ultimate knowledge. We do not know "what Being is . . . [we do not know] what knowledge is."

THE EMPIRICAL BASIS OF
TRANSCENDENT KNOWLEDGE

Against all metaphysical and correlated positivist constructions Nietzsche avers that we lack the needed (fore)knowledge of what counts as knowledge to begin with and that nevertheless, we posit it as given, demarcating what we name "knowledge" from error or illusion. This positioning of potentially totalizing or absolute, that is *true*, knowledge on the basis of our organic human perspective, proposing a univocal standard of what may count as knowledge, pretends to knowledge. Such pretensions have various expressions: all originate in organic error (Will to Power as interpretation), yet not all finish as classically metaphysical. The last qualifying consideration is an important one. For Nietzsche, we do have something very like transcendent knowledge which counts for us in everyday experience. The very "fact"—our language betrays us here—of experience provides a satisfying epistemic recourse in the current environment of scientific rationality.

At once, two points. First, it should be evident that when Nietzsche speaks of experience he is neither Lockean nor Hobbesian: he means not any immediate apprehension of a given world as an experience of what is true (this is the basis for traditional empirical or foundationalist epistemologies) but rather an interpretive projection of the Will to Power, understood in the context of organic error as a physiological reflex (i.e., ecophysiological). Second, following Hume's emphasis, experience for Nietzsche is anything to which we are already accustomed. So far from being a given of any kind, experience is contingent upon a preexisting structure or template or filtering apparatus or mapping mechanism.[62]

The structural metaphor is important. After Kant, nothing "experienced" *can* be experienced unless it clears or conforms to the conditions necessary for any experience at all. Only in this pre-formed sense does experience serve as the basis for further experience and so understanding: "What is 'knowing'? Referring something alien to the familiar, to the comfortable."[63] For Nietzsche the empirical reference in terms of the (already experienced) serves as a solid ground only in the limiting contextual sense of back- and foreground. Experience founds nothing—including what has already been experienced. Experience is nothing but interpretation. But if experience is only back- and foreground and not foundational knowledge, we need to ask what grounds the common interpretation of the "facts" of experience as ultimate facts.

Nietzsche holds that, as organic beings, we perceive on the basis of evaluative perspectives. In this, our structured/structuring expression of the world is not a given truth; it is much rather taken to be truth. C. S. Peirce once described the psychological motivation of inquiry as the desire to resolve doubt into (quiescent) belief; construed abductively, for Peirce, the successful explanatory procedure in this interest is thought to be true as well as comforting.[64] Nietzsche, however, refuses the possibility of any nonerring truth: the organic perspective is necessarily falsifying in itself. The illusory inventiveness of interpretation, as the character of the Will to Power, extends beyond human beings:

> The entirety of the organic world is the intertwining of entities [*Wesen*] with little worlds about them, where their energy, their desires, their habits are installed in the experiences external to them as their *outer world* . . . Naturally, of themselves they only have the same kind of false, invented, simplified representation.[65]

Interpretive perception is a projection, a construal from a particular perspective outwards, which then embraces the world: "The organic process constantly presupposes incessant interpretations."[66] This creative structural projection is the expression of the Will to Power.

This interpretative virtuosity is more than a human excellence, but it belongs, so says Nietzsche, to the essence of nature or what is Real even beyond the organic. The interpretational fiction functioning in perception goes backwards as well as forwards. Thus interpretation is fundamentally interactive. It is only on the basis of this same organic interpretive perspective that we can invoke the background world of "empirical experience" together with our foreground rationality to construct a unified science. This interpretive basis means that however great, however complex, however intricately reticulated it may be and be said to be, what the early Nietzsche named the "columbarium" of science in his essay "Beyond Truth and Lie" remains essentially perspectival and in this manner no more than a species-projective falsification. The perceptual organization of organic being (interpretive projection) establishes all higher knowledge, from perceptual consciousness to self-consciousness. Thus Nietzsche declares in a *Nachlaß* note: "It is our needs that interpret the world." Emphasizing the interpretive expression of the Will to Power, Nietzsche explains that "Every drive is a kind of lust to rule; each one has its perspective that it would like to compel all the other drives to accept as a norm."[67] It is this interest in overpowering that inspires the presumption of conceptual and perceptual knowledge—that is, what Nietzsche calls "perspectival estimates"—as proximate truth.

But this ideal in scientific understanding, presenting an absolute approximable through perceptual experience and conceptual reflection (back- and foreground perspective estimates) is only an illusion of truth (if truth is understood as "the adequate expression of all realities"). As human beings we remain within our own interpretive horizon. We do not express reality—much less anything like "all realities"—in a way that is "adequate" to any perception but our own. Nietzsche explains our organically inspired knowledge from experience on the basis of its very banality (in the sense of the Greek efficient, calculating, mechanical βάναυσία) in a way which recalls Peirce's analysis: "Regularity lulls the questioning (i.e, fear-inspiring) instincts to sleep."[68] For Nietzsche the banality of the procedure is a matter of adaptive technique and is thus deliberate, even if not a matter of individual or subjective decision. The sense of the Latin *ars sellularia* clarifies the connection of the banausic to the technical, and thereby to art, while retaining a reference to the effects of mechanical art or handicraft. What is important here is a representational technique which names the unknown or the unfamiliar in the guise of (association with) the known or the familiar. Nietzsche explains the process in terms of the physiology and psychology of adaptation (accomodation):

> First ground rule: what we have gotten *used* to no longer counts for us as a mystery, as a problem . . . everything that occurs *regularly* no longer appears questionable to us. Consequently, the first instinct of the knower is to *search for rules*, although naturally enough with the confirmation of a rule nothing is as yet "known"!—From this we get the superstition of the physicists: what can be preserved, that is, where the regularity of appearances permits the application of reductive formulas, has thereby been rendered *known*. They feel "secure": but behind this intellectual security stands the calming of frightfulness: *they want rules* because these strip the world of its fearsomeness.[69]

This "regularity," this *making* regular, is a regulative fiction in a literal and practical sense. Perceptual and conceptual, structural and focal, this fiction of regularity protects us from the unknowable, the "frightful and questionable character of existence." The rule of the world, the worldview of physics, of law-likeness in sum, protects us from the unknown.

It is essential to note again the *very efficacity* of such mechanically regulative fictions. For Nietzsche, social conventionalization of truth in language illustrates the life-preservative success of illusion as regularization (or perhaps better said: as symbolization or as formalization). We can recall that the standard ideal of both linguistic and logical truth is fixed by what the early Nietzsche names the semiotic equivalent of a *"Friedensschluss"*—or *"peace treaty."*[70] For Nietzsche, the process from experience to consciousness through language, and so on to logic and to science represents nothing more exotic than an organism's specific mechanism for self-preservation—"the *knowledge appararatus* as reductive [Verkleinerungs] *apparatus* . . . means of the nutritional apparatus."[71] The difference between human and animal perspectives lies in the superior plasticity of human interpretations: human perspectival fictions are proof against error. Human knowledge perspectives are not merely life-preservative errors; they are themselves further preserved in metaphysical, religious, or scientific concepts. Thus for Nietzsche, "Everything that sets man off from the animal depends upon this capacity to volatalize visual metaphors into a schema, thus to dissolve an image into a concept."[72]

This representational, conceptual metamorphosis is the crystallization of the past as past; the latter may be understood in the spirit of what Nietzsche regards as the tertiary critical historical perspective in the second of the *Untimely Meditations*.[73] With this human evolution, in the development of language and so of logic, interpretive fictions subsuming different experiences to similar experiences name them as the same. Likewise, similar objects are assumed to be, are taken as identical. To illustrate this process, Nietzsche considers the deliberately trivial example of a leaf in his essay "On Truth and Lie." A leaf is recognized as a *leaf* and thereby the particular leaf, the *individual* leaf is collapsed, so to speak, into the formal idea, the image, the imaginary projection of a leaf. In this way, the conceptualization of a "leaf" comes to a fancied existence apart from all experienced leaves. What is more important, through the utility of the concept or the name, all experienced leaves are regulated. One experiences not different leaves but, much rather, different instances of the same *thing*: the leaf. Naming, that is to say, language gives experience its conclusive character: we know experience—we have experience—through language. And in this way, a word, sound, idea, the name of the thing *becomes the thing*.

But this emphasis on language does not mean that Nietzsche is a nominalist, still less that he is an idealist. The very conventionality of language manifests the Will to Power as evaluation, that is, as interpretation. As value and expression, language reveals the cultural ecophysiological basis of knowledge. In this way, Nietzsche shows that objective empirical recourse, filtered through language, is itself semiotically grounded: "a uniformly valid and binding designation is invented for things, and this legislation of language likewise enacts the first laws of truth."[74] The linguistic or conceptual origin of the "truth" of experience in terms of its practical or ecophysiological exigency does not certify the accuracy, the adequacy that would be its truth as truth. In this early essay, and in his later reflections on related matters, Nietzsche observes that a certain efficiency of perception in recognizing similar events and objects (vis-à-vis selecting a course of nutrition or evasion) is essential to survival. But the value of such reductive regularity for life does not prove that perception offers an adequate evaluation of reality. There is no ὁμοίωσις between perception and representation, because conceptual representation is not and cannot be similar to its object. Human perception, for Nietzsche (as for Goethe), is inadequate to all realities because it does not (and it is self-adequational because it cannot) generate an accurate, mathetic understanding, however much it may be "like" or similar to such an understanding.

The question to be resolved here is the question of similarity. Nietzsche's take on knowledge begins with a rigorous examination of the idea of the *same*. For Nietzsche, any complete, adequate knowledge or understanding must accord ambiguity, that is, ambiguity understood as the sum of individual differences and distinction, its full expression in reality.[75] When we subsume perceptions under concepts (which is what we do in every act of human understanding), we achieve an effective estimation of similarities, rather than an affirmation of difference as difference or ambiguity. Even the perception of difference is possible only on the basis of an original presumption of similarity (or identity). For Nietzsche, there is no original formal identity that would ground this presumption: similar things are not similar in themselves but only similar for us. This is the unimpeachable ground of rationality, of *ratio* in its truest sense. We begin here and we remain here. Thus careful (scientific) attention to "reality" will never come to reveal the fiction of the general concept (say, to return to our example, a leaf): empirically observed differences will only confirm the conceptual framework (identity).[76] Efficient perception (and traditional conception) systematically overlooks (and must systematically overlook) certain aspects of experience. It is this inattention that Nietzsche refers to in describing the perceiving organism as "blindly playing around on the backs of things." The "playing around," the game in question comprises not only raw sense perception, "informed" by concepts but also the critical game of the empirical scientific knowledge project. For Nietzsche, "even science does what humanity has always done: it employs *what* it finds understandable, that it counts as *true*, for the purpose of explaining everything else: *humanization* in summa."[77]

Everything is perspectival and cannot be otherwise. Thus, Nietzsche explains our most presumptuous epistemic errors: "All our *organs and senses of knowledge* have been developed in accord with conditions of preservation and growth. *Trust* in reason and its categories, in dialectic, and therefore the *valuation* of logic, demonstrates only the experientially confirmed utility of the same for life, and *not* its 'truth.'"[78] As an ecophysiological expression, scientific knowledge assumes the basis of its validity in language (conceptual fiction). It is as a logical expression of the Will to Power that science works as an interpretive triumph, considered in its empirical efficacy. But for all that it is not true. We ought not forget Nietzsche's claim that the conditions of life might well be such as to (Nietzsche explicitly says that they can hardly be other than those that) require error: "Why could the world *insofar as it concerns us at all*—not be a fiction?"[79] Small differences (marginal errors from the scientific perspective) are discounted in favor of the recognizability of gross similarities. But it will be the small differences that must be brought into account for Nietzsche. Nietzsche's epistemological rigor lies in his refusal to accept the traditional margins between important and incidental differences. For these margins conceal a powerful ideology, the ideology of modernity and progress, or the enlightenment which favors the clear and the distinct at the expense of the ambiguous and the essentially ambivalent.

PERSPECTIVALISM AS EPISTEMOLOGY

By analyzing knowledge rather than the conditions of knowing, Nietzsche counters the traditional presumption of the possibility of knowledge with the proposition or the suspicion that knowledge might be not proper knowledge at all but no more than an illusion. His question challenges the sufficiency of the principle of sufficient reason.

What Kant posited as necessary, Schopenhauer for his part was able to uncover in the immediacy of the Will, that is, the intuition of the noumenal realm. More ambitious than Kant, Schopenhauer was able to find an indirect exhibition of the truth of the noumenal world in the direct experience of the Will. Thus he invented a kind of reflexive periscope—a means for looking round his own corner, or, as Nietzsche puts it—not unadmiringly—for effecting the fairy tale achievement of turning one's own eyes inward.[80] But it is clear that Nietzsche finds intuition, however magically direct, uncompelling. The fact and promise of intuition must be assumed (as Schopenhauer takes it from experienced Will) or posited (as Kant posits an ultimate ground as the foundational condition for knowledge, after showing the impossibility of knowing in any other way) or else taken as directly given as simply axiomatic by less scrupulous analytic thinkers.

According to a perspectivalist account, interpretive knowledge of the empirical world is necessarily based on selective experiences together with reports of past events and conceptual generalizations leading to more general predictions and explanations. Yet, it is not possible to exceed or embrace such generalizations with a higher truth. In this, Nietzsche's position affirms the intractability of the problem of induction and the problem of synthetic a priori knowledge. For if no transition is possible between empirical experience and regulative strictures or scientific laws, there is no precedent guarantee (as Hume observed) that accepted knowledge stands on anything but custom. As an alternative, to be sure, one can "make up" knowledge that is absolutely certain (having a subjective origin). Then scientific generalizations could be accepted as mere definitions or stipulations. Scientific efficacy and rationality would thus be explained without asserting any ultimate transcendent truth. In this pragmatic, conventional expression, sophisticated philosophy of science abandons the uncertain to the unknown and even to the unknowable. For Nietzsche, as we have seen, just this affirmation of nonknowledge presumes too much knowledge.

The critical epistemic question, addressing the range from metaphysical reflection to technological intervention, is Nietzsche's question. He cannot avoid concluding that (just as from a nonfoundationalist, anti-Platonic or constructivist perspective, mathematical knowledge is stipulated to be—already known to be—true, and does not depend on a world of existent mathematical entities for its "truth") the truth of empirical knowledge plays in hypothetical spheres. In balder, bolder terms, Nietzsche concludes that truth is an error—or better, a species of errors—and knowledge a fictionalization of the world. As one constructs in accord with method (stipulates) isosceles triangles, so one constructs or composes electrons, operas, and poems, as well as and, more ordinarily, tables, chairs, and spiders. Laws of nature, conceived as such fanciful constructions, work precisely as far as they are needed, that is, laws of nature *work* in all accounts and to all appearances. But, for Nietzsche, nothing in such accounts need be true and so the account is artifice—or only technically true, that is, insofar as it describes the world of our (scientific) human interest. The scientific world is as faithfully represented for our cultural interests as the spectrum of snow types represents part of a (especially a downhill or telarc) skier's world. The integrity of this representation from the scientific perspective is not at issue. What is to be questioned is whether these things—that is, scientific entities or appearances and the various kinds of snow—are "really" so, that is, adequately represented in all their realities. In a serious scientific culture the answer has to be (temporarily) affirmative. But for Nietzsche this affirmation, even as it is sophisticatedly qualified as potentially false, is necessarily self-serving. The philosophic perspective must (at least!) consider the possibility of the artifactual nature of scientific knowledge as a human interpretation.

From Nietzsche's critical epistemic position, true knowledge would have to be transcendental. "True knowledge" cannot be a limited or personal conviction—something I "know" to be true now or simply to be true for me.[81] But transcendental knowledge is precisely what is unavailable from an organic perspective. The question once again for Nietzsche is whether knowledge can ever be shown to be based upon anything other than faith. For Nietzsche the answer is no, but that in itself does not offer access to any kind of "negative" epistemology but is rather decisively an antiepistemology: "We simply lack any organ for knowledge, for 'truth': we 'know' (or believe or imagine) just as much as may be *useful* in the interests of the human herd, the species; and even what is here called 'utility' is ultimately also a mere belief."[82] In this way, Nietzsche's question concerning the basis of knowledge (including scientific knowledge) considers the nature (but by no means the fact) of what we take to be knowledge as belief.

For Nietzsche, the interpretive function of knowledge is its Will to Power. For human beings, the way of knowing is the way of securing continued existence. Such life-securing simplifications are what he considers to be falsifications:[83] "Knowledge is the *falsifying of the mutifarious and incalculable into the identical, similar, and calculable.*"[84]

If knowledge is falsification, if its truth is a lie, then since we as knowers do not (and this is Nietzsche's special point) recognize ourselves in this fiction, we are as liars, lying to ourselves. Our claim to knowledge is delusionary or self-deceptive. Nietzsche's analysis of the *innocent* yet inescapable duplicity of knowledge as it conflicts with "all realities," that is, in opposition to the Real, is subtle. To follow Nietzsche's thinking here demands that the reader be capable of thinking, of experimenting and taking risks. Nietzsche's doctrine is not only difficult to grasp but properly said *esoteric*. This doctrine is meant only for select readers: "In the last resort there exists an order of rank of states of soul with which the order of rank of problems accords; and the supreme problems repel without mercy everyone who ventures near them without being through the elevation and power of his spirituality, predestined to their solution."[85] The esoteric criterion in question works two ways. Differentiating between readers is only one of its interests and effects. It also defines the limits of the general benefit of the Nietzschean perspective.

As Nietzsche has it, humans are not designed for "knowledge." The pursuit of truth, (the truth of "all realities") when it is not a ruse in the service of life (which service, as we shall see later, has two opposed aspects) is deadly. And the pursuit of truth is always rooted in error, precisely because error is the condition of life.[86] Nietzsche writes in a note from the 1881 *Nachlaß*: "I recognize something true only in opposition to an actual living untruth. Thus truth comes into the world as a concept completely lacking power and first acquires power by inmixture with living errors. And for this reason, one must permit errors to flourish and acknowledge their dominion."[87] What is borne in this realm between error and truth is nothing less than the condition of all knowledge. "We must love and cultivate erring: it is the motherwomb of knowledge."[88]

What may earlier have seemed no more than a bare contentiousness in Nietzsche's rejection of both claims to knowledge and prudent reservations of Socratic nonknowing can now be seen in its proper significance. Whether one claims to know nothing or to have very little knowledge (as Nietzsche presents the opposition in the *Nachlaß* note), or to have some certain knowledge (as he publishes the opposition in *Beyond Good and Evil*), one claims what cannot be given when one claims that one does not err (the ideal of Cartesian certainty). Knowledge derives from error (as a specific interpretation). In this context, we can better consider the response Nietzsche recommends to the Cartesian-style pretender to privileged access to his own internal experience or idealistic knowledge of (a priori) "certain" truths:

> Whoever feels able to answer these metaphysical questions straightaway with an appeal to a sort of *intuitive* knowledge, as he does who says: "I think, and know that this at least is true, actual and certain"—will finds a philosopher today ready with a smile and two question marks. "My dear sir," the philosopher will perhaps give him to understand, "it is improbable that you are not mistaken: but why do you want the truth at all?"[89]

The philosopher's reply to the dogmatic thinker (here this thinker will be Descartes) is classically ironic: the specificity of this irony can only catch a certain reader's eye—and just as obviously, the Cartesian interlocutor so treated could only be baffled. Only the reader who knows something of Nietzsche's position on truth and experience, knowledge and certainty will be able to follow the sense of the answering question, and this sense is not a comic one. For the aesthetic possibility here lies beyond the arch deflection of the rational idealist's (or *mutatis mutandis*, an empiricist's) certain knowledge and may be discerned only on the broader plain of value. This value is a measure of life celebration: "To give existence an aesthetic meaning, *to increase our taste for it*, is the ground condition of every passion for knowledge."[90]

Again, Nietzsche's hierarchic elitism appears.[91] He writes for the reader who may be said to have lived, as did Nietzsche himself, "with a great question mark as his destiny."[92] Nietzsche's selective principle is significant for his general epistemology for the curious but essential reason that, as so emphasized and construed, such an elite vision would in fact lack popular appeal and (unlike most elite ideals representing a popular or vulgar elitism) making no appeal to popular subscription, this vision once properly construed would be in essence and generally unattractive. The appeal of Nietzsche's elite vision may be and has been compared to that of Aristotle's description of the great-souled man. Like Nietzsche's "embarrassingly political" ideology of the Will to Power, read in an exoteric domain, the esoteric dimension is likewise uncomfortably offensive.[93] This comparison is not fortuitous. The appropriately intrigued reply to an illustration of an idealized (elite) philosophic attitude might be the expression of admiration: How can we be like this (kind of person)? But the question here should really be: Would one want to be like this? It is hard to exaggerate the difficulty of conceiving a wholehearted affirmation of Nietzsche's requisite, rarified lifestyle.

Such an exceptional focus may well be intrinsically fascinating. But apart from the ordinary democratic enlightened drive to individual valorization, that is, apart from the banal desire to be thought select, to be "unique" or to stand out, the very nature of Nietzsche's exceptional, exoteric, or higher focus is itself extreme. The adventure (experiment) prescribed must lose its appeal just where the genuine danger and threat to the indivduality, the "uniqueness" of the average individual becomes apparent in its intrinsic threat to the average ideal of life. Nietzsche's experiment is least desirable (and we are still talking about *appeal* here) for those of us who value life (as we all do in the ascendant culture of subjectivity with its valuation of self-preservation as the highest ideal and concommitant valorization of comfort). Let us not fail to remember that Nietzsche was the philosopher who wrote against the commonplace conviction of the individual right to life (or economic well being): "one has no right, neither to life nor to work, nor indeed to happiness. It is no different for individual human beings than for the lowest worm."[94] Thus we read Nietzsche's description of an aristocratic culture: "A culture of exceptions, of attempts, of danger, of nuance as the consequence of a great abundance of Power."[95] This is not a culture, as it might be claimed, of sulking, disenfranchised intellectuals but a culture celebrating danger *and* the capacity for dangerous undertakings. Nietzsche's description is not meant to be lyrically romantic. Beyond those sceptics who can tolerate an abstract "rendezvous of questions and question-marks,"[96] Nietzsche's appeal is directed to those (remember here that these for Nietzsche are the ever as yet unencountered, still unreal, philosophically free spirits) who understand that "no philosopher hitherto has been in the right, and that a more praiseworthy veracity may lie in every little question-mark placed after your favourite words and favourite theories (and occasionally after yourselves)."[97] This is more than a common cynicism.[98] Neither a scepticism nor a relativism, perspectivalism is a *living* disposition: a deliberate position in thought and being in a "rendezvous of questions."[99] And, of course—and here is the value of the esoteric worldview in the present context—where there is room for questioning, science cannot be far behind. For Nietzsche, the selected seeker,

> ventures into a labyrinth, he multiplies by a thousand the dangers which life as such already brings with it, not the smallest of which is that no one can behold how and where he goes astray, is cut off from others, and is torn to pieces limb from limb by some cave-minotaur of conscience. If such a one is destroyed, it takes place so far from the understanding of men that they can neither feel it nor sympathize—and he can no longer go back![100]

The emphasis on difficulty, more precisely: on *suffering*, on what the early Nietzsche called the "tragic," and the later Nietzsche spoke of once again as the "Dionysian," is patent. Yet difficulty as such, the possibility (and perhaps the necessity) of failure, again, is not the point: "There are heights of the soul seen from which even tragedy ceases to be tragic."[101] To understand Nietzsche's position here recalls that he never denies the pessimistically construed epistemic stance. Although Nietzsche hopes to exceed nihilism, whether optimistically or pessimistically articulated, the aesthetic character of his resolution to the epistemic/moral quandary of existence can best be understood in the light of its origins as a pessimistic affirmation—that is, as an affirmative response to the working tragic insight, the effective illumination of "*art* in the light of life."

Perspectival knowledge looks to all possibilities, which never means that it encompasses all possibilities.[102] All perspectives, including supposed and trusted truths, are to play in questioning. This only means that they are to be thought. Yet we have seen that perspectivalism risks the consequences of absurd perspectives together with its own annihilation. The destructive risk of "going to ground" is accepted in good conscience because the aim is an enhanced—that is, artistically advanced, rather than merely preserved or maintained—life.

Thus understood aesthetically, perspectivalism's vulnerability to destruction does not render it a nihilism. Moved by the hermeneutic insight that knowledge is creative interpretation, Nietzsche's response is to try his hand at it. In this way, Nietzsche's critique of knowledge seeks to understand knowing in the manner of an artist. In this way Nietzsche represents art (here, knowledge) in the grand style as a means to life. Nietzsche's effort is to discover how far such deliberate aestheticism in life can be taken in the context of philosophical thought and, by extension, science. But this is not to say that Nietzsche's project champions a romantic beautification or subjective refinement of science. His guiding effort to uncover the possibilities for science (as art) in the light of life cannot be interpreted as a call to prettification, or even humanization, of science without serious and trivializing misunderstanding.[103] His question is rather the transformative perspective, the question of the grand style: Can we become the artistic creators of our own understanding, of our own science, and so our own existence? It will be in response to such a question that the teaching of the eternal return conditions the incorporation of such a creative aesthetic attitude toward life.

MULTIPLICITY AS INTERPRETATIONAL TRUTH:
THE METAPHYSICAL FICTION OF AN ABSOLUTE

Having thus explored Nietzsche's challenge to the basis of empirical, transcendental, and psychological knowledge, a provisional resumé may now be helpful. For Nietzsche, the practical end of utility is itself permeated by a fiction. This practical ideal is fictional in the constitutional or regulative sense, since it is a product of cultural, ecophysiological interpretation. But it is fictional in the traditional sense as well: as error generated from an illusion that is nevertheless presented as if it were unimpeachably true. For Nietzsche, practical and so scientific knowledge denies the complex character of the becoming chaos that is nature, or the Real. This complexity or ambiguity (*Vieldeutigkeit*) of the effectively unknowable Real is denied in favor of the effectively knowable rational-scientific categorization and organization—for Nietzsche: simplification—of the becoming of nature into a human-being-fitted world (outfitted according to whatever context, whether that involves religious morality, economic progress, or, indeed, radio telescopes, nuclear accelerators, or electron microscopes). We produce scientific conceptions, Nietzsche notes "to be able to live in a world."[104]

Nietzsche's Peirceian reflections on the conventionalism of knowledge explain the pragmatic organic interest of such simplifications. A world regulated by scientific laws is less threatening and less problematic than a world of chaos. To offer an example Nietzsche himself favored, such a limited focus is characteristic of the earthworm that has no consciousness of (and so from its limited perspective is therefore effectively, successfully able to ignore) the complex extension of the world about it. This ignorance works until the worm is unfortunate enough to encounter some extrusion of that world upon it, for example, when some beak pokes at it or foot treads upon it. Then, Nietzsche says, *contracting*, it shows its appreciation of the "reality" of this foreign intrusion. Its self-retroactive response to that world-intrusive assault is conservatively *interpretative*. In this example, we see what Nietzsche understands by power-expressive interpretation. By turning, the worm diminishes the likelihood of being stepped on again. The worm's shrinking response is effective: *self*-contraction works to *effectively*, *actually* contain or shrink the *world*.

The limit upon any critical knowledge venture remains the

utility of interpretive falsification (its "truth"): one must be able
to live with whatever "truth" is proposed. As the earthworm
manifests the range of its perspectival conviction in its full or contracted extension, so the scientific interpretive accounts reach
their own rational level within the constraints of life. Ergo, "This
subtler honesty and skepticism came into being wherever two
contradictory sentences appeared *applicable* to life because both
were compatible with basic errors . . . also wherever new propositions, though not useful for life, were also not evidently harmful
to life.[105] The closer look at "the flowers of the hedgerow" that
marks the botantist's perspective, separating it from the less-dis-
cerning selection of the bee or butterfly, has its origin, "as the
expression of an intellectual play impulse."[106] But there is nothing
that is truly neutral in this playfulness. The ideal of a "botanic
benevolence" or any other neutral scientific good will may not be
affirmed because it does not exist. What starts as play becomes its
own reason for being—the play *becomes* the thing, and it then
works in its own right as a perspective over and against other possibilities. We are ever, for Nietzsche, concerned with the expression of the Will to Power when we speak of interpretation, even
when we speak of play:

> Gradually such judgments and convictions filled the human
> brain, and a ferment, struggle, and lust for power developed
> within this tangle. Not only utility and delight but every kind of
> impulse took sides in the battle over these "truths." The intel-
> lectual battle became a task, a stimulus, a profession, a duty,
> something dignified—and eventually knowledge and striving for
> the true took their place as a need among other needs.[107]

This means that "knowledge collided with those primeval basic
powers"[108] Such a collision has only one outcome. Knowledge
works on the basis, and is the further expression, of the old
organic errors. They must be considered together: "two lives, two
powers, both in the same human being."[109] When Nietzsche asks
then, "To what extent can truth endure incorporation?"[110] what
he means to question is the subtlety and free play of the will to
truth; such a will must be distinguished from institutionalized scientific truth and its nihilistic metaphysics. Again, what is involved
here is a discriminating critique and beyond that an interest in a
higher possibility for knowledge (in art), and so, for life.

A NOTE ON THE TYPOLOGY OF SCIENCE
AND PHILOSOPHY: THE WILL TO POWER

Nietzsche understands the scientific illusion of truth as a reactive manifestation of the human Will to Power expressed as a drive to dominate (i.e., to fully appropriate or exploit) the earth.[111] Despite the circumstantial fact that the reflective expression of science is permeated with conditional claims, qualifications, and assertions of nonknowing, that is, despite science's role in confirming Nietzsche's "hyperrealism"—the world in its ambiguity and constant flux as a chaos[112]—the scientific Will to Power is an absolute drive.

The casual experience Nietzsche reports expressing the difference between a distant landscape and the immediacy of the countryside itself as it strikes the approaching walker, as the difference between the near and the far, the object for consideration and the immediacy of a country world, can be repeated by anyone who chooses to investigate, as does Merleau-Ponty, the "phenomenology of perception" or, indeed, by any child watching battalions of ants crawling through the high grass forest of a moss bed or a picnic luncher leaning against a tree and noticing with fascination (or revulsion according to temperament) the literal, crawling "buzz of life" upon its bark. This interest proves Hollingdale's assertion that "Nietzsche wanted to be a phenomenalist."[113] But it is more illuminating to speak of Nietzsche's *hyperrealism* than of his "phenomenalism" where Nietzsche writes, "Walking in the outside world, I am always amazed to think how everything strikes us with such wonderful precision: the woods, thus and so, and the mountain, thus and so, and all with no hint whatsoever of confusion, error, or faltering with reference to any of our sensations."[114] For Nietzsche, again, later notes the tendency to overlook things as they are in the helter-skelter, shimmering Jamesian or Ur-Joycean world of blooming, buzzing confusion must betray even a critical perspective and reveals "the hatred of becoming, against the careful observation of becoming."[115]

It has been suggested that the disingenuousness of scientific or what is often called "Popperian fallibilism"—the modern sophistication of science and scientific philosophy—is obvious in the persistence of the ideal of scientific progress. This ideal may be represented as a knowing unknowing: science is an approximation of what is ultimately an objectively true account of its subject matter.

In the context of the absolute dimension that is implied by the progressive ideal in science, it should be acknowledged that Nietzsche has been read by many authors, most notoriously by Heidegger, as affirming just this very "earth mastery" as part of his own philosophic ideal. Often proposed in support of this reading is Nietzsche's prescriptive/descriptive comment: "*Actual philosophers, however, are commanders and law-givers.*"[116] Yet the reading Heidegger uses for his own separate purposes, taking this particular text as an expression of Will to Power, is misleading. Nietzsche's concern is not to justify world domination as an ideal but rather to distinguish between ordinary scholars or scientists and philosophers. Philosophers may require scientific training, but that can only be preparatory, never a philosophic goal as distinct from the scientific.[117] While philosophy is properly interested in the inestimable, the goal of science is the calculation of the objective world.

For Nietzsche, it is no more than an "illusion that something is *known* when we possess a mathematical formula for an event: it is only *designated, described*; nothing more!"[118] The rational structure provided in the mathematical expressions or theoretical formulas of natural science is a constitutive imposition or fanciful composition. For Nietzsche, the natural conformity to law uncovered by scientific investigation in a rational image repeats for its part only the "model of a thoroughgoing *fiction* . . . In this a way of thinking can be *fabricated* . . . [and thereby] all affects, feeling, and willing are thought away."[119] Of course, the elimination of all affinities, all subjective feelings and desires, conforms to nothing less than the (ascetic) ideal of objective scientific rationality. Nietzsche argues that the scheme projected by objective scientific rationality is fictional (an interpretive scheme, rather than an exhaustive expression of all realities):

> This kind of thing never happens in reality, which is unutterably, differently, complicated. Inasmuch as we impose these fictions as *schemas*, in thinking the factual event we thereby simultaneously *filter* it through a simplifying apparatus. We thus relegate it to *symbolic expression* and to *communicability* and to the *remarkability* of logical procedures . . . There is no logical thinking in reality and no axiom of either arithmetic or geometry can be deduced from it, just because it isn't there.[120]

Because it is an interpretive projection, science cannot "explain" but is condemned to mere constructive description: "We operate with nothing but things that do not exist, with lines, planes, bodies, atoms, divisible time spans, divisible spaces."[121]

It could be said that Nietzsche does not accept the rule of a science that invents axioms and works with (for example) ideal gases and then takes these descriptions for (literally and so-called) absolutes, because the supreme value of integrity, that is, *Redlichkeit*, which requires a kind of truth (*Treue*) to the ambiguity of nature. In this openness to "all realities" (perhaps including the possibility of nonideal protean gases) Nietzsche's philosophy of science ultimately moves beyond science—to art and life. Jean Granier expresses the rational (metaphysical) ground of the ostensive connection between what is known through human mathematico-scientific efforts and what is "real":

> As this is defined in philosophic speculation, science has never thought through the relation between man and being on its own account. Science implies the postulate of affinity between "being" and the essence of man where that involves the possession of ratio. The systematization of laws, their reduction to the smallest possible number, the conviction that the progress of knowledge conditions the psychological and moral blossoming of humanity, only makes sense in terms of the tacit belief according to which everything occurs as if there were a plan to the universe, a plan which itself redounds to a divine wisdom combining in itself absolute intelligence with the most abundant goodness.[122]

The measure of evaluation in question is the logical standard of mathematics and the ultimacy and precision of the ideal logos that has guided Western metaphysics since Plato. For Nietzsche, this ideal logos is itself equivalent to the Christian logos as a divinely ordained and so preordained guide to human knowing. The formal and the spiritual ideals of logical rationality permeate science as the "latest and noblest form"[123] of the ascetic ideal in Western metaphysics. Granier offers the relevant distinction that talk of metaphysics is not a matter of reiterating "to the point of banality, [the thesis] according to which science always necessarily implies *a* metaphysics. It is rather a matter of establishing that, according to Nietzsche, science is metaphysics itself. . . . [Nietzsche] *challenges science in order to overcome metaphysics alto-*

gether."[124] The position I take regarding Nietzsche's evaluation of science as a metaphysical (fundamentally Socratic) project has a clear parallel in Granier's assertion of the Heideggerian perspective that "the metaphysical understanding of 'Being' as an ideal automatically entails some definition of the criterion of truth."[125]

BEYOND TRUTH AND LIE

We have seen that, for Nietzsche, there are many truths because there are many circumstances, many capacities, many interests, all proliferating a plurality of perspectives and culminating in a dizzying abundance of possibilities.[126] But if there are many "truths," there is no truth as such, that is, there is nothing that is in itself, for itself true: "There are many kinds of eyes. Even the sphinx has eyes—and consequently there are many kinds of 'truths,' and consequently there is no truth."[127]

As we have seen, the statement that there is no truth does not present the ideal of metaphysical or transcendent knowledge: it opposes that ideal because everything is interpretation (or Will to Power). The goal and rule of valid logical argument is its conservative assurance of truth (justification), if truth is to be had. Only where the premises are given as true to begin with may one validly proceed to (similarly true) conclusions. The logical effort is necessarily tautological: it assumes truth to preserve it.

But if there is no truth, only interpretation, then anything can be concluded from whatever is ordinarily taken to be true. Hence Nietzsche's fascination with the Assassin's motto, "Nothing is true: everything is allowed," does not blindly presume the principles of ordinary logic but is rather concinnously employed. For the benefit of believers in ordinary logic, A and -A can be posited as equivalent. If it is false that nothing is true or false, anything follows. The value of truth is elided in the ambivalence of insistent contradiction. This notional dissonance is characterized by the fluid terms, contradictions, aphorisms, metaphors, hyperbole, and the tropings and the concinnous invocations and so on making up the conceptual challenge of Nietzsche's style.

With respect to truth, logic, and lie, what may (loosely) be called Nietzsche's "logic" is an antilogic. Nietzsche's "logic" is not Parmenides' or Aristotle's any more than it is Russell's or Tarski's. While the interest of traditional logic is the preservation of truth (which requires, as Nietzsche says, and standard logic does not contest this, that "truth" is to be assumed in advance), Nietzsche for his part, wishes to express the multiple truths of human understanding together with the ambivalent, ever-ambiguous self-manifestation of nature. Thus Nietzsche can speak of "truth" as he does in the fourth maxim/arrow in *Twilight of the Idols*: "'All truth is simple.'—Is that not a compound lie?"[128] In the *Nachlaß* note expanding this observation, Nietzsche coordinates the true with the Real, as seen in Lacanian terms, that is, as opposed to the imaginary truth ideal of simplicity: "All truth is simple: that is a dyadic [*zweifache*] lie. Everything that is simple is plain [*Bloß*] imaginary, it is not 'true.' However what is real, what is true, is neither simple, nor indeed ultimately reducible to singularity."[129] Although consistently and by definition non-truth-conservative, Nietzsche's antiepistemology concerns our interaction with the world as the Real.[130] It is because there is no world apart from that interaction for us that Nietzsche's antiepistemology survives the ambiguity and indeterminacy of the empirical world in the small and in the large (it affirms just this variational perspective). The perspectivalist's point for a philosophy of science is that a logic that opposes formal-analytic logic is likely to be more appropriate to a "scientific" description of reality, once given Nietzsche's standards of rigor and the ambivalence of the real world, as nature or like the Lacanian Real, construed according to the full, ambiguous, chaotic sense Nietzsche gives to it.[131]

The logico-empirical view of the world is amenable either to a static, nonstochastic or stochastic descriptive logic. Contemporary Western science has successfully employed the former view, and more recently the latter has come into vogue in its project. This practical success, however, is not inconsistent with perspectivalism. Nietzsche claims, then, that "'contrary to the scientific presumption of regularity, 'Things' do not behave regularly, according to a *rule*: there are no things (—they are fictions invented by us); just as little do they follow the constraints of necessity."[132] He makes this claim because "There is no law: every power draws its ultimate consequence at every moment. Calculability exists because things are unable to be other than what they are (*keine mezzo termine*)."[133] From a perspectival position such as ours, "We need units to be able to calculate: but that does not mean we must suppose that such units exist."[134] The process of interactive calculation is possible because our own knowing is an instance of the Will to Power. Where the character of the world is Will-to-Power-and-nothing-else, Nietzsche reminds us, "We belong to the character of the world. There is no doubt of *that!*"[135] This relational world-character, understood in a reciprocal exchange, shows the radical value of what it means to employ interpretive perspectives:

> the world, apart from the conditions of life in it, the world which we have not reduced to our being, our logic and our psychological prejudices, does *not* exist as the "world in itself." The world is essentially relational, presenting under certain conditions a *different aspect* from each vantage point; from each point its being is essentially different.[136]

To say that our conditions for knowing determine what can be known is not a relativism since we are ourselves interpretive expressions of the Will to Power. But because "Logic and mechanics are only applicable to the most superficial . . . [a] domination of the manifold through an art of expression,"[137] contemporary scientific culture lacks aesthetic taste or an artist's sense. For this reason Nietzsche takes care to warn the logical gentlemen of physics, thereby obliquely warning us, we philosophers, that science may not measure "up to" Nietzsche's canon of *good taste*, "the taste of reverence for everything that lies beyond your [our] horizon."[138] It is for this reason that Nietzsche can speak of science as one of the stupidest descriptions possible. More critically, he claims that science is "anti-aesthetic." In other words, beyond stupidity, thickness, bad taste, or presumption, science is an aesthetic expression that lacks artistic sensibility.

This is a special point, and we need to tarry here. To speak of Nietzsche's "hyperrealism," as I previously expressed his perspectival attention to detail, devolves upon a minute consideration of differences in nature (the Real) and the fuzziness of perception and conception, not as objections to the perceptual/conceptual achievement, but as a description of this achievement as an interpretive self- and world-structuring expression of the Will to Power. From such an interactional perspective, science can be named a tasteless and even stupid world-interpretation. It is science's stupidity that is at stake, where stupidity is an inadequate understanding that fails to comprehend this very inadequacy.

But is this so? Is science "stupid" or, better, less offensively said, *inadequate*? If we reflect upon the example considered earlier concerning (the trivial) differences between leaves, or if we simply call to mind the cliché celebrating the variety to be found among individual snowflakes, fingerprints, or inkblots, it is obvious that variation, mutability, and ambiguity characterize a certain level of reality and should perhaps (strictly taken) function in the descriptive calculus, the "logic" and the "science" describing this level of reality.[139] Such attention to the Real corresponds to Nietzsche's conviction that simplicity rather than being a royal road to the heart of things is purely ideal, a fiction. Reality is, he says, unspeakably complex because reality is composed of individual real things not simple natures. Science does not render the details of this reality, because the aim of science is universal not individual, because such an aim requires selection. This selectivity guarantees scientific precision even as it leaves it open to Nietzsche's critiques. The individual must be subsumed under the rubric of the token, the general; the individual as such can only be marginal and spurious. It is thus significant that even the recent affection for fractals and chaos theory reflects only a search for ordering structures. The project or end of this discipline is precisely as named in the title of the well-known book by Prigogine and Stengers, *Order out of Chaos*. Thus the "new alliance" with nature is contracted according to the ancient terms of dominion. The point here is not to criticize this brave new alliance but to push its claim to the point of reformulation and legitimation. Here Erwin Chargaff goes a bit further, if he is also more diffuse, in Nietzsche's direction, highlighting details and other distinctions that made in excess lead not to clarity but to ambiguity.[140] One wants in Nietzsche's spirit to count in all the little mistakes, the marginalities, the fuzzy details previously disregarded ("annihilated" or "damned") by science.

In this limited way, science may be described as inadequate or, as Nietzsche says, "stupid." But science regards this kind of inadequacy as *unimportant*. Science is effectively, in practice, in theory in "attitude," insensitive to an adduced inadequacy (its own inadequacy to the not merely "soft" but quite literally fluid or [but not unstable] items such as snowflakes and inkblots). Such an insensitivity to such soft details is exactly the scientific "tough"-mindedness explicitly praised by pragmatic philosophy and tacitly endorsed by philosophers of science. For Nietzsche, the excess marginality of the realities so ignored outlines not inconsequential details but the abyss that threatens and defines existence. Just because Nietzsche's perspectivalism projects a multivalued logic along with its wild "truths," Nietzsche's vision necessarily has more affinity for the marginal (spurious but still) phenomenal world than other (rationalist scientific) conceptions.

Nietzsche's basic point against Aristotelian or formal logic as we have seen is that where truth is not given to begin with, it can hardly be preserved. Instead of logical form, what may now be called the "logicality"[141] or consistency of perspectivalism takes truth in interpretive terms. In Nietzsche's generous attention to the detail of life and the inconsistencies that are hardly visible from certain (alternate logical) perspectives hardly visible, "truth" (from the perspective of the Will to Power) is various and impermanent. Thus Nietzsche writes, "It is wonderful that the assumptions of mechanics are adequate for our needs (machines, bridges, etc). They are after all very big needs, and the 'little mistakes' don't enter into consideration."[142]

With reference to the mariginalities within the Real, Nietzsche's multivalued perspectivalism affirms the statistically insignificant possibility, the ambivalent possibility, and hence the impossible possibility. We have noted that a Nietzschean philosophy of science would reflect the heretofore neglected, smoothed-out, smoothed-over "little mistakes." Beyond the calculative complexity entailed by such an attention to marginal errors, it is this openness to reality as it is a human projection and as it is Real (unknowable beyond our cultural project and perspective) that recommends the critical and reflexive power of Nietzsche's thought on the nature of human knowledge and the scientific endeavor as a new line in the philosophy of science.

The phenomenalist advantage of Nietzsche's perspectivalism is evident: changing one's aesthetic valuation or descriptive focus according to the relative value of events in connection with other events—moving from a perspective directed to an invariant or invariants over a variable range to one open to variability *per se*—affords the philosopher both another description of reality as well as another attitude vis-à-vis human knowing and, so, possibilities for human being. Perspectivalism, in sum, affirms becoming, ambiguity, and ambivalence in place of the static, absolutely defined, or univocal.

Nietzsche's perspectivalism might thus offer the basis for interpretational (higher-order) logic beyond affirmation/negation.

As a deliberately, contrivedly trivial (like Nietzsche's leaf) example as an illustration of this possibility for scientific understanding, let us consider the (observational, matter of fact) question, Is it raining? The corresponding responses yes/no are mutually exclusive alternatives, easily arranged in two-tier truth tables. However, either answer, that is, either yes or no, might be empirically justified (perspectivally) at any time. That is also to say: at the same time *and* under the same conditions both responses may be offered with empirical justification, depending on, to use an experimental metaphor, one's sampling parameters.

The justification for this claim depends upon the affirmation of the continuum that is the Real process of (natural) becoming. For Nietzsche, what appears to best human discernment as a (delimitable, regular) continuum is a chaos rendered continuous by the discerning process itself. This chaos is the contrastive basis for the limit conceptual/perceptual image which Nietzsche reads as *"ein Blitzbild aus dem ewigen Flusse."*[143]

Given our human affection for the extension of such a *Blitzbild* (in generalization), together with our banal human terror of or animal blindness to the "eternal flux," the answer to the question, Is it raining? will depend upon the definition (or the relevant *Blitzbild*) of rain. But, with this convention we are in the domain of stipulation, that is, what Nietzsche calls "tautology." Because perspectivalism promotes different descriptions of natural states than those currently enjoying a legitimated application in contemporary science, Nietzsche's perspectivalist flux-proposition concerning the continuous nature of the real world and its manifestations might prove an advantage for a philosophy of science.

Rain is at once a sprinkling or a shower, a drizzle or a steady rain or a driving rain, a downpour or a drencher, a soaker, and so on, all the way to a deluge, a cataclysm, or an inundation. (Doubtless other selective interests would generate some other scale of rain terms.) Then, to this catalogue, one might wish to add the difficult to classify: the odd rain drop from nowhere, for example,—and these are important considerations—with/without overhead air-conditioners, passing birds, in warm or cold weather, and so on. To fix our selective parameters we may ask how many drops it takes to make it "rain": two drops, three drops, four—indeed, what precisely counts as a drop? just a drop of water? must it be a drop of rain? Is there a difference? What distinguishing details are important, viewed from the differing perspectives of a meteorologist, a special-effects technician, or indeed, the research scientist *simulating* rain (causing rain for the purposes of a meteorological study or the effects of rain-absorption by plants)? Is simulated rain involving a shower of water more *rainlike* than the visual effect of slashed sheets of mylar waved across the stage to the audio accompaniment of falling raindrops in a theater production or even the pure simulacrum of the video-effect addition of shimmering or fuzzy electronic rain "falling" on the video-monitor "before" or "on" TV-studio/video performers as safe from rain as they are from sun or cold?

On the terms of our everyday illustration, let me note that, in the English language, an array of descriptive terms for rain corresponds to the (now denounced as overrated when not exactly adduced as a proof of a general conception or comprehension) versatility of Eskimo snow perceptions or else, more plausibly, the concept of snow represented in the worldview of cross-country, downhill, and telarc skiers. *Snow* is a multifarious constant in these and other contexts. The same name may be given to all types of snow but, the professional or the weekend skier experiences a variety of "snows." Snow will vary in the experience and representation for the sidewalk snow-shoveller as for the child making angels or rolling snow for a snowman or making snowballs.

At issue here is the old problem of the relation between universals and particulars. Nietzsche would emphasize that we do not merely gain in conceptual power when we use universal terms but also lose our ultimate connection with the individual. The universal is indequate to any representation of the individual in its individuality. Particular truth is sacrificed to the general ideal of truth, so much so that we have no individual understanding of truth. In this example, the varying descriptive terms for the same physical or meteorological phenomenon are ordered according to some (necessarily subjective) standard of intensity or significance, some aesthetic or experimental criterion. But, however ordered, the qualitative/quantitative dimension of rain is relative to our own interpretational perspective. There is no more rain, no enhanced "raininess" in a downpour than in a drizzle, or a single raindrop except of course—and this would be Nietzsche's point— from the viewpoint of the individual actually in the rain: watching it fall, singing in a musical number or fantasy, or merely tramping stolidly home. For Nietzsche, rain like everything else is a relational event, and language describes it that way. For Nietzsche, the relational report is what our language is fit for and language cannot exceed this report. Thus he writes, "The demand for an *adequate means of expression* is *senseless*: it lies in the essence of a language, a medium of expression, to express no more than a basic relation"[144]

The example of rain or attention to particular individuals (leaves) may seem irrelevant to a scientific context where what is wanted is precisely universal knowledge. But the point can be extended—and Nietzsche makes this very extension—to the difference between "identical" chemical compounds, or to the variety present in what we take to be a particular element: "Nothing happens twice: the oxygen atom is without an identical match. In truth the assumption that there are countless identities *suffices* for us."[145]

Oxygen is an unchangeable substance (that is its chemical definition; its molecular weight and subatomic composition are "metaphysical" concepts: "There is nothing invariant in chemistry—that is only appearance, merely a scholastic prejudice. We have imported the invariant, my dear physicists, from metaphysics as always. Read off from the surface alone, it is completely naive to assert that diamond, graphite, and coal are identical. Why so?"[146] This claim contradicts traditional scientific doxa concerning the indiscernibility and hence identity of not only oxygen atoms but also electrons (and other particles).[147] Nietzsche's relational perspectival position testifies to his openness. It is plain that he seeks to avoid denial where he simply does not know. Instead he emphasizes the creative functional construal (the interpretive character) of knowledge and the senses. For Nietzsche, "the 'logical,' the 'artistic,' is our constant engagement. What made this power as sovereign as it is? Obviously that given the pellmell of impressions, no living being could endure without it."[148] For psycho-biologists and psychologists, talk of species-specific perceptual selectivity says nothing specifically new. Indeed biologically grounded theories of perception incorporating cognitive theory but concerned with the issues of ecological theories of perception (Gibson, Marr) and bioevolutionary theories of knowledge reflect a growing and exciting field of investigation.[149] Nietzsche's own interest is however broader than the differences and parallels to be noted between frog, cat, and human worlds of vision: Nietzsche's specific interest is the aesthetic perspective that is the topic to be explained in subsequent chapters in terms of the question of temporality and of life in the grand style.

NOTES

1. "Leben ist die Bedingung des Erkennens. Irren die Bedingung des Lebens und zwar im tiefsten Grunde Irren. Wissen um das Irren hebt es nicht auf! Das ist nichts Bitteres!" (KSA IX, p. 504).

2. "Was unterscheidet den wahren und den falschen Glauben?" (KSA XII, p. 265).

3. "Das πρῶτον ψεῦδος: wie ist die Thatsache der Erkenntniß möglich? ist die Erkenntnis überhaupt eine Tatsache? was ist Erkenntniß? Wenn wir nicht *wissen*, was Erkenntniß ist, können wir unmöglich die Frage beantworten, ob es Erkenntniß giebt. Sehr schön! Aber wenn ich nicht schon 'weiss', ob es Erkenntniß giebt, geben kann, kann ich die Frage 'Was ist Erkenntniß' gar nicht vernunftigerweise stellen" (KSA XII, p. 264), WP 520.

4. "Hätten wir nicht die Künste gut geheissen und diese Art von Cultus des Unwahren erfunden: so wäre die Einsicht in die allgemeine Unwahrheit und Verlogenheit, die uns jetzt durch die Wissenschaft gegeben wird—die Einsicht in den Wahn und Irrthum als in eine Bedingungen des erkennenden und empfindenden Daseins—gar nicht auszuhalten" (FW 107, KSA III, p. 464), GS, p. 163.

5. Cf. JGB 1; and above JGB 16.

6. FW 354.

7. Any projected Nietzschean epistemology must be accounted a nihilist (or "negative") epistemology. Cf. Kaufmann, Löwith, and Danto. In a recent article, J. Wilcox dubs the position of nihilist(ic) epistemology "Nietzsche's negativity," seeking a theological parallel ("Nietzsche Scholarship and 'The Correspondence Theory of Truth': The Danto Case," p. 347).

8. "Eine systematisch, ausgearbeitete und in sich widerspruchsfrei Erkenntnistheorie hat Nietzsche nicht vorgelegt, hätte dies doch seinem philosophischen Selbstverständnis radikal widersprochen" (Kirchhoff, "Zum Problem der Erkenntnis bei Nietzsche," p. 16).

9. Clark, *Nietzsche on Truth and Philosophy*, p. 1.

10. Ibid., p. 21.

11. By speaking of a *ecophysiological* perspective I mean a perspective determined not only by the physiological or physical constitution of the interpreting perspective but also its ecology or relative world-circumstance. By such a coinage, I mean to express Nietzsche's valuation not only of the empiricist's physiological or sensual perspective but also with the unfolding of will to power as world (ecological) interpretive expression. In other words, paralleling an Aristotelian intentionality, the ecophysiological perspective is reciprocally determined by a physiology and an ecology that are not opposed but continuous.

12. Clark, *Nietzsche on Truth and Philosophy*, p. 27.

13. Explaining Heidegger's phenomenological historical dialogue of thoughtful meditation as embracing both negative and positive moments, William J. Richardson affirms this to be the endeavor "to comprehend and express not what another thinker thought/said, but what he did not think/say, could not think/say and why he could not think/say it" (*Heidegger: Through Phenomenology to Thought*, p. 22). Heidegger explains the project of foundational thought as eliciting what is possible for thought in what is (to use Richardson's apposition) "thought/said" because thought thinks beyond the thinker's power to limn its movement: thought thinks in advance of itself, i.e., in the possibilities an original thought opens up for thinking. This means that a genuine thinker's thought is rich beyond itself so far as it is true to thought: "Das höchste denkerische Sagen besteht darin, im Sagen das eigentlich zu Sagende . . . so zu sagen, daß es im Nichtsagen genannt wird" (*Nietzsche I*, p. 471). The wealth of thought can be revealed not by examining what is complementary to the thought but rather (in a movement through the schema of original thought) by considering what is possible or sketched out for thought. Even as one must exceed what a thinker "thought/said," one remains true to the thought. Inasmuch as the epistemological critique takes its origins from the very questionableness of knowledge as a possibility, this claim is epistemological.

14. As opposed to the Davidsonian generosity vis-à-vis the "very idea" of conceptual schemes, Nietzsche's perspectival view is different. Davidson's generosity works on what appears to be a Tarskian metalevel. It presents the issue of truth on my side and on your side from one perspective. Nietzsche's perspectivalism attends precisely to the positionality of the imposition of such convictions.

15. Hollingdale, *Nietzsche*, p. 1.

16. Hollingdale, *Nietzsche*, p. 131.

17. Grimm, *Nietzsche's Theory of Knowledge*, p. 100.

18. For example, Warnock, "Nietzsche's Conception of Truth," and, reflecting Jaspers's influence, Pfeffer, *Nietzsche: Disciple of Dionysus*; and the always interesting Morgan, *What Nietzsche Means*.

19. Useful texts on the topic include, in German, those by Kirchhoff, Kaulbach, Abel, Gebhard, as well as Stegmaier, "Nietzsches Neubestimmung der Wahrheit" and Bittner "Nietzsches Begriff der Wahrheit"; and, in French, Granier, *Le probléme de la vérité dans la philosophie de Nietzsche* and Andler, *Nietzsche: Sa vie et sa pensée* (esp. vols. 2 and 3), as well as Deleuze and Philonenko.

20. "Decken sich die Bezeichnungen und die Dinge? Ist die Sprache der adäquate Ausdruck aller Realitäten?" (UWL, KSA I, p. 878), TL, 81.

21. We will need to return to a consideration of the structure of

what counts as knowledge in the technological scientific age, the incarnation of a *deus ex machina*—in an emphatically different sense from its dramatic origin.

22. "Der Forscher nach solchen Wahrheiten sucht im Grunde nur die Metamorphose der Welt in den Menschen . . . Sein Verfahren ist: den Menschen als Maass an alle Dinge zu halten, wobei er aber von dem Irrthume ausgeht, zu glauben, er habe diese Dinge unmittelbar als reine Objekte vor sich. Er vergisst also die originalen Anschauungsmetaphern als Metaphern und nimmt sie als die Dinge selbst" (UWL, KSA I, p. 883), TL, p. 86.

23. "Wir haben uns eine Welt zurecht gemacht, in der wir leben können" (FW 121, KSA III, pp. 447–48), GS, p. 177.

24. "Mit der Annahme von Körpern, Linien, Flächen, Ursachen und Wirkungen, Bewegung und Ruhen, Gestalt und Inhalt: ohne diese Glaubensartikel hielte es jetzt keiner aus zu Leben!" (KSA III, pp. 477–78), GS, p. 171.

25. "Deutlicher geredet und grob und gründlich: synthetische Urtheile a priori sollten gar nicht 'möglich sein': wir haben kein Recht auf sie, in unserm Munde sind es lauter falsche Urtheile" (KSA V, pp. 25–26), cf. WP 530.

26. Henke finds this reference positively essential in his review "Gedanken zum Umgang mit dem Grundstrukturen und Elementen Quasi-Naturwissenschaftlichen Bezüge im Werk Nietzsches" (*Nietzsche-Studien* 18 [1989]. Henke refers to Klaus Spiekermann on the indispensable reference to the eternal return in a "scientific" or merely "quasi-"scientific connection (K. Spiekermann, "Nietzsches Beweise für die Ewige Wiederkehr," *Nietzsche-Studien* 17 [1988]). For my reading, against these and other commentators, what Nietzsche has to say about science and epistemology exceeds the (cosmological, cosmophysical) issue of the Eternal Return.

27. See Michel Serres for the converse view, a visionary conception with regard to the Franco-Belgian "new" scientific alliance with nature ("Commencements," *Le Monde* 4:1 [1980]).

28. B. Magnus, "Nietzsche and the Project of Bringing Philosophy to an End," *Journal of the British Society for Phenomenology* 14: 315 (1983).

29. KSA XII, p. 191, my emphasis. On limited truth (tautology): cf. UWL 1, KSA I, p. 878; KSA VII, pp. 500, 493, u.a.

30. "Dass es dauernde Dinge gebe, dass es gleiche Dinge gebe, dass es Dinge, Stoffe, Körper gebe, dass ein Ding Das sei, als was es erscheine, dass unser Wollen frei sei, dass was für mich gut ist, auch an und für sich gut sei" (FW 110, KSA III, p. 469), GS, p. 169.

31. TI. "Die Wirklichkeit zeigt uns einen einzückenden Reichthum der Typen, die Üppigkeit eines verschwenderischen Formenspiels und -

wechsels" (GD, KSA VI, p. 86). This should be considered together with
Nietzsche's rebuke to the natural sciences: "Man soll es [Dasein] vor
Allem nicht seines vieldeutigen Charakters entkleiden wollen: das
Fördert der gute Geschmack, meine Herren, der Geschmack der
Ehrfurcht vor Allem was über euren Horizont geht" (FW 373, KSA III,
p. 625; cf. GS, Preface 4).

32. "Solche irrthümliche Glaubensätze . . . wurden selbst innerhalb
der Erkenntniss zu den Normen, nach denen man "wahr" und 'unwahr'
bemass—bis hinein in die entlegensten Gegenden der reinen Logik" (FW
110, KSA III, p. 469).

33. " . . . Eine willkürliche Festsetzung des Begriffs 'wahr', die sich
nicht beweisen läßt! Es ist ein einfaches 'das *soll* als wahr gelten, soll
"wahr" heißen!' Im Hintergrunde steht der Nützen einer solchen Geltung des Begriffs 'wahr' . . . Das *bedeutet also*: 'was sich beweisen läßt,
ist wahr' setzt bereits *Wahrheiten als gegeben voraus*" (KSA XII, p.
191).

34. "Denn das Beweisbare appellirt an das Gemeinsamste in den
Köpfen (an die Logik): weshalb es natürlich nicht mehr ist als ein Nüt-
zlichkeits-Maaßstab im interesse der Meisten" (ibid.).

35. "Die größte Fabelei ist die von Erkenntniß" (KSA XII, p. 141),
WP 555.

36. "Man müßte *wissen*, was Sein *ist*, um zu *entscheiden*, ob dies
und jenes real ist (z.b. 'die thatsachen des Bewußtseins'); ebenso was
Gewißheit ist, was *Erkenntnis* ist und dergleichen.—Da wir das aber
nicht wissen, so ist eine Kritik des Erkenntnißvermögens unsinnig: wie
sollte das Werkzeug sich selber kritisiren können, wenn es eben nur *sich*
zur Kritik gebrauchen kann? Es kann nicht einmal sich selbst definiren!"
(KSA XII, pp. 104–5). See note 49 below.

37. "Mein Auge, wie stark oder schwach es nun ist, sieht nur ein
Stück weit, und in diesem Stück webe und lebe ich, diese Horizont-Linie
ist mein nächstes grosses und kleines Verhängniss, dem ich nicht ent-
laufen kann. Um jedes Wesen legt sich derart ein concentrischer Kreis,
der einen Mittelpunct hat und der ihm eigenthümlich ist. Ähnlich
schliesst uns das Ohr in einen kleinen Raum ein, ähnlich das Getast" (M
117, KSA III, p. 110), D, p. 117.

38. "Ihre Empfindung führt nirgends in die Wahrheit, sondern beg-
nügt sich Reize zu empfangen" (UWL, KSA I, p. 876), TL, p. 80.

39. "Die Gewohnheiten unserer Sinne haben uns in Lug und Trug
der Empfindung eingesponnen: diese wieder sind die Grundlagen aller
unserer Urtheile und 'Erkenntnisse',—es giebt durchaus kein Entrinnen,
keine Schlupf- und Schleichwege in die wirkliche Welt! Wir sind in
unserem Netze, wir Spinnen, und was wir auch darin fangen, wir können gar nichts fangen, als was sich eben in unserem Netze fangen lässt"
(KSA I, p. 876), TL, p. 247.

40. "Die adäquate Ausdruck aller Realitäten" (UWL, KSA I, p. 876).

41. "Die angenehmen, Leben erhaltenden Folgen der Wahrheit: gegen die reine folgenlose Erknenntniss ist er gleichgültig, gegen die vielleicht schädlichen und zerstörenden Wahrheiten sogar feindlich gestimmt" (UWL, KSA I, p. 878), TL, p. 81.

42. "Ähnliche Qualitäten, sollten wir sagen, statt 'gleich' . . . und 'ähnlich' für uns. Es kommt nichts zweimal vor, das Sauerstoff-atom ist ohne seines Gleichen, in Wahrheit, für uns *genügt* die Annahme das es unzählig gleiche giebt" (KSA IX, p. 531).

43. "Das Ähnliche is kein Grad des Gleichen: sondern etwas vom Gleichen völlig Verschiedenes" (KSA IX, p. 505).

44. "Das Übersehen des Individuellen und Wirklichen giebt uns den Begriffe, wie es uns auch die Form giebt, wohingegen die Natur keine Formen und Begriff, also auch keine Gattungen kennt, sondern nur ein für uns unzugängliches und undefinirbares X" (UWL, KSA I, p. 880), TL, p. 83.

45. Cf., for example, FW 354; JGB 16; WP 530, KSA XII; etc., and see below. We can note here this is the nexus of a possible correspondence with the later Wittgenstein and, in another sense, with the conceptually schematic precisions of Quine and, to an even closer degree, Davidson. Yet Gunter Abel, for one, finds this last connection particularly compelling, I disagree with Abel's perception of such analytic convergence. Below I try to demonstrate that where the logical reticence of the thinking of Davidson and perhaps even more (because of its rhetorical effect apart from all ironic niceties) the writings of Wittgenstein inevitably *end*, Nietzsche *begins*.

46. Cf. KSA XIII, p. 193.

47. Ibid., pp. 271, 371, 373.

48. "Denn auch unser Gegensatz von Individuum und Gattung ist anthropomorphisch und entstammt nicht dem Wesen der Dinge, wenn wir auch nicht zu sagen wagen, dass er ihm nicht entspricht: das wäre nämlich eine dogmatische Behauptung und als solche ebenso unerweislich wie ihr Gegentheil" (UWL, KSA I, p. 880), TL, pp. 83–84.

49. "Um eine solche Unterscheidung machen zu können, müßte man sich unsern Intellekt mit einem widerspruchsvollen Charakter behaftet denken: einmal, eingerichtet auf das perspektivische Sehen, wie dies noth thut, damit gerade Wesen unsrer Art sich im Dasein erhalten können, andrerseits zugleich mit einem Vermögen, eben dieses perspektivische Sehen als perspektivisches, die Erscheinung als Erscheinung zu begreifen" (ibid., XII, p. 241).

50. "Wer zum Beispiel das 'Gleiche' nicht oft genug aufzufinden wusste, in Betreff der Nahrung oder in Betreff der ihm feindlichen Thiere, wer also zu langsam subsumirte, zu vorsichtig in der Subsump-

tion war, hatte nur geringere Wahrscheinlichkleit des Fortlebens als Der, welcher bei allem Aehnlichen sofort auf Gleichheit rieth. . . . die nicht genau sehenden Wesen hatten einen Vorsprung vor denen, welche Alles 'im Flusse' sahen" (FW 111, KSA III, pp. 471–72), FW, p. 171.

51. Thus he prefaces his comment that physical science be considered nothing but a (very uncautious) interpretation of the world with the following excuse/apology: "Man vergebe es mir als einem alten Philogen, der von der Bosheit nicht lassen kann, auf schlechte Interpretations-Künste den Finger zu legen" (JGB 22, KSA V, p. 37). But see too R. Schmidt's discussion of Nietzsches's "Blitzbild" noted below.

52. "Unser Denken ist oberflächlich und zufrieden mit der Oberfläche, ja es merkt sie nicht" (M 125, KSA III, p. 116), D, p. 76.

53. This is the spirit that must be added to the conclusion cited above in the quotation from the 1885–86 *Nachlaß*: "wie sollte das Werkzeug sich selber critisiren können, wenn es eben nur sich zur Kritik gebrauchen kann?" (KSA XII, p. 105).

54. "Versuchen wir den Spiegel an sich zu betrachten, so entdecken wir endlich Nichts, als die Dinge auf ihm. Wollen wir die Dinge fassen, so kommen wir zuletzt wieder auf Nichts, als auf den Spiegel.—Diess ist die allgemeinste Geschichte der Erkenntniss" (M 243, KSA III, pp. 202–3), D, p. 141.

55. "Diess ist der eigentliche Phänomenalismus und Perspektivismus, wie ich ihn verstehe: die Natur des thierischen Bewusstseins bringt es mit sich, dass die Welt, deren wir bewusst werden können, nur eine Oberflächen- und Zeichenwelt ist, eine verallgemeinerte, eine vergemeinerte Welt,—dass Alles was bewusst wird, ebendamit flach, dünn, relativ-dumm, generell, Zeichen, Heerden-Merkzeichen wird . . . Es ist, wie man erräth, nicht der Gegensatz von Subjekt und Objekt, der mich hier angeht: diese Unterscheidung überlasse ich den Erkenntnistheoretikern, welche in den Schlingen der Grammatik (der Volks-Metaphysik) hängen geblieben sind. Es ist erst recht nicht der Gegensatz von 'Ding an sich' und Erscheinung: denn wir 'erkennen' bei weitem nicht genug, um auch nur so scheiden zu dürfen. Wir haben eben gar kein Organ für das *Erkennen*, für die 'Wahrheit'"(FW 354, KSA III, p. 593).

56. JGB 13.

57. "Es macht mir wenig aus, ob sich heute Einer mit der Bescheidenheit der philosophischen Skepsis oder mit religiöser Ergebung sagt: 'das Wesen der Dinge ist mir unbekannt' oder ein Andrer, Muthigerer, der noch nicht genug Kritik und Mißtrauen gelernt hat: 'Das Wesen der Dinge ist mir zu einem guten Theile unbekannt.' Beiden gegenüber halte ich aufrecht, daß sie unter allen Umständen noch viel zu viel zu wissen vorgeben, zu wissen sich einbilden" (KSA XII, pp. 240–41; cf. KSA XII, p. 136).

58. "Das will sagen: ausgestattet mit einem Glauben an die 'Real-

ität', wie als ob sie die einzige wäre, und wiederum auch mit der Einsicht über diesen Glauben, daß er nämlich nur eine perspektivische Beschränktheit sei in Hinsicht auf eine wahre Realität. Ein Glaube aber, mit dieser Einsicht angeschaut, ist nicht mehr Glaube, ist als Glaube aufgelöst. Kurz, wir dürfen uns unsern Intellekt nicht dergestalt widerspruchsvoll denken, daß er ein Glaube ist und zugleich ein Wissen um diesen Glauben als Glauben" (ibid.).

59. This must be compared with Clark's discussion of just this expression of Nietzsche's claims.

60. "Es ist genug, die Wissenschaft als möglichst getreue Anmenschlichung der Dinge zu betrachten, wir lernen immer genauer uns selber beschreiben, indem wir die Dinge und ihr Nacheinander beschreiben. Ursache und Wirkung: . . . in Wahrheit steht ein continuum vor uns . . . Die Plötzlichkeit, mit der sich viele Wirkungen abheben, führt uns irre, es is aber nur eine Plötzlichkeit für uns" (FW 112, KSA III, p. 473).

61. "wie das Insekt oder der Vogel eine ganz andere Welt percipiren als der Mensch, und dass die Frage, welche von beiden Weltperceptionen richtiger ist, eine ganz sinnlose ist, da hierzu bereits mit dem Maassstabe der *richtigen Perception* d.h, mit einem *nicht vorhandenen* Maassstabe gemessen werden müsste" (UWL, KSA I, p. 884), TL, p. 86.

62. In this schematic ascription, Nietzsche is as Kantian as most modern perceptual psychology.

63. "Was ist 'erkennen'? Zurückführen von etwas Fremden auf etwas Bekanntes, Vertrautes" (KSA XII, p. 187; cf. FW 355), WP 575, 608.

64. Peirce, "The Fixation of Belief."

65. "Das Ganze der organischen Welt ist die Aneinanderfädelung von Wesen mit erdichteten kleinen Welten um sich: indem sie ihre Kraft, ihre Begierden, ihre Gewohnheiten in den Erfahrungen außer sich heraus setzen, als ihre *Außenwelt* . . . von sich selber haben sie natürlich ebenfalls nur eine solche falsche erdichtete vereinfachte Vorstellung" (KSA XI, p. 503).

66. "Der organische Prozeß setzt fortwährendes Interpretiren voraus" (KSA XII, p. 140), WP 643.

67. "Unsre Bedürfnisse sind es, die die Welt auslegen . . . Jeder Trieb ist eine Art Herrsucht, jeder hat seine Perspective, welche er als Norm allen übrigen Trieben aufzwingen möchte" (KSA XII, p. 315), WP 481.

68. "Die Regelmäßigkeit schläfert den fragenden (d.h. fürchtenden) Instinkten ein" (KSA XII, p. 188).

69. "Erster Grundsatz: das woran wir uns *gewöhnt* haben gilt uns nicht mehr als Räthsel, als Problem. . . . alles, was *regelmäßig* geschieht, scheint uns nicht mehr fragwürdig. Deshalb ist die *Regelsuchen* der erste Instinkt des Erkennenden: während natürlich mit der Feststellung der

Regel noch gar nichts 'erkannt' ist!—Daher der Aberglaube der Physiker: wo sie verharren können d.h. wo die Regelmäßigkeit der Erscheinungen die Anwendung von abkürzenden Formeln erlaubt, meinen sie, sei erkannt worden. Sie fühlen "Sicherheit": aber hinter dieser intellektuellen Sicherheit steht die Beruhigung der Furchtsamkeit: *sie wollen die Regel*, weil sie die Welt der Furchtbarkeit entkleidet" (KSA XII, pp. 187–88).

70. KSA I, p. 877; TL, p. 81.

71. "Der *Erkenntniß-Apparat* als Verkleinerungsapparat . . . Als Mittel des Ernährungs-Apparates" (KSA XI, p. 111).

72. "Alles, was den Menschen gegen das Thier abhebt, hängt von dieser Fähigkeit ab, die anschaulichen Metaphern zu einem Schema zu verflüchtigen, also ein Bild in einen Begriff aufzulösen" (UWL, KSA I, p. 881), TL, p. 84.

73. Section 5.

74. "Es wird eine gleichmäßig gültige und verbindliche Bezeichnung der Dinge erfunden und die Gesetzgebung der Sprache giebt auch die ersten Gesetz der Wahrheit" (UWL, KSA I, p. 877), TI, p. 81.

75. Reality understood as "multiple realities" or the Real.

76. Thus Nietzsche may be said to express an extreme "phenomenalism" (as Hollingdale has it) or what I name a "hyperrealism" in his account of Nature as the Real: as unremitting chaos. If this is held to imply a correspondence theory of truth (as Wilcox and Danto hold, to different degrees), where a "true" account is to be one that stands in accord with the Real, it must be affirmed that this is so if one, first, admits a completely incoherent— because necessarily various—notion of correspondence. Second, this can only be reserved against Nietzsche's notion of truth as fiction, that is, either as difference-nihilating or life-enhancing abstraction. It must be emphasized that the nature of naming as such makes a Nietzschean correspondence theory an absurdity: language would have to be remade in accordance with the ambiguity or multiplicity of the world, and that would mean, the iterative multiplicity of a world we may not experience because of the generalizing effects of our senses. So construed, a Nietzschean account would require a language assigning unique names to all things, without any abstractions or judgments of similarities: i.e., without collective nouns, without names for "things" of certain kinds. Nietzsche attacks the fiction of such collective concepts or things or values in themselves.

77. "Auch die Wissenschaft thut, was der Mensch immer gethan: *etwas* von sich, das ihm als verständlich, als *wahr* gilt, zur Erklärung benutzen alles Anderen—*Vermenschlichung* in summa" (KSA XI, p. 132).

78. "Alle unsere *Erkenntnissorgane und -Sinne* sind nur entwickelt in Hinsicht auf Erhaltungs- und Wachsthums Bedingungen/Das *Ver-*

trauen zur Vernunft und ihren Kategorien, zur Dialektik, also die *Werth-schätzung* der Logik beweist nur die durch Erfahrung bewiesene *Nützlichkeit* derselben für das Leben: nicht deren '*Wahrheit*'" (KSA XII, p. 352).

79. "Warum dürfte die Welt, *die uns etwas angeht*—nicht eine Fiktion sein?" (JGB 34, KSA V, p. 54), BGE, p. 47.

80. This leads to the presumptuous ideal of pure, objective knowledge: "hier wird immer ein Auge zu denken verlangt, das durchaus keine Richtung haben soll"(GM III:12, KSA V, p. 365), GT 5.

81. For Nietzsche, far from any subscription to relativism, "internal" facts have no special privilege, as immediate knowledge. Nietzsche applies the same criticisms to our knowledge of the inner as of the external world. JGB 16, 17, 19, 21, among others.

82. "Wir haben eben gar kein Organ für das Erkennen, für die 'Wahrheit': wir 'wissen' (oder glauben oder bilden uns ein) gerade so viel als es im Interesse der Menschen-Heerde, der Gattung, nützlich sein mag: und selbst was hier 'Nützlichkeit' gennant wird, ist zuletzt auch nur ein Glaube" (FW 354, KSA III, p. 593), GS, p. 300.

83. Cf. JGB, KSA V, pp. 24, 34. A. Nehamas, in his *Life as Literature*, has recently drawn attention to the fact that simplification as such is not falsification. His claim in this would be more interesting if (1) Nietzsche had not said that that simplification is falsification, and not only with that single verbal formula; and (2) if Nehamas did not then (in subsequent paragraphs) retract his charge and show how one could understand how Nietzsche could (justifiably) say what he said: that simplification falsifies, namely, when a simplified position is conjoined with a claim to render an absolute account. But this "simplified" position reaches to the roots of idealism, formalism, and a good deal of the analytic enterprise, not to mention science itself. For literalists or for those who have never read Nietzsche's own account, Nehamas's version rewards consideration.

84. "*Erkenntniß ist Fälschung des Vielartigen und Unzählbaren zum Gleichen, Ähnlichen, Abzählbaren*" (KSA XI, 506). Note that Nietzsche here maintains, with the very word, the so-called falsification thesis that M. Clark denies, on his behalf, as error.

85. "Es giebt zuletzt eine Rangordnung seelischer Zustände, welcher die Rangordnung der Probleme gemäss ist; und die höchsten Probleme stossen ohne Gnade Jeden zurück, der ihnen zu nahen wagt, ohne durch Höhe und Macht seiner Geistigkeit zu ihrer Lösung vorherbestimmt zu sein" (JGB 213, KSA V, p. 148), BGE, p. 126.

86. JGB 4.

87. "Ich erkenne etwas Wahres nur als Gegensatz zu einem wirklich lebendigen Unwahren. So kommt das Wahre ganz kraftlos, als Begriff, zur Welt und muß sich durch *Verschmelzung mit Lebendigen*

Irrthümern erst Kräfte geben! Und darum muß man die Irrthümer leben lassen und ihnen ein großes Reich zugestehen" (KSA IX, p. 506).

88. "Wir müssen das Irren lieben und pflegen, es ist der Mutterschooß des Erkennens" (KSA IX, p. 504).

89. "Welcher sagt, 'ich denke, und weiss, dass dies wenigstens wahr, wirklich gewiss ist'—der wird bei einem Philosophen heute ein Läheln und zwei Fragezeichnen bereit finden. 'Mein Herr, wird der Philosoph vielleicht ihm zu verstehen geben, es ist unwahrscheinlich, dass Sie sich nicht irren: aber warum durchaus Wahrheit?'" (JGB 16, KSA V, p. 30).

90. "Dem Dasein eine äthetische Bedeutung geben, *unseren Geschmack an ihm mehren*, ist Grundbedingung aller Leidenschaft der Erkenntniß" (KSA IX, p. 504; cf. 505; JGB 24, 56).

91. It is likewise no accident that comments such as "Die Heerden-Menschen und die *selbsteignen* Menschen" and "*Heerdenmenschen* und *Sondermenschen*!" appear in the same notebook (KSA IX, pp. 516 and 524, respectively).

92. "Für Jeden, der mit einem großen Fragezeichnen wie mit seinem Schicksale zusammen gelebt hat" (KSA XII, p. 157).

93. For the characterization of Nietzsche's "embarrassing" politics see Sokel, "The Political Uses and Abuses of Nietzsche in Walter Kaufmann's Image of Nietzsche." For a further discussion of this embarrassing connection in express terms of the very utility of the "political Nietzsche" see Ansell-Pearson, "The Significance of Michel Foucault's Reading of Nietzsche," especially pp. 268–70. See also Minson, *Genealogies of Morals* for a completely, disastrously "embarrassed" reading of Nietzsche, a reading which painfully dates (a recent 1985!) and disfigures an otherwise valuable book. And for a more cavilling account of the connection between Foucault's reading of Nietzsche see Pizer, "The Use and Abuse of 'Ursprung.'"

94. "Man hat kein Recht, weder auf Dasein, noch auf Arbeit, noch gar auf 'Glück': es steht mit dem einzelnen Menschen nicht anders als mit dem niedersten Wurm" (KSA XIII, p. 98). Cf., "dass du aber ein *Recht* auf Existenz habest, dünkt dich unwiderleglich!" (FW 335, KSA III, p. 561). See also chapter 6.

95. "Eine Cultur der Ausnahme, des Versuchs, der Gefahr, der Nüance als Folge eines großen Kräfte-Reichtums" (KSA XII, p. 41).

96. "Stelldichein, wie es scheint, von Fragen und Fragezeichen" (JGB 1, KSA V, p. 15), BGE, p. 15.

97. "Das bisher noch kein Philosoph Recht behalten hat, und dass eine preiswürdigere Wahrhaftigkeit in jedem kleinen Fragezeichnen liegen dürfe, welches ihr hinter eure Leibworte und Lieblingslehren (und gelegentlich hinter euch selbst) setzt" (JGB 25, KSA V, p. 42), BGE, p. 38.

98. Cf. JGB 26.

99. JGB 1.

100. "Begiebt sich in ein Labyrinth, er vertausendfältigt die Gefahren, welche das Leben an sich schon mit sich bringt; von denen es nicht die kleinste ist, dass Keiner mit Augen sieht, wie und wo er sich verirrt, vereinsammt, und stückweise von irgend einem Höhlen-Minotaurus des Gewissens zerrissen wird. Gesetzt, ein solcher geht zu Grunde, so geschieht es so ferne vom Verstandniss der Menschen, dass sie es nicht fühlen und mitfühlen:—und er kann nicht mehr zurück!" (JGB 29, KSA V, p. 48), BGE, p. 42.

101. "Es giebt Höhen der Seele, von wo aus gesehen selbst die Tragödie aufhört, tragisch zu wirken" (JGB 30, KSA V, p. 48), BGE, p. 43.

102. Nietzsche writes that the text allows infinite interpretations: "insofern wir die Möglichkeit nicht abweisen können, dass sie unendliche Interpretationen in sich schliesst" (FW 374, KSA III, p. 627). He does not claim (contra Nehamas and others) that there are infinite interpretations. His conclusion from his premises is more modest: there is no (one) correct interpretation.

103. See Heidegger, *Nietzsche I*, pp. 243–54.

104. FW 121, KSA III, p. 477. Cf. MA 33. "Wir haben eine Conception gemacht, um in einer Welt leben zu können, um gerade genug zu percipiren, daß wir noch es *aushalten*" (KSA XIII, p. 271).

105. "Jene feinere Redlichkeit und Skepsis hätte überall dort ihre Entstehung wo zwei engegengesetzte Sätze auf das Leben *anwendbar* ershienen, weil sich beide mit den Grundirrthümern vertrugen . . . wo neue Sätze sich dem Leben zwar nicht nützlich, aber wenigstens auch nicht schädlich zeigten" (FW 110, KSA III, p. 470), GS, p. 170.

106. GS 110, p. 170.

107. GS, p. 170. "Allmählich füllte sich das menschliche Gehirn mit solchen Urtheilen und Überzeugungen, so entstand in diesem Knäuel Gährung, Kampf und Machtgelüst. Nützlichkeit und Lust nicht nur, sondern jede Art von Trieben nahm Partei in dem Kampf um die 'Wahrheiten'; der intellectuelle Kampf wurde Beschäftigung, Reiz, Beruf, Pflicht, Würde—: das Erkennen und das Streben nach dem Wahren ordnete sich endlich als Bedürfniss in die anderen Bedürfnisse ein" (FW 110, KSA III, p. 470).

108. "Die Erkenntnisse und jene uralten Grundirrthümer auf einander stiessen" (FW 110, KSA III, p. 470).

109. "Beide als Leben, beide als Macht, beide in dem selben Menschen" (FW 110, KSA III, p. 470).

110. "Inwieweit verträgt die Wahrheit die Einverliebung?" (FW 110, KSA III, p. 470), GS, p. 170.

111. Note that this is not identical with Heidegger's expression of

this same notion. See chapter 8 and my "Heidegger's Reading of Nietzsche and Technology: A Musical Retrieve of the Range of Value and Light Feet," in *The Fate of the New Nietzsche*, ed. Keith Ansell Pearson and Howard Caygill.

112. Cf. KSA IX, p. 537.

113. Hollingdale, *Nietzsche*, p 134. Nietzsche, however, offers the counter of an odd, isolated gloss: "Phenomeno-manie" (KSA XII, p. 239). In this context, Nietzsche's condemnations of what he names the *"hyperbolische Naivetät"* (KSA XIII, p. 49) of human beings must be considered.

114. "Ich bin immer erstaunt, ins Freie tretend zu denken, mit welcher herrlichen Bestimmtheit alles auf uns wirkt, der Wald so und der Berg so und daß gar kein Wirrwarr und Versehen und Zögern in uns ist in Bezug auf alle Empfindungen" (KSA IX, p. 537).

115. "Der Haß gegen das Werden, gegen die sorgfältige Betrachtung des Werdens" (KSA XIII, p. 444).

116. *"Die eigentlichen Philosophen aber sind Befehlende und Gesetzgeber"* (JGB 211, KSA V, p. 145), BGE, p. 122.

117. The conclusion that Nietzsche seeks a kind of "world-mastery" is not an impossible one: given Nietzsche's exhortation of self- and world-overcoming as the ultimate project for free spirits. But one would do well to avoid this conclusion: its aesthetic sense is elusive and mistaking that sense is potentially dangerous.

118. "Illusion daß etwas *erkannt* sei, wo wir eine mathematische Formel fur das Geschehen haben: es ist nur *bezeichnet, beschrieben*: nichts mehr!" (KSA XII, p. 105), WP 628, p. 335.

119. "Das Muster einer vollständigen *Fiktion* . . . Hier wird ein Denken *erdichtet* . . . alle Affekte, alles Fühlen, und Wollen wird hinweg gedacht" (KSA XI, p. 505).

120. "Es kommt dergleichen in der Wirklichkeit nicht vor: diese ist unsäglich anders complicirt. Dadurch daß wir jene Fiktion als *Schema* anlegen, also das thatsächliche Geschehen beim Denken gleichsam durch einen Simplifikations-Apparat *filtriren*: bringen wir es zu einer *Zeichenschrift* und *Mittheilbarkeit* und *Merkbarkeit* der logischen Vorgänge . . . In der Wirklichkeit giebt es kein logisches Denken, und kein Satz der Arithmetik und Geometrie kann aus ihr genommen sein, weil er gar nicht vorkommt" (ibid.).

121. "Wie könnten wir auch erklären! Wir operiren mit lauter Dingen, die es nicht giebt, mit Linien, Flächen, Körpern, Atomen, theilbaren Zeiten, theilbaren Raumen" (FW 112, KSA III, p. 473).

122. "C'est que la science n'a jamais repensé pour son propre compte le rapport de l'homme et de 'l'être' tel qu'il avait été défini par la spéculation philosophique. Science implique le postulat de l'affinité de 'l'être' et de l'essence de l'homme, pour autant que celui-ci enveloppe la

possession de la *ratio*. La systématisation des lois, leur réduction au plus petit nombre possible, la certitude que le progrès du savoir conditionne l'épanouissement psychologique et moral de l'humanité, n'ont de sens que par la croyance tacite selon laquelle tout se passe comme s'il y avait un plan de l'univers, plan qui lui-même renverrait à une sagesse divine alliant en elle l'intelligence absolue à la bonté la plus généreuse" (Granier, *Le problème de la vérité dans la philosophie de Nietzsche*, p. 84).

123. GM 3; see further, chapter 6 below.

124. "Fastidieuse à force de banalité, selon laquelle la science implique toujours nécessairement *une* métaphysique. Il s'agit d'établir que la science est, selon Nietzsche, *la* Métaphysique elle-même . . . [Nietzsche] *veut contester la science pour surmonter la Métaphysique tout court*" (Granier, *Le probléme de la vérité dans la philosophie de Nietzsche*, p. 73).

125. "La compréhension métaphysique de 'l'être' comme Idéal entraîne automatiquement une certain définition du critère de la vérité" (ibid., p. 60).

126. The notion of freedom suggested here should be defined as necessary and perspectival.

127. "Es giebt vielerlei Augen. Auch die Sphinx hat Augen: und folglich giebt es vielerlei 'Wahrheiten', und folglich giebt es keine 'Wahrheit'" (KSA XI, p. 498), WP 540.

128. "'Alle Wahrheit ist einfach'—Ist das nicht zweifach eine Lüge?" (GD, KSA VI, p. 59: TI, p. 33).

129. "'Alle Wahrheit ist einfach: das ist eine zweifache Lüge. Alles, was einfach ist, ist bloß imaginär, ist nicht 'wahr'. Was aber wirklich, was wahr ist, ist weder Eins, noch auch nur reduzierbar auf eins" (KSA XIII, pp. 478–79).

130. Cf. KSA XII, p. 141; KSA XIII, p. 271.

131. Ambivalence is the affective interest varying with and within perspectives. Ambivalence reflects the ineluctable obliquity of perspective. In Heideggerian language, of course, this may be spoken of as "unconcealment." Thus Nietzsche's "phenomenological" interest is fundamentally ambivalent: "Die Welt ist Erscheinung—aber nicht *wir* allein sind Ursache daß sie erscheint. Noch von einer anderen Seite her ist sie unreal" (KSA VII, p. 495).

132. "'Die Dinge' betragen sich nicht regelmäßig, nicht nach einer *Regel*: es giebt keine Dinge (—das ist unsere Fiktion) sie betragen sich ebensowenig unter einem Zwang von Nothwendigkeit" (KSA XIII, p. 257), WP 634.

133. "Es giebt kein Gesetz: jede Macht zieht in jedem Augenblick ihre letzte Consequenz. Gerade daß es keine mezzo termine giebt, darauf beruht die Berechenbarkeit" (KSA XIII, p. 258), WP 634.

134. "Wir haben Einheiten nöthig, um rechnen zu können: deshalb

ist nicht anzunehmen daß es solche Einheiten giebt" (KSA XIII, p. 248), WP 635.

135. "Wir gehören zum Charakter der Welt, das ist kein Zweifel! (KSA XIII, p. 33).

136. "Die Welt, abgesehen von unserer Bedingungen, in ihr zu leben, die Welt, die wir *nicht* auf unser Sein, unsere Logik, und psychologischen Vorurtheile reduzirt haben existirt nicht als "Welt an sich" sie ist essentiell Relations-Welt: sie hat, unter Umständen, von jedem Punkt aus ihr *verschiedenes Gesicht*; ihr Sein ist essentiell an jedem Punkt anders" (ibid.; WP 568).

137. "Logik und Mechanik sind nur auf das *Oberflächlichste* anwendbar . . . ein Bewältigung der Vielheit durch eine Kunst des Ausdrucks" (ibid., XII, p. 190; cf. FW 373).

138. "Der Geschmack der Ehrfurcht vor Allem, was über eurem Horizont geht!" (FW 373, KSA III, p. 625). Cf. KSA XII, p. 257).

139. This attention to little, to contradictory detail, the detail of the world we actually "live in" is what atracts such polymath philosophers (mathematicians) as M. Serres to science. And he is not wrong to find this appealing.

140. See Chargaff, *The New Heraclitean Fire*, in addition to his many other books on medical and natural philosophic themes.

141. G. Abel analyzes the value of Nietzsche's discursive claims in terms of "logicality." See Abel, *Nietzsche*, esp. chapters 6, 8.

142. "Es ist wunderbar, daß für unsere Bedürfnisse (Maschinen, Brücken usw.) die Annahmen der Mechanik ausreichen, es sind eben sehr grobe Bedürfnisse, und die 'kleinen Fehler' kommen nicht in Betracht" (KSA XI, p. 531; cf. FW 46).

143. KSA IX, p. 502. Rüdiger Schmidt finds the source of this metaphor in the young Nietzsche's experience of landscape images glimpsed through the windows of a railroad car. See Schmidt, *"Ein Text ohne Ende für den Denkenden." Studien zu Nietzsche*, p. 19. It is irrelevant whether one traces the image to this early experience or reads out of it as Schmidt does, a kind of semiotic of loss and of passing. For the point here is that the train image with its window-framed access to the world inspired not only Nietzsche but also Ingmar Bergmann and Jacques Lacan, captures the sight of the human being, the knower in the world. The resonance with Heraclitus, while less psychoanalytically revealing and less psychologistically appealing (we find ourselves more easily on the train with its "Cinematic Blitzbilder" than we can represent ourselves to ourselves in the "ewigen Flusse" as such), is also doubtless more philosophically viable.

144. "Die Forderung einer *adäquaten Ausdrucksweise ist unsinnig*: es liegt im Wesen einer Sprache, eines Ausdrucksmittels, eine bloße Relation auszudrücken" (KSA XIII, p. 303).

145. "Es kommt nichts zweimal vor, das Sauerstoff-atom ist ohne seines Gleichen, in Wahrheit, für uns *genügt* die Annahme, daß es unzählige gleiche giebt" (ibid., IX, pp. 531–32; cf. KSA XI, p. 455).

146. "Es giebt nichts Unveränderliches in der Chemie, das ist nur Schein, ein bloßes Schulvorurtheil. Wir haben das Unveränderliche *eingeschleppt*, immer noch aus der Metaphysik, meine Herren Physiker. Es ist ganz naiv von der Oberfläche abgelesen, zu behaupten, daß der Diamant, der Graphit, und der Kohle identisch sind. Warum?" And he answers, "Bloß weil man keinen Substanz-Verlust durch die Wage constatieren kann!" (ibid., XIII, p. 374).

147. Cf. ibid., IX, pp. 530, 531; XIII, p. 374. On the other hand, current research in high-energy physics indicates by means of its proliferation of particles (often onetime observtion or "Mayfly" particles) that atomic and subatomic consistency may be no more than an epiphenomenon of wild flux and difference on a quark (subquark) level. Atomism, however, is a very old way of seeing the world.

148. "Diese 'logische', diese 'kunstlerische' ist unsere Fortwährende Thatigkeit. Was hat dieser Kraft so souverän gemacht? Offenbar, daß ohne sie, vor Wirrwarr der Eindrücke, kein lebendes Wesen lebte" (ibid., XI, pp. 435–36).

149. In addition to clasical English works on the eco-psychology of perception by Gibson and Marr see the more philosophical accounts included in Wuketis et. al. on evolutionary epistemology.

CHAPTER 4

Under the Optics of Art and Life: Nietzsche and Science

RESUMÉ: THE ECOPHYSIOLOGICAL GROUND OF KNOWLEDGE

As seen in the last chapter, Nietzsche's basic epistemological question is the question concerning the possibility of saying what counts as knowledge. Thus, as we have seen, Nietzsche asks, "How is the fact of knowledge possible? Is knowledge a fact at all? what is knowledge? If we do not know what knowledge is, we cannot possibly answer the question whether there is knowledge."[1] But if the ideal of universal knowledge is problematic for Nietzsche, what is to count as truth? More particularly, how is science to proceed without the conceptual tools of generalization? It would seem that Nietzsche's (anti-) epistemic standard requires that science be reduced to an anecdotal, wholly inductive affair, eschewing even general conclusions on the pain of being "inadequate" to the complexity of nature. How express a "perspectival optic" that might be true to this same complexity? What advantage is gained by Nietzsche's critical epistemology of body and world for an expression of the perspectival optics of art and life?

For Nietzsche, the evolutionary force of knowledge must favor those who categorize things and events more rather than less quickly and more rather than less roughly. The results of this quick and rough approximation of things are not only categorial simplifications but communicational convenience and survival: those who attended to the little differences as if they made a difference perished. For Nietzsche, then, "whatever becomes consious *becomes* by the same token shallow, thin, relatively stupid, general, sign, herd signal."[2]

As organic beings we perceive on the basis of evaluative perspectives. This means that we cannot think of knowledge in the

Platonic way as μάθησις or as eternal truth because we take for absolute truth what is attained merely by way of organic short-cuts in the service of what both Peirce and Nietzsche named "quiesent belief." This organic disposition of our knowing apparatus and self-reference of our epistemic standards is what Nietzsche means when he speaks of science in general and our knowledge of nature in particular as "humanization in summa."[3] In other words, for Nietzsche, "our needs that interpret the world."[4] Thus one may not speak of knowing "all realities" in a way "adequate" to any perception or set of interests other than our own. Yet it is clear that the traditional ideal of knowledge (ὁμοίωσις) requires nothing less. Nietzsche's Copernican revolution is a conservative theory of knowledge that denies traditional knowledge claims where they go too far. Given the complexity, fluidity, and extent of the world to be known, and given our own perspective upon that world and the built in importunity of that perspective *as* perspective, we "simply lack any organ for *knowledge*, for 'truth.'"[5] Knowledge is thus born of error in the interest of power. Accordingly, "Life is the condition of knowing. Error, the condition of life and, *indeed*, erring in the deepest way."[6] What we lack is the metaphysically necessary given of a pure connection, or coordination between knower and known. Thus analytic rigor may still be able to pose as a demarcational criterion for truth, in Nietzsche's expression of such a criterion—"what can be proven is true." But Nietzsche counters that this demand of itself "presumes truth as given in advance."[7] For Nietzsche's part, without a given, without the grounding of truth in advance, we turn to consider the character of the world of which we are a part, which world for Nietzsche does not obey laws, if it is also plain that "things are unable to be other than what they are."[8] The world is itself interpretive (Will to Power), and the issue for Nietzsche that brings us to truth and life again is the aesthetic expression of that interpretive dynamic.

SCIENCE AND NIHILISM

Despite a radicality we have already seen to be problematic in its complexity, it is useful to state Nietzsche's position on the problem of science and the question of nihilism as these bear upon or reflect one another. For Nietzsche, the Western cultural ideal of science is inevitably a manifestation of reactive being. This in turn means that the culture of science, which Nietzsche will call the "latest and noblest form of the ascetic ideal," is in fact the insidiously vital representation of a culture organized against the essence of life. In this way, the culture of science is, in Nietzche's eyes, the culture of nihilism, if nihilism is also a complicated issue to be discussed like pessimism in terms of strength and weakness, or like art, in terms of lack and abundance. But positive, or negatively construed science is nihilistic as it is organized against the nonabsolute facts of life and death, against the ambiguity of change and decay.

Nietzsche evaluates what he calls the ascetic ideal in Western culture as inherently if productively (positively) nihilistic. Yet the primary ascetic ideal, that of Western Christianity appears to have lost much of its importance. The *metarécits* of the enlightenment, humanism, Marxism, and so on appear to have lost much of their influence. How much more diminished then is the *metarécit* of the "Church," of "Christendom," which was itself once superceded by these fading *récits?* What Nietzsche shows, and this is key if one would understand the persistence of the modern in the dynamics of the postmodern movement, that is, the survival of the modern despite all challenges, is the very ascetic mechanic of insistence. In the dissembling aesthetic of asceticism, an ideal persists in another guise, indeed as we shall see, it can persist through under any guise whatever. Thus as we shall see below, the mechanic of *Ressentiment* adopts what it resents, as what it resists, as its own and in its own name. The survival of an ideal or ethos in another guise represents what Nietzsche called "reactive values." Science thus perpetuates the ascetic ideal of reactive religion by other means. Thus although Nietzsche focusses upon Judeo-Christian culture, his analysis of the working of the ascetic ideal entails an equipollent and for our purposes illuminating focus on what is today distinguished from Judeo-Christian culture as science as such. What is more, in its relevance for contemporary sensibilities, the viability of Nietzsche's analysis of ressentiment and the nihilistic manifestation of its ascetic ideal must be sought within a scientific schema.

Like antecedent expressions of the ascetic ideal, the notion of science is fundamentally ambivalent. Despite the necessity of scientific knowledge for the preservation of life in our culture, its inspirational drive is fundamentally nihilistic. Nietzsche suggests that our confidence in the efficiency of (scientific) knowledge grounds the value of science. Hence taken as objective evidence of scientific progress, scientific technologies are effective (working) demonstrations of science's success as knowledge. Indeed, science is far and away the most successful kind of knowledge possible. We have already seen that this efficient interest—reading efficacy as (progressive) truth—can be taken to reveal the nihilism of science in the exclusive perspective of its ideal.[9] As in Nietzsche's exposition of Judeo-Christian moral value schemes, the trajectory of scientific progress outlines the direction of nihilism: "Our natural science is bent upon *destruction* (*Untergang*) in its pursuit of knowledge."[10] Note that this destructive process is not, for Nietzsche, a conscious or deliberate effort in our culture. It is not Nietzsche's claim that anyone pursues science or scientific goals seeking annihilation. Rather what is key here is implicit: the roots of science are to be found in the illusions needed to preserve life:

> Science only *continues* the process *constituted* by the essence of the species, which makes belief in certain things endemic, and which separates and exterminates unbelief. The *similarity* of sensation attained (concerning spatial or temporal sensation or the feeling of large and small) becomes a condition for the existence of the species, [although having nothing to do with the truth].[11]

The aim of science, then, so far from being directed toward destruction, is power over nature: "The whole knowledge apparatus is an apparatus for abstraction and simplification—aimed not at knowledge but at power over things."[12] Nietzsche criticizes the power-economy, the politics of knowledge later in the same collection of notes from 1884: "Science—transforming nature into concept for the sake of the mastery of nature."[13] In accord with this power-drive, Nietzsche finds the preludes for science in magic and alchemy. These latter practices sponsor and reflect "a thirst, a hunger, a taste for *hidden* and *forbidden* powers."[14] This taste for forbidden power is not an expression of what Nietzsche considers higher capacities but is tailored to what he calls the *"plebianism of the spirit."*[15] The interest of emancipated thinkers in science and philosophy, the purposes such *"hidden* and *forbidden* powers" are meant to serve, is human progress or comfort: "the universal green pasture happiness of the herd, with security, safety, comfort, and an easier life for all."[16] The meaning of the nihilism involved in this can make sense only in the light of Nietzsche's understanding of the Will to Power and the magnification of life in the thought of the Eternal Return. Where the expression of the Will to Power is directed to life diminution (as is the case where what is to be secured is no more than the least demanding level of life, and that at any price) the ultimate consequence can only be nihilism. But, once again, this nihilistic circumstance is the effective but not the intended result. Echoing the emphasis in "On the Advantage and Disadvantage of History for Life," Nietzsche writes in a note from the 1872–73 *Nachlaß*: "If I speak of the frightful possibility that knowledge leads to decline, I am far from wishing to pay the current generation a complement: it has no such tendencies of itself. Yet if one views the process of science since the fifteenth century, this very power and possibility [of decline] is obvious."[17] The nihilistic possibility for scientific progress is ambivalent, taking the directionality of this ambivalence in the full literality of the word: the business of going under is either to be in the service of life-enhancement or an incidental, practical consequence of a technological life-style, which, last for Nietzsche, can only diminish the possibility for life in the grand style.

Nietzsche's interest in tracing the ambivalence of the scientific ideal of progress is less that of mapping out the possibilities of a technological disaster or the dangers of a large scale war than the coordinate consequence of reduced opportunities for the articulation and relevance of the life of genius and art in a therapeutically bureaucratic and democratic society oriented to the advancement of commonness.[18] Thus he writes, "Tremendous counter-forces have to be called upon to cross this natural, all too natural progressus in simile, the continuing development of mankind into the similar, ordinary, average, herdlike—into the *common!*"[19] The implications and significance of the nihilism of contemporary culture must always be considered in such an hierarchical perspective. Thus Nietzsche mourns the lost possibility of the life of genius and art. This concern does not mean that Nietzsche is a romantic thinker. Nietzsche's appeal is hardly broad enough for that. The "lost" possibility of an artistic life has only Nietzsche himself and perhaps those who might understand him as rightful mourners. But even this company of free and like spirits cannot claim to be contenders for the life of genius and art. Most of us must remain unmoved by both this possibility as well as its eclipse.[20] Assuming his irrecusable esotericism, Nietzsche's position should seem clearer when he writes (paralleling the conclusion of *On the Genealogy of Morals*):

> given: that we live as a result of an error: what can the "will to truth" be? Isn't it rather a "will to death"—could the efforts of philosophers and men of science [*wissenschaftlichen Menschen*] perhaps be a symptom of decadent, dying-out life, a kind of life-excess of life itself? *Quaeritur?* and here one really could become reflective.[21]

It is not art then but emancipated scientific culture (by and large) that affirms the self-preservation and the democratic exaltation of the smaller self: the little or mass human being. Where, for a perspectivalist philosophy, all knowledge is belief, there is a fundamental difference between ornamental (or aesthetically maintained) belief and the desperate (or nihilistic) faith in one's beliefs. The latter structure embodies a self-preservative project. What Nietzsche criticizes in distinguishing between life-affirmative and life-denying beliefs is the impotent institution of the latter as representative of absolute realities (or "provisional" scientific truths) rather than as invented, compensating projections of an artistic vision. An artistic conscience constitutes the difference between the illusion of truth and the truth of an illusion and, that is to say, between the epistemology of nihilism and the nihilism of the scientific knowledge ideal.

REALITY AND TRUTH: THE DOMINATION OF TRUTH

For Nietzsche, as we have seen, the scientific project is the discovery of the Real as calculable. The *objectivity* of this ratiocinative, calculative interest is not equivalent to innocence; nor does such an objective focus (alone) justify the scientific project.[22] For Nietzsche, "Objectivity and justice have nothing to do with one another."[23] There is no interest-free knowledge project, and science itself, in its logicizing and mathematizing (in the interest of basic natural comprehension—that is, as Nietzsche would say, in the interest of mastery over nature) has a special interest in a fixed image of the world. The fact that mastery over nature entails interference with natural processes as such is not denied in science: the utility of science is, after Kant's regulative analysis, the business of putting (effective) questions to nature. This inquisatorial constraint is hermeneutically suspect (so far as it denies its inherent circularity), but at the same time it is hermeneutically productive (that is, and on its own terms, the tautology of a vicious circle has excellent logical credentials). As Heidegger expresses this classic phenomenological observation: "Science always encounters only what *its* kind of representation has admitted beforehand as an object possible for science . . . scientific knowledge is compelling."[24]

Assuming that nature answers those questions that express the interests of our culture, the philosopher can undertake to ask the objective question: what is so marvelous about nature that it responds to our questions? or, queried from the subjective side: what excellence characterizes our cultural interests that nature turns out to have such a responsive affinity for just the questions we ask? We have seen that Nietzsche's answer would detail the following insights: we are of the world and in such a way as to interpret that world after our own image (the subjective perspective), where everything in the world works in the same interpretive fashion with regard to everything else (the objective perspective).

For the first subjective interpretation, consider Nietzsche's analysis of the scientific interest of the "type of the theoretical man." This type "finds the highest object of his pleasure in the process of an ever happy uncovering that succeeds through his own efforts."[25] The logic of this unveiling requires its interminable suspension. This perpetual deflection or extension of the scientific project is its promise, or, if one likes, its "promissory" dimension.

So conceived, the process of discovery expresses a pathological desire.[26] If the goal of scientific inquiry were nothing more than a simple uncovering or revelation of the facts, its interest would quickly be either satiated or abandoned: there are any number of facts and no ultimate terminus to the project of their discovery. The phenomenon of physiological satiation illustrates this point. While the satisfaction of desire is the death of desire, a deferred or partial satisfaction can be its quickening.[27] As the scientific process progresses toward a distant limit, some indirection must be assumed to sustain its joy or interest in discovery. As it is dependent on what I will call "artifice" or "experimental stimulation" in this way, science is far from being a pure or innocent pursuit of truth. To the contrary, science's pursuit of the truth is like the pathology of voyeurism: the eye of science is like a hypertrophied, unblinking organ.[28]

Speaking of science metaphorically, the objects of scientific research may then be called "fetishes."[29] Nietzsche suggests that failing an artificial structure, the project of science would have to be relinquished, for the pure project of obtaining complete (not to speak of unconditioned) knowledge is (if it is not impossible) endless.[30] Describing the relationship of the man of science to the unveiling of the goddess *Truth*, Nietzsche concludes: "There would be no science if it were concerned only with that *one* nude goddess and with nothing else. For in that case her devotees would have to feel like men who wanted to dig a hole straight through the earth."[31] That truth should be represented as a female deity is not new in the philosophic tradition; that the mysteries of this goddess should be closer to those of Baubo or Sais is surprising, shocking. With this metaphorical reading of the traditional metaphor of truth, Nietzsche effects a crude transformation of the essence of this convention. In this rude context, the image of digging a hole straight through the earth expresses the problem of a pure desire for truth: the game is infinite (from any perspective).

What is capital here is the search itself; *process* is the key to the progress ideal. For Nietzsche's point is that without presupposing a Promethean titanic temperament, what mortal being would ever deliberately pursue such a Sisyphean project of infinite effort and infinitely deferred return? The possibility of such a positive engagement would depend upon whether one had learned "to be satisfied with finding precious stones or discovering laws of nature?"[32] The key to delayed satisfaction is found in the points of contact confirming desire and maintaining its tension. For the digger in question, of course, these contact points are the jewels signifying the worth of continued burrowing. The practicability of science corresponds to its project of crafting or carving out the laws of nature (or facts) in its investigation of nature. The tactics of scientific continuity or self-perpetuation express nothing less than the secret or essence of science. Lessing (for Nietzsche "the most honest theoretical man")[33] articulates this same secret in his over-frank admission "that he cared more for the search after truth than for truth itself."[34]

The generation of laws and facts permits the examination and experimental expression of scientific nature; this productivity can be made the subject of further (nihilistic) reflection. Like the caged monkey, so disturbed by confinement that it meticulously examines itself for vermin to the point of self-mutilation, the logic of science once turned in upon itself confronts its own limits. It is at this point that science and logic itself contemplate with appropriate "horror how logic coils up at these boundaries and finally bites its own tail."[35] One might take this reflexive prediction to suggest that Nietzsche extrapolates the ultimate end of science as a self-annihilation in a tangle of self-contradictions gotten from its own nihilistic logical ground. This is not the case: by speaking of science as a perverse pursuit of truth, the point to be made is that it is autodynamic: it is nothing less than a perpetual motion machine, the holy grail of the mechanical sciences.

Nietzsche holds that scientific nihilism may be represented as manifesting a chronic cultural sickness. The nihilism of a metaphysical culture is not the same as the death of that culture: a culture of nihilism is merely a degenerate culture. And such degeneration is not a matter of recent decline, for Nietzsche traces it back to Socrates, the herald of the brightest flower of philosophy and the optimistic promise of happiness through reason. Socrates taught that knowledge should be sought at any price,[36] and he made this claim in the wake of Greek tragic art and its willed illusion.[37] At the time, Nietzsche tells us, this was a needed therapy: the Greeks were on the verge of losing the strength (Nietzschean "nobility") needed for the tragic worldview. The sickness in question was a culture of ascendant weakness, that is, a democratic or reactive slave culture.[38] The cultivation of the Socratic knowledge ideal preserved, as therapy can, the culture in its sickness. Today, the same culture of degenerate life is (pre)served by the scientific ideal of truth.

Originally, the ideal of (decadent) truth was advanced in the face of life against a tragic aesthetic (and hence as an antiaesthetic ideal). The aesthetic ideal of life is tragic just because it inevitably entails the (innocent) dissolution of the individual—innocent or not. But as life expression—even full expression—is a matter of context, not every expression will be utter or "to the death." The *incomplete* expression of power is only ennervating: rather than giving itself out utterly, it gives itself out to an enfeebled remainder. And a reactive, inadequate, or impotent will can be supreme where its ultimate achievement converts the vulnerable convalescent to an irreversible decadent. Alternatively, the tragic aesthetic is an artistic affirmation of life requiring an abundance of power.

To understand the meaning of abundant power, affirmation, or excess, two further correlative observations must be made, drawing out two terms that have special significance in this context but might otherwise appear utterly innocuous: *self-preservation* and *happiness*.

The lack of power fated in this way to be ultimately triumphant forces an antiaesthetic countermove in the interest of *self-preservation*. Lacking an abundance of power, a democratic culture seeks nothing but its own preservation, its therapeutic life-sustaining comfort, and so its *general happiness*. In the service of these democratic goals, the project of science seeks to understand and ultimately to control nature.[39] The decadent sickness supported by Socratism—and the science (or scientism) it makes possible—has its "symptoms":[40]

> it combats Dionysian wisdom and art, it seeks to dissolve myth, it substitutes for a metaphysical comfort an earthly consonance, in fact, a *deus ex machina* of its own, the god of machines and crucibles, that is, the powers of the spirits of nature recognized and employed in the service of a higher egoism; it believes that it can correct the world by knowledge, guide life by science, and actually confine the individual within a limited sphere of solvable problems.[41]

These "symptoms" of modern science are not recognized as signs of any illness. Rather have they come to be taken as the supreme signs of a healthy culture. Thus currently reviewed, myth has been dissolved (sophisticated rationality) and metaphysical (or religious) comforts have given way to earthly ones (technological advance). This is scientific progress. If the new faith in a scientifically improved world and the optimism of a scientifically regulated life happen to be less idealistic today than their Alexandrian counterparts, that is simply because they have been effectively achieved. The efficacy of such an achievement depends upon an optimistic inattention to the intractable dimensions of a complete knowledge of the Real (regarded abstractly)[42] and a rational restraint of human demands (viewed practically): one must desire what the artifices of science (and technology) can offer. This, again, is to say that there are no unimaginable technological possibilities, that is the meaning of the point suggested under the rubric of a limited (*einen allerengsten*) sphere of solvable problems."[43] Because the scientific project is as effective as it is, the nihilistic aspect Nietzsche outlines in his symptomatology of science is dangerously difficult to perceive.[44] If the conservative ideal denies life, it must be admittted that this denial preserves life (at its lowest level). The desire to live at any cost pushes the value of life to an extreme.

Against this reduced standard and artificial love of life, Nietzsche preferred the pessimism of a Schopenhauer (or even a Kant) to the disguised danger of Socratic culture. A pessimistic artifice is better (has more potential) than (Socratic) optimism's equal or parallel artifice. So Nietzsche's suggestion that a deep optimism may be possible beyond pessimism may be read as an apology for Schopenhauer's vision. Thus Nietzsche suggests that Schopenhauer "perhaps by [the logic of his pessimism], and without really intending to, may have had his eyes opened to the opposite ideal."[45] Both pessimism and optimism articulate compensatory structures corresponding to societal and individual needs. But to use Nietzsche's distinction between sickness and health, the pessimist's inspiration is healthier one and it presupposes health. Pessimism reflects a "truer" conception of reality—whereby and for the first time an attention to the "limitless periphery of science" may be brought to awareness. Such a healthy conception of reality is preliminary (as a kind of *Fragezeichnen*) to a critical investigation of human circumstances and beliefs and so (possibly) to an artistic basis for science. Opposed to this possibility, on the other hand, scientific or Socratic optimism ignores the consequences of its own "limitless periphery" (the scientific project can go on forever), and, inattentive to reality in the small and in the large (i.e., to reality as it is Real: incalculable and incomprehensible), it perpetually seeks an encompassing structure for reality. For Nietzsche, as we have seen, this practical scientific procedure neither begins from nor results in knowledge:

> Not to know but to schematize—to impose upon chaos as much regularity and form as our practical needs require. In the formation of reason, logic, the categories, it was *need* that was authoritative; the need, not to "know," but to subsume, to schematize, for the purpose of intellligibility and calculation.[46]

By containing the Real (natural "Chaos") within the structure of a symbolic model, the project of science is able to work out the rule(s) of that containment: specifying or stipulating what is to count as real. In a word, the project of reactive science is an articulation of reality. Nietzsche expresses this scientific articulation of reality as a kind of structuralist's cat-and-mouse game or if one prefers another analogy, the reality compensation/symbolization of Freud's little grandson's *Fort-Da* exercise: "One finds in things nothing more than what one has oneself inserted in them (*hineingesteckt*): this child's game, which as such I am far from belittling, is called science?"[47]

Regarded in this way, the interpretive, inventive scientific project is a creative construal of reality, very like a child's game. Like a child, the scientist has forgotten (and it must be emphasized that the successful scientist needs to forget) that he or she is playing a game. For Nietzsche, the structure of logic and natural scientific law remains an artifice or an illusion (*Irrthum*).[48]

It is not my intention, nor was it Nietzsche's, to denounce the illusion that is the scientific project. For, according to Nietzsche, illusion is necessary for life itself: "there would be no life at all if not on the basis of perspective evaluations and appearances."[49] Logic simply offers one aesthetic possibility or perspectival valuation among others. So, as Günter Abel observes with a different emphasis in another context, if one may consider logic (or science) and art as comparable interpretive schemes, then the opposition traditionally asserted to hold between them is hard to support.[50] The question to be asked is whether the aesthetic character of the illusion is active or reactive, that is, whether "hunger" or "abundance" is productive in today's science.[51] How distinguish the optimistic illusion of truth (reactive lack) from the possibility of a science operating as a creative (active fullness) interpretation of nature and (so) ourselves? The answer will be the focal topic of the concluding chapter: previewed here, it can be said that creative (and that also means a poetic or artistically conceived) science must match the achievements of the tragic myth in a creative manifestation expressing both nature and humanity in the perspective of the artistic exaltation of life.

SCIENCE: REALITY AND ILLUSION

The scientific world is a human construction. Scientific reality is an illusion, it is not the Real as such, and it is not true: "We have *created* the world that possesses values. Knowing this we know, too, that reverence *for truth* is already the consequence of an illusion."[52] This illusion of scientific truth, for Nietzsche, is that one can explain reality when one has merely given it a mathematical and logical (theoretically scientific) description. And in fact such a description of reality, according to many sober and contemporary standards, is paradigmatically, rigorously scientific. But, for Nietzsche, our scientific projection of the world represents only the symbolic structures of our own understanding.[53] In an early text, he articulates this insight with an almost burlesque metaphor: "We behold all things through the human head and cannot cut off this head; while the question nonetheless remains what of the world would still be there if one had cut it off."[54]

The traditional objectivity of number (mathematics and logic) does not survive Nietzsche's critical perspective. As he criticizes the surreptitious moral impulse of Kant's critique of pure reason, so he criticizes our habit of inventing rational structures and applying them to nature with all the noises of lucky discovery.[55] Nature, as Nietzsche conceives it, "does not by any means strive to imitate man."[56]

Without the metaphysical conviction that the ideal objects of mathematical calculation match the objects of the real world in a formal way, Nietzsche declares that mathematics "would certainly not have come into existence if one had known from the beginning that there was in nature no exactly straight line, no real circle, no absolute magnitude."[57] This assertion has a particular vehemence. As long as science works with numbers (or concepts in general, that is, as long as science follows an exclusively rationalist ideal), it works with reality-falsifying simplifications (idealisations, in the scientific, nonartistic, sense): "The invention of the law of numbers was made on the basis of the error which has been dominant even from the earliest times: that there are identical things (but in fact nothing is identical with anything else); at least that there are things (but there is no 'thing')."[58] The scientific vision denies the wild and subtly various chaos that Nietzsche takes to be a more (perspectivally) accurate account of nature. This underlines the delusory character of science where, for Nietzsche, numbers and the objects they describe "are valid only in the human world." Thus, in all seriousness, he can write in a later text: "it will do to consider science as an attempt to humanize things as faithfully as possible; as we describe things and their one after another, we learn how to describe ourselves more and more precisely."[59] Regarded in this anthropomorphic way, however, scientific reality is arbitrary. Made in our own image, science cannot claim to be an ultimate representation of the world: "The question is whether there could not be many other ways of creating such *an apparent* world—and whether this creating, logicizing, adapting, falsifying is not itself the best-guaranteed *reality*; in short, whether that which 'posits things' is not the sole reality."[60] The possibility of such a positive plurality of perspectives and the circumstance of our own involvement with and organic dependence upon the world means that we have every interest in as efficient a vision of the world as possible. Yet this efficiency can be attained without at the same time advancing a "true" vision. Why would human beings limited to their own "corner," their own perspectives, Nietzsche wonders, have any interest in the world at large? Responding to this question of sufficient reason, Nietzsche continues, if an organism does in fact develop a drive to know, ought one to ask if this drive satisfies a particular organic need or life-interest (including that of cultural life)? This is Nietzsche's reflection: "It is essential to *determine* what concepts and formulas must be: means for comprehensibility and calculability. *Practical application* is the goal, that man be able to help himself to nature . . . Science: the conquest of nature for the ends of man."[61] As the drive to mastery over nature for human purposes in the scientific project, the drive to know can be expressed only in accordance with the physical or organic limits of the human perspective. As human beings we are caught in the prison of our own perceptual/conceptual construction, a projective horizon that in turn is determined by our ecophysiological needs and constraints. Although we cannot exceed our own projective, that is, already self-exceeding, interests (each knowledge increment is determined as it is the horizonal unfolding, or expression, of its original transcendence), the project of objective knowledge (science) seeks a reality beyond what comes into appearance as such in its own project. One might think of science in this way as the project of the project, or the project of epistemological excess.

The interest of science thus (as objective knowledge) goes beyond its transcendental structure as an historical perspective and seeks the ultimate ground of its own ground. This project of the project could be taken to be identical with the fundamental project of philosophic questioning, except that the scientific (and so too the traditionally philosophic) approach endeavors to find the ultimate basis of its project on the basis of that project alone. This is a metaphysical, logical endeavor and cannot succeed. One cannot remain on the level of science to think the project of science, just as one cannot reveal the meaning of Being on the level of beings. Beyond the duplicity of the ideal of reflexive science, Nietzsche doubts whether an aesthetic justification (not merely a justification for limited interests, but a justification in the interest of the grand style) of the "project of the project" could ever be given. For Nietzsche, "reliance on reason and its categories, on dialectics, hence the *estimation* of logic, proves no more than its experientially confirmed *utility* for life and *not* its truth."[62] Our claim to objective knowledge may be no more than a convenient fiction, and it is certainly not an absolute truth: "Life is grounded upon the presumption of a belief in enduring and regularly recurrent events. The more powerful life is, the broader is the calculable simultaneously constituted world. Logicizing, rationalizing, systematizing as aids to life."[63] The necessity of objective knowledge for life, the notion of a correspondence (for Nietzsche, an aesthetic or appropriate adequacy between our interest and ways of thinking and what is real) is, as it were, made and not found. Human knowledge of its objective interest (in whatever historical manifestation) is projected or filtered, it is not purely received. Nietzsche laughs at the notion of an "immaculate perception." Thus the "arbitrary" character of reality reveals the creative possibility of the knowing project: "Truth is therefore not something true, that might be found or discovered—but something *that must be created.*"[64] In this creative possibility, Nietzsche assumes that our purposes and the world of our interests do coincide. But the necessity of this coincidence does not establish a similarly necessary connection with things in themselves: there are no such things. Instead this coincidence is aesthetically construed. The value of this aesthetic, its creative or merely receptive orientation, determines its value for life. It is as an ambivalently aesthetic claim that the claim to objective knowledge is to be understood as arbitrary. Nietzsche's claim in this is not so much that the world really is chaos (although he suggests this possibility, in a parallel with our external and internal sensitivity and sensibilities) as that one has no other option than to conceive knowledge through the glass metaphor of an optic.[65] In distinguishing between knowledge and belief, Nietzsche wishes to distinguish our understanding of the world and the world as we cannot know it: "the world of 'phenomena' is the rectified [*zurechtgemachte*] world that we *perceive as real.* 'Reality' consists in the

constant recurrence in their *logicized character* as similar, recogniz-
able, related things, all in the belief that here we can reckon, we can
calculate."[66] Logic, then, works to render a world comprehensible
and, so, calculable. Its scientific utility is to be found in its consistency. But this is a projective fiction: "We need units in order to
count, but it may not be assumed that such units [of measure]
exist."[67] The production of measurement units—not only kilos and
joules are so designated but species, stars, colors, and so forth are
included as well—as identifiable, reidentifiable things, is based
upon the (likewise fictitious) concept of the self (Ego) as a subject:
that is, our concept of the subject as a substratum, or in-dividuum
that persists as an identifiable self through time. On this ground,
we understand things and their qualities (predications), causes and
effects. Yet Nietzsche observes, "Mechanics formulates the appearance of resultants withal semiotically in the sense of sensory and
psychological means of expression—all without approaching the
causal force."[68] What Nietzsche means by the possibility of knowledge corresponds then to the logical capacity (schematising function) of our senses understood as the creative expression of our
senses. Science and art have the same root, and they do not differ
by virtue of the superior claim of the first to represent the logical
order actual in the world among the "things themselves." Indeed,
if advantage is to be had here, Nietzsche gives the palm to art but
only as the "good will to appearance."[69]

For Nietzsche, "Thinghood was invented by us from the
first."[70] Reality as we take it to be in itself and the understanding
we have of reality necessarily correspond insofar as this is trivially
insured by the reciprocity of any relational construction.[71] This
kind of interpretive correspondence is the communication or dialogue noted by Prigogine and Stengers: "Whatever we call reality,
it is revealed to us only through the active construction in which
we participate."[72] Nietzsche emphasizes this same interpretive
dynamic construction as the "interpretive character of all
occurences. There is no event-in-itself. What transpires is a group
of appearances *selected out* and organized [zusammengefaßt] by
an interpreting being."[73] But this interpretive character is not limited to the human subject: what is characterized in this expression
of "the interpretive character of all coming to pass" is reality as it
happens: "*Es giebt kein Ereignis an sich.*"

But we must ask how, limited to his own perspective and
acknowledging this very limitation as he so explicitly does, Nietz-

sche nevertheless goes on to claim that *"there is no event in itself."* How can he *know* this? We have seen that his project is not that of describing the absolute structure of reality: Nietzsche does not simply assert that there *is* no event in itself. Instead, he says that every event occurs as an eventuating irruption among beings: all happening is relational or perspectival. Nietzsche's perspectivalism thus "de-privileges" the subject for the benefit of the object—"the world seen from within . . . would be will to power and nothing else"[74]—but this means: for the benefit of the dynamic event or happening. In the full characterization supplied through the *Nachlaß*, the coming into being of an event is not in-itself but fundamentally phenomenal: it is exposed and comprehended by an interpretive essence. What happens is never (it is nonsense to speak of it) an isolated phenomenon; it comes to pass among other events (*inter pres*) that structure it, circumscribe it, or yield to its presence: "all events, all motion, all becoming as a determination of degrees and relations of force, as a *struggle*."[75] The phenomenal character of an event bespeaks its relational and expositional character vis-à-vis other interpretive phenomena, and not for human interpretive perspectives alone. Among these interpretive occasions, the human perspective is supreme not because of any ontic excellence but rather just because it is our perspective. Because of the cosmic interpretive dynamic, "we may not ask: 'who then interprets?' for the interpretation itself is a form of the will to power, . . . as an affect."[76] The *world*, it is well to recall here, is *Will to Power*.[77]

In this, we see again that what differentiates Nietzsche's perspectivalism from any relativistic approach and from subjectivism in particular is its insistence that "interpretation" (as the event of the world) or objective constitution is not the limited prerogative of the human perceiver. There is an interpretive dynamic operative in all energetic or position/temporal interactions: "every center of force adopts a perspective toward the entire remainder . . . Reality consists precisely in this particular action and reaction of every individual part toward the whole."[78] That is to say that every node in the cosmic flux has its own—is its own—interpretive relation to that All.[79] To pursue the cosmological implications of this point of view, one needs a critical and so challenging orientation toward the contemporary scientific belief structure.[80] We hear again Nietzsche's perspectivalist critique of the rationalist presumption:

It is no different with the faith with which so many materialistic natural scientists rest content nowadays, the faith in a world that is supposed to have its equivalent and its measure in human thought and human evaluations—a "world of truth" that can be mastered completely and forever with the aid of our square little reason. What?[81]

THE MEANING OF NATURE AND CHAOS: A NOTE ON NIETZSCHE'S "CHAOS SIVE NATURA"[82]

In a passage from The Gay Science usually invoked in connection with the idea of the Will to Power, Nietzsche declares that despite the relative order of the solar system and the organic life made possible in it, the "total character of the world . . . is in all eternity chaos."[83] Niezsche's definition of the world as a chaos ("to all eternity") makes no sense unless the word "chaos" can be understood in its original significance as a disrupted original order. As it is defined today, chaos is an utter lack of order, or disorganization. In this connotation (English as well as German, French, and even Greek) the expression of chaos rather than signifying a Hesiodic primordial state of affairs (that simply came to be), represents the failing of a primary (referential) order. In this way, chaos is an evolutionary rather than an originary term. What is currently called "chaos theory" reverses this expression, but even here the Newtonian thermodynamic scheme seems to prevail, and entropy represents the destruction, not the beginning of the cosmos.

Nietzsche's reading of chaos (or, as he says, nature) brings us not to modern physics as some of his commentators have suggested but precisely, incisively back to the Greeks.[84] As the potential raw chance of nature, the chaos of original becoming is formed or interpreted from the perspective of inorganic or organic interests. From a human perspective, the resulting physical/metaphysical arrangement is the cosmos. Thus projected, the notion of cosmos varies with the relevant aesthetic cultural composition. In at least one such conception, the two terms chaos/cosmos can perhaps be given an independent definition (as Nietzsche's definition seems to require and as Heidegger argues both for and against). Yet to do this, one would have to accept what is (primarily) a broadly Semitic notion of chaos as the lifeless void or yawning, impotent nonvital, and ultimately unreal feminine or watery abyss prior to the masculine initiative of divine movement creating the world. This is in opposition to *any* kind of primordial archaic Greek disorder, as articulated by Hesiod or even more distinctively and philosophically suggestively (if also gnomically) as implied by Anaximander's concept of ἄπειρον, akin to what Lacan speaks of as the Real. It is important to stress that Nietzsche's understanding of chaos is not empty nothingness but of a *plenary* if still horrifying nothingness.[85] I will later wish to suggest that the Semitic denotation of a feminine void-chaos has effectively been "mixed" with the still quasi-feminine but self-productive, and thus potent, archaic Greek expression of chaos in Hesiod's conception of a pristinely productive pell-mell. This cross offers a metonymically resonant convergence with the Greek sense of chaos as a generative fullness (as the Theogonic primordial χάος, the Anaximandrian ἄπειρον or Plato's Timean χώρα). This reading may well approximate the genesis of the notion as Nietzsche employed it.

In a study of Nietzsche's notion of chaos, Walter Gebhard notes that the origin of the Greek term χάος is nonsubstantive but must be found in the verb, χανεῖν. An opposition can then be supposed, presenting the ordering process, χοσμέω, as primordial.[86] This ordering process becomes the pristinely determining worlddorder or cosmos. Despite its surface consonance with tradition and traditional usage, Gebhard's interpretation clashes with Nietzsche's perspectival account and trivializes his philological precision as well.[87] In Nietzsche's favor (and against Gebhard's commentary) Nietzsche's reading of χάος meshes with (even with its Eddic and Semitic resonances) the (spirit of the) Greek sense of (the origins of) chaos. Thus Heidegger, although superficially in accord with Gebhard on the non-Greek origins of Nietzsche use of chaos, recalls the primordial communality between mythic traditions resonant in his reading of the classical meaning of the term: "Chaos, χάος, χαίνω signifes the yawning, that which yawns, the gaping out of itself. We comprehend χάος *in its closest connection with an original exposition of the essence of* ἀλήθειᾰ as the self-opening of the abyss."[88] These not only Greek mythic resonances heighten Heidegger's Nietzschean reading of Ur-Genesis that sounds in Hesiod's *Theogony*: "Ἦ τοι μὲν πρώτιστα χάος γένετ' (And in the very beginning Chaos came to be)."[89]

In these accounts of cosmic origins, it is chaos, the void, or formless waste (waters or ice) rather than the world ordering process of cosmos that is generatively primordial. Ordering is born out of chaos. The primordiality of the void, understood as raw possibility, that is, as becoming recurs in Nietzsche's notion of the world taken as chaos *to*, and thus and in this wise: *from* all eternity. Heidegger's explanation of this characterization of the world as eternal and necessary chaos as an expression of the Will to Power in the terms of the Eternal Return is true to Nietzsche's thought (preserving Nietzsche's own harmonies), illuminating the complex question of Nietzsche's understanding of the world in terms of what he named his most difficult thought, the Eternal Return.[90]

In a recent commentary on Nietzsche's idea of chaos, J. Granier takes up the significance of Heidegger's penetrating insight, "Becoming, i.e., here at once the character of the being of the world entire as the eternal chaos of necessity is the eternal return of the same."[91] From this expression of the nature of becoming, Granier explains Nietzsche's association of chaos and necessity (and this is an important contribution) as an interpretive phenomenon. It is only as an interpretive phenomenon that the basic character of all beings understood as the Will to Power (the essence of all beings) can be expressed in the Eternal Return of the same (as the necessary mode of existence of all beings). For Granier, "being is text."[92] Thus he can explain that "as primitive text, nature is the chaos-being that manifests itself as a signifiying process, the figures of which trace neither a system nor a chaos but merely a veil."[93] For Granier, "*Das Gleiche*—being that recurs in the Eternal Return, renders copresent nature and veil in the equivocal unity of the text."[94] This text is (as raw nature and as veil) the basis for interpretation. Yet this is not to claim a trivial phenomenalism in Nietzsche's name. The process of interpretation is not separate from the text as such. We are hardly speaking of Platonic illumination here—much rather of Heideggerian or, indeed, pre-Platonic revelation. For this reason: "The Same, designating the being of the phenomenon links nature and the scrawls of interpretation in such a fashion that the text remains, for all interpreters, enigmatic."[95] There is no real or true nature, because "l'être chaos ... est ohne Grund,—ein Abgrund, un abïme,"[96] or, as Granier deliberately puts it, risking the claim to articulate what cannot be articulated: "The real is a chaos with absolutely no common measure with human exigency."[97]

It is understood as an abyss that chaos serves to inspire the thought of the Eternal Return of the same, where Heidegger writes, "The *thought* of the eternal return of the same *fixates* by determining *how* the world essentially is the necessitous chaos of perpetual becoming."[98] And any instant, each moment "each in its own way is a resonance of the whole and a harmonious entry into the whole."[99] Only for such a view of necessity does it become clear how "in truth the thought of thoughts [the Eternal Return] grants supreme lucidity and decisiveness to beings at every moment."[100]

This Heideggerian emphasis brings out the unique possibility of the thought of the Eternal Return: everything recurs eternally. This possibility may not be compared with its alternative, which would read: everything once only. "Once only"—that is, merely once—is no investiture weighing an action in the thought of eternity. What is eternal in this sense is not a matter of infinitely recurring identical cases: indiscernibility is not its watchword but becoming, in its innocence as the same.[101] A. Philonenko's observation is important here: "plainly the Eternal Return could not be the Eternal Return of [temporal] contents."[102]

Above I noted that in the provenance of the Nordic cosomology of the Eddas, chaos (gap) is what comes to be independently of gods. Terminating in a final consummately destructive *ragnarok* or twilight of the gods, the image of chaos again offers the alternative scheme that permits Nietzsche to conceive the chaos/cosmos opposition apart from a cosmos privileging logos or ἀρχή. By denying the cosmos its archic originative and ruling as well as its telic or guiding and ending principles, that is, by understanding reality as multiple realities, Nietzsche recovers the principle significance of the primordial Greek term κόσμος just insofar as it may be thought as a (purely) aesthetic concept. For the archaic Greek, we know, to speak of a κόσμος is to refer to a (proper) cosmetic and aesthetic arrangement. For Nietzsche, this is to say that organization or schematizing interpretation in Heidegger's sense is essential to any cosmos.

The (interpretive) aesthetic dimension recalls the connotations of the archaic Greek κόσμος as a fitting (and as it ultimately came to mean: ornamental) arrangement or harmony. C. Kahn points out that Homer's use of "κόσμος, κοσμέω . . . denote in general any arrangement or disposition of parts which is appropriate, well-disposed, effective."[103] The transfer from the original significance, Kahn observes, of a "'neat arrangement', is an easy one to the wider decorative sense of κόσμος as 'finery, rich adornment.'"[104] Granier notes that this "adornment" presents an interpretation of the primordial text. The primordial text has nothing in common with the traditional, medieval notion of a book of nature, because the cosmos is nothing but a veil in this sense: reading "text" metonymically as texture, textile. This interpretation of the veil of nature, whether construed as Schopenhauer's *Maya* or Lacan's *Real*, holds all the way up to the most objective level of contemporary science, because as Granier expresses Nietzsche's explicit perspective, "In the final analysis, logic is no more than a way of appropriating the real, deforming it on the basis of our determined needs, which, for human beings, fix even the orientation of our perspective."[105] Beyond this veil, which has for its part a divine appeal—and is thought to conceal the same—lies the horror of the ambiguous, contingent world-being, that is: Chaos. This Being is horrible, is terrifying, because its ambiguity and contingency is *necessary*.

This insight into necessary contingence is what freezes Hamlet's action, but it is this fatal prepossession that must be counted as Dionysus's last secret. In connection with the expression of the Will to Power, we leave the somber realm of law to enter the world of play, what Heidegger refers to as the "possible." This is the possible that corresponds to one's essential, one's utmost *possible* possibilities: the fully, ineluctably possible, rather than the merely partly, or irresolutely possible. Granier explains this sense of possibility/play as "the duplicity of Being: the play of mortal truth and illusion-values, according to an infinity of interpretations—the play of chaos and life."[106] It can be misleading to speak of possibility and play, because the contemporary image of physics following the indeterminacy principle (and in the wake of popularising accounts) comes all too easily to mind. This connection, even on the terms of a Nietzschean philosophy of science, is not to be desired. Yet I do not mean to dismiss this connection without a distinguishing word in its favor.

In the spirit of this philological excursus, the very notion of theoretical probability in physics or chemistry cannot play as lightly as Nietzsche would like. Because the finitude of probability repeats its cosmetic origin in its function in modern cosmology, it is meant to produce usable approximations, but as a means of calculation it is far from being an affirmation of chaos.[107] Echoing Mallarmé, Deleuze suggests the different kind of dancing or "light-footed" probability ("playing") appropriate for Nietzsche's vision: "Not a probability distributed over several throws but all chance at once; not a final, desired, willed combination, but the fatal combination, fatal and loved, *amor fati.*"[108] It is as an affirmation of chaos (or dancing) that the theory or application of probability affirms in its very possibility, although not its actual expression, Nietzsche's understanding of the most "scientific" comportment: *amor fati.* It is as an affirmation of chaos (or dancing) that the theory or application of probability affirms in its very possibility, although not its actual expression, Nietzsche's understanding of the most "scientific" comportment: *amor fati.* Thus, to twist the image of probability to illuminate Nietzsche's special vision, the pattern that a dancer might leave on a dusted stage should be contrasted with the drunkard's walk. The latter is the image used in stochastic discussions of nonregular series. The drunkard's walk is opposed to the dancing step. It is not only the deliberate aesthetic (or the physical dance-steps read as the indeliberate sign of the aesthetic movement of the dance) that converts the image; the interpretive contexts, the performance, the ecstatic *play* appropriate in the dance, induce the extended work of effective contrast. The stochastics expert knows that the dancer will "throw" [off] all calculation. Aesthetically the dance performance although choreographed is incalculable because, like any performance, it joins spontaneity and necessity. A free, expressive dance is determinately unpredictable: the occasion, the event of a necessary chaos, it is the very and saccharine image of chance.

Although it is an interpretive, cosmetic image like any other (i.e., like alchemy, religion, and myth), today's science takes itself too seriously, failing the good taste of conscience that would acknowledge its origin in its cultural interests, as in a prejudice.[109] Nietzsche's fundamental critique of science is that scientific cosmology is taken to be truth, where it is no more than an interpretive perspective, or illusion: a projection of given interests. For Nietzsche, the cosmic order of modern cosmology is nothing but a human perspective imposed on an unending chaotic flux. One knows this flux only interpretively, on both rational and unconscious levels. Deleuze says, "For the dice throw is multiple affirmation, the affirmation of the one and the many. But all parts, all the fragments, are cast in one throw; all of chance, all at once. This power, not of suppression of multiplicity but of affirmation of it all at once, is like fire."[110] The fire it resembles (and the game it plays with itself) is easy to recognize as Heraclitus's fire. Borrowed from Heraclitus too is the notion of multiplicity and the affirmation of the many, "all at once," in the order of measure that is the operation of the mutual determination and interpretation of the world playing as the manifestation of constantly metamorphosing Wills to Power.

REALITY AND ILLUSION: THE INTERPRETIVE DYNAMIC

The Real is not the reality envisioned by human conceptual schemes but "the formless, unformulizable world of chaos-sensations—that is, *another kind* of phenomenal world, one 'unknowable' for us."[111] So far from opposing the idea of a chaos world, our constitutional capacity in its ecophysiological sensual and conceptual expression makes the assumption (together with the regularity) of experience possible. It is important to note that Nietzsche's special interest in the sciences of his day suggested that things are more complicated than the senses communicated them to be and that this complication is effective on the same directly perceptual level. So Nietzsche speaks of the simplified world of everyday and theoretically mediated perceptions as a matter of "refinement." Just as perceptual judgments, scientific laws, and scientific entities are possible upon the basis of flattened differences (*"und die kleinen Fehler kommen nicht in betracht"*—that is, "margins of error," "standard deviation," and the theoretical finesse, "curve fitting," are ways of *not counting* these "little" mistakes), excluding potential exceptions. And yet science hardly denies what it thus discards thereby, as it were, having it "both ways."

Nietzsche's attention to the duplicity of science addresses the limits of reflective knowledge. We are not as successful as Lewis Carroll's heroine, for we are unable to pass "through the looking-glass" into our very selves, as the reflective ground of what we know: "When we try to examine the mirror in itself we discover in the end nothing but things upon it. If we want to grasp the things we finally get hold of nothing but the mirror.—This, in the most general terms, is the history of knowledge."[112] In a preface to the same text written half a decade later, Nietzsche would speculate on what it could mean for Kant to reflect critically on reflection: "come to think of it, was it not somewhat peculiar to demand of an instrument that it should criticize its own useful-ness and suitability? that the intellect itself should 'know' its own value, its own capacity, its own limitations? was it not even a little absurd?"[113] In the *Nachlaß*, he then concluded, "The intellect cannot criticize itself, . . . in order to criticize the intellect, we would have to be higher beings with 'absolute knowledge.'"[114]

Thus Nietzsche denigrates the possible powers of our "foursquare little human rationality" while yet remaining himself a seeker of truth. It is this search that leads him to "the problem of the value of truth."[115] Perhaps things could be seen from differ-ent perspectives, perhaps we should try to incorporate or embody these perspectives; perhaps these foreign perspectives include the most absurd, the most unpleasant, the most difficult of perspec-tives; perhaps our choice of viewpoint determines what is to be seen, that is, determines what is there to be seen, brings it as such into the open to be seen? Perhaps this revelatory seeing counts as much as the passive object, the object in itself? Perhaps, indeed, there is no such passive object? In this vein, he writes, "The world is *not* thus and thus: and living things see it as it appears to them. Rather, the world is made up of such living things and for each of them there is a small corner from which it measures, observes, sees and fails to see. The 'essence' is lacking: The 'becoming' or 'phenomenal' is the only kind of being."[116] Nietzsche's insight here challenges the tendentious grammatical distinction between subject and object, active and passive,[117] organic and inorganic, and even life and death.[118] Thus when he asks, "if perspectivism is inherent in all things," he can answer, "This would be possible if everything in existence were somehow essentially capable of per-ception."[119] In this light, we can read a subsequent reflection: "Given, however, that we install particular values into things, when we then forget that we were the instigators, these values work upon us in turn."[120] There is a reciprocity between interpre-tations and perspectives: "every center of force adopts a perspec-tive toward the entire remainder."[121] The transformation of a "chaos" of sensations into a cosmos suitable for human under-standing may be an arbitrary fantasy but it is nonetheless an effec-tive symbolic construction of the world: "the 'world' is only a word for the collective play of these actions."[122]

On the basis of the relational interaction of the world con-ceived both in its subjective and objective dimensions as Will to Power and in its ultimate interactive involvement with itself ("The world seen from within, the world described and defined according to its 'intelligible character'—it would be 'will to power' and nothing else.")[123] Nietzsche can write: "A quantum of power is determined by the effects that it exercises and resists. As conceived in itself, the *adiaphorie* does not exist. Essentially there is a will to violation and defense against violation. Not self-preservation: every atom has an impact on the whole of being."[124] To conceive the world—"from within" as Nietzsche says—not from the assumption of our own interest but rather as the world might be seen according to its interest is, in effect, to construct the object subjectively. But this is not a "noumenal" perspective, and it is not an anthropocentric vision. We have seen that the perspectivalist position adopts an interpretive perspective beyond the subjective and extending beyond the objective as well. To say that everything is interpretation is to affirm that our own view of the world effects only our own interpretive transformation of the world: "As a science of *motion*, mechanics is already a translation into the sensory language of human beings."[125]

The assertion that science is a sensual/conceptual human "translation" of the world does not deny that science "works." The point is to show the subjective interest of science's (today less) vaunted "objectivity" together with the subjective interests of its "objects." Science works with illusions. But we have seen that Nietzsche goes on to make the stronger claim about science: "It is a profound and fundamental good fortune that scientific discoveries stand up under examination and furnish the basis, again and again, for further discoveries. After all, this could be otherwise."[126] Scientific facts are not only nonabsolute but also a matter of chance: they could have been otherwise!

What does this still-strange claim mean—given the subjective (interpretive) character of scientific objectivity and the subjective (interpretive) perspective of the scientific object, that is, given the uselessness of the distinction between subject and object in a world understood as the expression of mutually interpretational Wills to Power? We can recall that for Nietzsche, "the total character of the world, however, is in all eternity, chaos."[127] This world is "das Reich der Zufälle," the kingdom of chance and "cosmic stupidity." In an earlier text, Nietzsche observed: "The Greeks called this realm of the incalculable and of sublime eternal narrow-mindedness *Moira*."[128] To censure *Moira's* unfathomable eternal realm as "narrowminded" (*Bornirtheit*) is to refer to its insensitivity to human projects as regarded from a *human* perspective.

Now, the human conception of an insensitive nature is a systematic expression of the cultural illusion of science in the West and the enlightenment's reverse mythological vision of a mechanical nature revealed through the sobriety of scientific sophistication. Ergo, a deliberate self-diminution may well advance the (seemingly contrary) aims of exaggerated egoism.[129] The implications of this vain (that is, this presumptuous and affected) modesty will be the subject of the following chapter. Here, we need only note that the privilege of the subject and its ideal of objectivity are errors (delusions): "Once you know that there are no purposes, you also know that there is no accident."[130] In Nietzsche's vision of the world, necessity and purpose are dissociated. But in the ideal order of nature's conformity to law, what opposes purpose is accident. Without this ideal, without purpose, there is no accident and only chance with its necessity can remain. Thus, it is no accident that the coin comes up heads or the dice roll gives a six and not a seven.[131] The child's game that the universe plays is an eternal dice game; its necessity is the luck of the throw: "Those iron hands of necessity which shake the dice-box of chance play their game for an infinite length of time: so that there *have* to be throws which exactly resemble purposiveness and rationality of every degree."[132] If we look at the world from our own perspective, as we must, it is still possible to catch glimpses of the Real: "Judged from the point of view of our reason, unsuccessful attempts are by all odds the rule, the exceptions are not the secret aim, and the whole musical box repeats eternally its tune which may never be called a melody—and ultimately even the phrase 'unsuccessful attempt' is too anthropomorphic and reproachful."[133] And Nietzsche reminds us that from an artist's perspective—although the philologist and even the philosopher may count as the standard for rigor here—"Nature . . . is no model. It exaggerates, it abstracts, it leaves gaps. Nature is *chance*."[134] But the prospects from this perspective for the philosopher, the lover of wisdom, are by no means discouraging. If it is impossible to determine "How far the perspective character of existence extends or indeed whether existence without interpretation, without 'sense,' does not become 'nonsense'; whether, on the other hand, all existence is not essentially engaged in *interpretation*."[135] We have a wider possibility for knowledge once we are able to admit

that there may well be "infinite interpretations" in the world. And
taken from the proper perspective of an artist, we have to consider
this "new infinite" a new image of infinity. This perspective is elusive. Still more damning as we have already implied, such an artistic perspective is ambiguous. Even the philosopher after Nietzsche's new ideal considers him or herself merely "*contemplative
by nature and thereby overlooks that he himself is really the poet
who keeps creating this life.*"[136] This unawareness must apply to
all human beings, insofar as they are actively conscious (thus the
via contemplativa determines a Nietzschean *via activa*),[137] that is,
insofar as they are artistic beings. The human being is "a testimo-
nial to what gigantic forces can be set in motion by a little being
with a multifarious disposition." For Nietzsche we are "*Beings
who play with stars.*"[138] Thus, "In man, *creature* and *creator* are
united: in man, there is matter, fragment, excess, clay, mud, mad-
ness, chaos; but in man there is also creator, sculptor, the hardness
of the hammer, the divine spectator and the seventh day."[139] It is
on the terms of this boldly ambivalent affirmation—and with the
qualification that not all human creative potentialities can be
equally, deliberately artistic—that a creative aesthetic style may
become a scientific, as indeed a philosophic, possibility. And
again, we read that "Only we have created the world *that concerns man!*—But precisely this knowledge we lack, and when we
occasionally catch it for a fleeting moment we always forget it
again immediately; we fail to recognnize our best power and
underestimate ourselves, the contemplatives, just a little. We are
neither as proud nor as happy as we might be."[140] Nietzsche
attends to the ever eclipsing, barely perceptible, dangerously
ambiguous life of genius and art. In this a selective solicitation
resounds from Nietzsche's text: as a temptation to certain readers,
those of an artistic, those of a scientific, philosophic bent.

NOTES

1. WP 530.
2. "Dass Alles, was bewußt wird, ebendamit flach, dünn, relativ-
dumm, generell, zeichen, Heerdenmerkzeichen *wird*" (FW 354, KSA III,
p. 543).
3. "Vermenschlichung in summa" (KSA XI, p. 132).
4. WP 481.

5. "Wir haben eben gar kein Organ für des *Erkennen*, für die 'Wahrheit'" (FW 354, KSA III, p. 543).

6. "Leben ist die Bedingung des Erkennens. Irren die Bedingung des Lebens und zwar im tiefsten Grunde Irren" (KSA IX, p. 504).

7. "'Was sich beweisen läßt, ist wahr' setzt bereits Wahrheiten als gegeben voraus" (ibid. XII, p. 191).

8. WP 634.

9. The establishment of such scientific ideals obscures any sense of the immensity of the ambivalent chaos that is nature and the contingency and ambiguity that is organic life.

10. "Unsre Naturwissenschaft geht auf den *Untergang*, im Ziele der Erkenntniß hin" (KSA VII, p. 480).

11. "Die Wissenschaft setzt also den Prozeß nur *fort*, der das Wesen der Gattung *constituirt* hat, den Glauben an gewisse Dinge endemisch zu machen und den Nichtglauben auszuscheiden und absterben zu lassen. Die erreichte *Ähnlichkeit* der Empfinding (über den Raum, oder das Zeitgefühl oder das Grob- und Kleingefühl) ist eine Existenzbedingung der Gattung geworden, aber mit der Wahrheit hat es nichts zu thun" (ibid. IX, p. 501).

12. "Der ganze Erkenntniß-Apparat ist ein Abstraktions -und Simplifikations-Apparat—nicht auf Erkenntniß gerichtet, sondern auf Bemächtigung der Dinge" (KSA XI, p. 164), WP 508.

13. "Wissenschaft—Umwandlung der Natur in Begriff zum Zweck der Beherrschung der Natur" (KSA XI, p. 194). Cf KSA XI, pp. 114, 158, 209.

14. "Durst, Hunger und Wohlgeschmack an *verborgenen* und *verbotenen* Mächten" (FW 300, KSA III, p. 539).

15. "*Plebejismus des Geistes*" (FW 358, KSA III, p. 605).

16. "Das allgemeine grüne Weide-Glück der Heerde, mit Sicherheit, Ungefährlichkeit, Behagen, Erleichterung des Lebens für Jedermann" (JGB 44, KSA V, p. 61).

17. "Wenn ich von der furchtbaren Möglichkeit rede, daß die Erkenntniß zum Untergang treibt so bin ich am wenigsten gewillt, der jetzt lebenden Generation ein Compliment zu machen: von solchen Tendenzen hat sie nichts an sich. Aber wenn man den Gang der Wissenschaft seit dem 15ten Jahrhundert sieht, so offenbart sich allerdings eine solche Macht und Möglichkeit" (ibid. VII, p. 482). Nietzsche further writes "Unsre Naturwissenschaft geht auf den *Untergang*, im Ziele der Erkenntnis hin" (ibid., p. 480). This kind of *decline* leads to self-overcoming.

18. That is, the promotion of the universally unique.

19. "Man muss ungeheure Gegenkräfte anrufen, um diesen natürlichen, allzunatürlichen progressus in simile, die Fortbildung des Menschen in's . . . Gemeine!—zu kreuzen" (JGB 268, KSA V, p. 222).

20. Nietzsche claims that a certain nobility is required for even this last, apocalyptic perspective, that is for those who are "not noble enough to see the abysmal disparity in order of rank and abysm of rank between man and man" (BGE 62); "Um die abgründlich verschiedene Rangordnung und Rangkluft zwischen Mensch und Mensch zu sehen" (KSA V, p. 83).

21. "Gesetzt, wir leben in Folge des Irrthums, was kann denn da der 'Wille zur Wahrheit' sein? Sollte er nicht ein 'Wille zum Tode' sein müssen? —Wäre das Bestreben der Philosophen und wissenschaftlichen Menschen vielleicht ein Symptom entartenden absterbenden Lebens, eine Art Lebens-Überdruß des Lebens selber? Quaeritur: und man könnte hier wirklich nachdenklich werden" (KSA XI, p. 649).

22. "Ist es wahr, dass jene Objectivität in einem gesteigerten Bedürfniss und Verlangen nach Gerechtigkeit ihren Ursprung hat?" (UM II:6, KSA I, p. 285).

23. "Objektivität und Gerechtigkeit haben nichts miteindander zu thun" (UB II:6, KSA I, p. 290), UM, p. 91.

24. Heidegger, "Die Wissenschaft trifft immer nur auf das, was ihre Art des Vorstellens im Vorhinein als den für sie möglichen Gegenstand zugelassen hat. Man sagt, das Wissen der Wissenschaft sei zwingend" ("Das Ding," *Vorträge und Aufsätze*, p. 162).

25. "Sein höchstes Lustziel in dem Prozess einer immer glücklichen, durch eigene Kraft gelingenden Enthüllung" (GT 15, KSA I, p. 98).

26. The term *desire* is not proposed as a technical (psychoanalytic, or philosophical, or even literary) term but comprehends everyday usage.

27. Thus for Nietzsche, the eros of the scientific desire to know requires its own frustration: this is obvious in the formula of an "unknowing knowing" used to characterize the ideal of science in prior chapters. A "knowing knowing" would be the full accomplishment of science according to the naive objective realist ideal; a "knowing unknowing" recognizes the finitude or limits of knowledge. The idea of an unknowing knowing is essential to an objective realist epistemology; but it is nonetheless a chimaera which assumes as its own the absolute knowledge (even as what it does not know) from the perspective of the second. We have encountered this tactic before in the scheme of basic relativism.

28. GT 14; cf. UB III:6 (KSA I, p. 393ff.). This emphasis on the pathology of the scientific world-orientation was clear even before Nietzsche's extended analysis of the "Selbstsucht der Wissenschaften" in the third of the UB.

29. Nietzsche uses the very same term (in a related context—but concerning a higher analytic level) in one of his last books: "wir kommen in ein grobes Fetischwesen hinein, wenn wir uns die Grundvoraus-

setzungen der Sprach-Metaphysik, auf deutsch: der Vernunft, zum Bewusstsein bringen"(G-D, KSA VI, p. 77).

30. Sade's writings demonstrate that the distinguishing feature of (de)natured desire is the constant challenge of its own perpetuation.

31. "Es gäbe keine Wissenschaft, wenn ihr nur um eine nackt Göttin und um nichts Anderes zu thun wäre. Denn dann müsste es ihren Jüngern zu Müthe sein, wie Solchen, die ein Loch gerade durch die Erde graben wollten" (GT 15, KSA I, p. 98).

32. "Dass er sich . . . genügen lasse, edles Gestein zu finden, oder Naturgesetze zu entdecken" (KSA I, pp. 98–99).

33. "Der ehrlichste theoretische Mensch." Note that this is to say a good deal for the man who charged himself with a pious response to a hypothetical divine temptation (if God were to offer the truth with one hand and the search after the other, Lessing declared that he would choose the Platonic course). Nietzsche conceived the theoretical human being as a thoroughly duplicitous human being. There are, of course, both less and more charming kinds of duplicity, both honest and dishonest dramatic imagery.

34. "Dass ihm mehr am Suchen der Wahrheit als an ihr selbst gelegen sei" (KSA I, pp. 98–99).

35. "Die Logik sich an diesen Grenzen um sich selbst ringelt und endlich sich in den Schwanz beisst" (GT 15, KSA I, p. 101), BT, p. 98.

36. Cf. C-D, KSA VI, p. 72.

37. Cf. GT 13.

38. Cf. G-D, KSA VI, pp. 71, 72; GT 18 and "Vorrede 4": I; cf. KSA IX, p. 274. See chapter 6 below.

39. The nihilistic "fruits of optimism," so Nietzsche writes, can ripen: "wenn der Glaube an das Erdenglück Aller, wenn der Glaube an die Möglichkeit einer solchen allgemeinen Wissenscultur allmählich in die drohende Forderung eines solchen alexandrischen Erdenglücks, in die Beschwörung eine Euripideischen deus ex machina umschlägt" (GT 18, KSA I, p. 117). We have seen precisely this demand (and its satisfaction) accelerate in our own culture. Nietzsche understood the appeal of such mechanical (now technological) "gods." See Z, "Prologue," on the topic of the "letzten Menschen"; part IV may be read ironically as suggesting that the "ultimate" men are not very far from these last men.

40. But it may not be forgotten that if the goal of science answers a bland comfort, it could also express an artistic challenge: "Actually, *science* can promote either goal." "In der That kann man mit der Wissenschaft das eine wie das andere Ziel fördern!" (FW 12, KSA III, p. 384), GS, p. 86.

41. "Dass sie die dionysische Weisheit und Kunst bekämpft, dass sie den Mythus aufzulösen trachtet, dass sie an Stelle eine Metaphysis-

chen Trostes eine irdische Consonanz, ja einen eigenen deus ex machina setzt, nämlich den Gott der Maschinen und Schmelztiegel, d.h. die im Dienste des höheren Egoismus erkannten und verwendeten Kräfte der Naturgeister, dass sie an eine Correctur der Welt durch das Wissen, an ein durch die Wissenschaft gegleitetes Leben glaubt und auch wirklich im Stande ist den einzelnen Menschen in einen allerengsten Kreis von lösbaren Aufgaben zu bannen" (GT 17, KSA I, p. 115), BT, p. 109.

42. This Real of nature is chaos to all eternity. Here, Nietzsche defines the Real (quite piously) as "Das Unheil im Wesen der Dinge" (GT 9, KSA I, p. 69).

43. Ibid. Marcuse makes this same point without however losing his own Marxist commitment to a technological utopianism throughout his *One-Dimensional Man*.

44. What Nietzsche first describes as "der unumschränkt sich wähnende Optimismus!" expresses in an early form the apocalyptic vision of a willed nothingness. Cf. GM III:28, KSA V.

45. "Der hat vielleicht ebendamit, ohne dass er es eigentlich wollte, sich die Augen für das umgekehrte Ideal aufgemacht" (JGB 56, KSA V, pp. 74–75).

46. "Nicht 'erkennen', sondern schematisiren, dem Chaos so viel Regularität und Formen auferlegen, als es unserem praktischen Bedurfniß genug thut . . . zum Zweck der Verständigung, der Berechnung" (KSA XIII, pp. 333–34), WP 515.

47. "Man findet in den Dingen nichts wieder als was man nicht selbst hineingesteckt hat: dies Kinderspiel, von dem ich nicht gering denken möchte, heißt sich Wissenschaft?" (KSA XII, pp. 153–54; cf. G-D, KSA VI, p. 91).

48. See G-D, KSA VI, pp. 90–91.

49. "Es bestünde gar kein Leben, wenn nicht auf dem Grunde perspektivischer Schätzungen und Scheinbarkeiten" (JGB 34, KSA V, p. 53).

50. See Abel, *Nietzsche*, esp. pp. 173ff.

51. Cf. GS 370. Elsewhere, Nietzsche employs the (regrettable but not misleading) term: *Weibs-Aesthetic*, to denote the essentially reactive, receptive, spectator's aesthetic he opposed; see KSA XIII, p. 357. See below: chapters 6 and 7.

52. "Wir haben die Welt, welche Werthe hat geschaffen! Dies erkennend erkennen wir auch, das die Verehrung der Wahrheit schon die Folge einer Illusion ist" (KSA XI, p. 146), WP 602. Cf. BGE 34.

53. This historical projection is the sober antidote to the presumption of a marvelous correspondence between our human understanding and the world, which Nietzsche finds too lavish an hypothesis. Although language analysis of Anglo-American philosophy, after Wittgenstein, finds its inspiration in the legacy of William of Ockham, a parallel should not be too closely drawn. The interest of the latter is nominalism;

Nietzsche responds simply with an ego-cutting edge in the thought: "keine rücklaufigen Hypothesen!" (KSA XI, p. 170).

54. "Wir sehen alle Dinge durch den Menschenkopf an und können diesen Kopf nicht abschneiden; während doch die Frage übrig bleibt, was von der Welt noch da wäre, wenn man ihn doch abgeschnitten hätte" (MA I:9, KSA II, p. 29), H I:9, p. 15.

55. "Das theologische Vorurtheil bei Kant, sein unbewußter Dogmatismus, seine moralistische Perspektive als herrschend, lenkend, befehlend. Das πρῶτον ψεῦδος : wie ist die Thatsache der Erkenntniß möglich?" (KSA XII, p. 264).

56. "Strebt durchaus nicht darnach, den Menschen nachzuahmen!" (KSA III, p. 468), GS 109, p. 168.

57. "Es in der Natur keine exakt gerade Linie, keinen wirklichen Kreis, kein absolutes Grössensmaass gebe" (KSA II, p. 31), H, p. 16.

58. "Die Erfindung der Gesetze der Zahlen ist auf Grund des ursprünglich schon herrschenden Irrthums gemacht, dass es mehrere gleiche Dinge gebe (aber thatsächlich giebt es nichts Gleiches) mindestens dass es Dinge gebe (aber es giebt kein 'Ding')" (KSA II, p. 40), H 19, pp. 22–73.

59. "Es ist genug, die Wissenschaft als möglichst getreu Anmenschlichung der Dinge zu betrachten, wir lernen immer genauer uns selber beschreiben, indem wir die Dinge und ihr Nacheinander beschreiben" (FW 112, KSA III, p. 473), GS 112, pp. 172–73.

60. "Die Frage ist, ob es nicht noch viele Art(en) geben könnte, eine Solche *scheinbare* Welt zu schaffen—und ob nicht dieses Schaffen, Logisiren, Zurechtmachen, Fälschen die best-garantirte *Realität* selbst ist" (KSA XII, p. 396).

61. "Man muß . . . *festhalten:* was Begriffe und Formeln nur sein können: Mittel der Verständlichung und Berechenbarkeit, die *praktische Anwendbarkeit* ist Ziel daß der Mensch sich der Natur bedienen könne. Wissenschaft: die Bemächtigung der Natur zu Zwecken des Menschen" (ibid. XI, p. 91).

62. "Das *Vertrauen* zur Vernunft und ihren Kategorien, zur Dialektik, also die *Werthschätzung* der Logik beweist nur die durch Erfahrung bewiesene *Nützlichkeit* derselben für das Leben: *nicht* deren 'Wahrheit'" (ibid. XII, p. 352).

63. "Leben ist auf die Voraussetzung eines Glaubens, an Dauerndes und Regulär-Wiederkehrendes gegründet; je mächtiger das Leben, um so breiter muß die errathbare, gleichsam *seiend gemachte* Welt sein. Logisirung, Rationalisirung, Systematisirung als Hülfsmittel des Lebens" (ibid. XII, p. 385).

64. "Wahrheit ist somit nicht etwas, was da wäre und was aufzufinden, zu entdecken wäre,—sondern etwas, *das zu schaffen ist*" (KSA XII, p. 385), WP 552.

65. We are circumstantial, organic, finite beings; it would be marvelous indeed if our perspectival vantage had nothing to do with what we can know. It is of interest to add that Nietzsche's position on the wondrous (miraculous) character of knowledge is reversed in most realist philosophy of science and in H. Putnam's Internal Idealism: for Putnam, the idea that the striking "success" of science be possible without objective correspondence requires a miraculous series of accidents. To this degree, consideration of our perspectival limits is obviated by the external test of scientific success.

66. "Die Welt der 'Phänomene' ist die zurechtgemachte Welt, die wir *als real empfinden*. Die 'Realität' liegt in den beständigen wiederkommen gleicher, bekannter, verwandter Dinge in ihrem *logisirten Charakter*, im Glauben, daß wir hier rechnen, berechnen können" (KSA XII, pp. 395–56).

67. "Wir haben Einheiten nöthig, um rechnen zu können: deshalb ist nicht anzunehmen daß es solche Einheiten giebt" (KSA. XIII, p. 258).

68. "Die Mechanik formulirt Folgeerscheinungen noch dazu semiotische in sinnlichen und psychologischen Ausdrucksmitteln, sie berührt die ursächliche Kraft nicht" (KSA XIII, p. 259).

69. FW 107.

70. "Die 'Dingheit' ist erst von uns geschaffen" (KSA XII, p. 396).

71. This suggests a transformed correspondence theory of truth (or as Danto has it: a pragmatic conception (*Nietzsche as Philosopher*, see final chapters). But the idea of even a pragmatically effective mapping process—still less a real structure of correspondence—is less relevant to Nietzsche's thought than the fundamental notion of interchange or interpretational exchange.

72. Prigogine and Stengers, *Order out of Chaos,* 229.

73. "Der interpretive Charakter alles Geschehens. Es giebt kein Ereigniß an sich. Was geschieht, ist ein Gruppe von Erscheinungen *ausgelesen* und zusammengefaßt von einem interpretirenden Wesen" (KSA XII, p. 38).

74. "Die Welt von innen gesehen ... sie wäre eben 'Wille zur Macht' und nights ausserdem" (JGB 36, KSA V, p. 55).

75. "Alles Geschehen, alle Bewegung, alles Werden als ein Feststellen von Grad und Kraftverhältnissen als ein Kampf" (KSA XII, p. 385), WP 552.

76. "Man darf nicht fragen: 'wer interpretirt denn?' sondern das Interpretiren selbst, als eine Form des Willens zur Macht, hat Dasein ... als ein Affekt" (KSA XII, p. 140), WP 556. Cf. JGB 36.

77. JGB 36. Cf. KSA XI, p. 611.

78. "Jedes Kraftcentrum hat für den ganzen *Rest* seine *Perspektive* Die *Realität* besteht exakt in dieser Partikulär-Aktion und Reaktion jedes Einzelnen gegen das Ganze ... " (KSA XIII, p. 371), WP 567.

79. " . . . jedes Kraftcentrum—und nicht nur der Mensch—von sich aus die ganz übrige Welt construirt" (KSA XIII, p. 371). On the basis of this generalized interpretation/interaction structure, a number of writers in the German tradition of Nietzsche scholarship find parallels between Nietzsche and Leibniz (here, in connection with the Monadology and Occasionalism, though this association is naturally extended to appearance of the Pythagorean self-reflexive account of the Eternal Return in the Theodicy). Nietzsche's perspectivalist view is far stronger than Leibniz's monadology: his version of occasionalism in the interactive tension of the cosmos does not permit an external rule or law, for this reason, despite the separate direction of the notion of the Eternal Return, this inter-knotted flux cannot be thought as ultimately or completely predictable. The affinity between Leibniz and Laplace does not extend to Nietzsche on the same line. However, the association between Nietzsche and Leibniz is suggestive on any number of levels, provided one avoids mechanistic readings. On this topic, noteworthy authors include, in German, G. Abel, W. Gebhard, and F. Kaulbach; and, in English, G. Stack and A. Moles.

80. Yet in support of this same challenge the direction of many areas of contemporary research could be cited (e.g., quantum mechanics in physics, catastrophe/probability theory in biochemical/biological growth models, as well as the (organic) chemical descriptions Nietzsche favoured. See Kirchhoff and Abel for representation of the first, Löw and Maturana for some limited consideration of the second, and, of course, Mittasch for the third.

81. "So viele materialistische Naturforscher zufrieden geben, [mit] dem Glauben an eine Welt, welche im menschlichen Denken, in menschlichen Werthbegriffen ihr Äquivalent und Maass haben soll, an eine Welt der Wahrheit," der man mit Hülfe unsrer viereckigen kleinen Menschenvernunft letztgültig beizukommen vermöchte—wie?" (FW 373, KSA III, p. 625), GS 373, p. 335.

82. KSA IX, p. 515. Reference is to Book I of *Zarathustra*, composed as Nietzsche put it, "in the style of the Ninth Symphony."

83. "Der Gesammt-Charakter der Welt ist dagegen in alle Ewigkeit Chaos" (FW 109, KSA III, p. 468), GS 109, p. 168.

84. See Moles, *Nietzsche's Philosophy of Nature and Cosmology* and Klaus Spiekermann's recent *Naturwissenschaft als subjektlose Macht? Nietzsches Kritik physikalischer Grundkonzepte.*

85. On the notion of plenary origin, George Steiner makes the grand claim that "for every German thinker after Herder, the word Ursprung is resonant. It signifies not only 'source,' 'fount,' 'origin,' but also that primal leap (Sprung) into being which at once reveals and determines the unfolding structure, the central dynamics of form in an organic or spiritual phenomenon" (Introduction to *The Origin of Ger-*

man Tragic Drama, by Walter Benjamin, trans. J. Osborne [London: NLB, 1977], 15–16).

86. W. Gebhard, "Erkennen und Entsetzen. Zur Tradition der Chaos-Annahmen im Denken Friedrich Nietzsches," 20ff.

87. Indeed, this analysis of chaos would have more bite as a proposed criticism if Nietzsche were merely a philologist or if philology were as limited a discipline in Nietzsche's practice as it has come to be in the United States. Nietzsche's background interests and competence exceeded a simple attention to the Greeks (even with regard to a Greek term) and must be thought to reflect Semitic and Eddic accounts and resonances, together with his own celebrated and criticized intuition or genius.

88. "Chaos, χάος, χαίνω bedeutet das Gähnen, Gähnende, Auseinanderklaffende. Wir begreifen χάος im engsten Zusammenhang mit einer ursprünglichen Auslegung des Wesens der ἀλήθεια als den sich öffnenden Abgrund (vgl. Hesiod, *Theogonie*)" (Heidegger *Nietzsche I*, p. 350). Heidegger's account also includes a specific restriction of Nietzsche's understanding of chaos that would place it between classic and contemporary accounts.

89. Cf. Heidegger, *Nietzsche II*, trans. D. Krell. After a brief analysis of the Hesiod fragment, Krell proposes that Heidegger's reading be interpreted in the sense of the interpretation offered in *Introduction to Metaphysics* as the "Timaean *chora*, the receptacle of space . . . the open region in which all beings can first appear and be in being" (Heidegger, *Nietzsche II* [trans. Krell], 92; Heidegger, *Nietzsche I*, 350).

90. Heidegger, *Nietzsche I*, 339–56ff. Chaos conceived in this connection is necessary becoming.

91. "Der Werde, d.h., hier zugleich der Seinscharakter des Weltganzen als das ewigen Chaos der Nothwendigkeit ist die ewige Wiederkunft des Gleichen" (ibid., 371).

92. Granier, "La pensée nietzschienne du chaos," 132. (L'être est texte.)

93. "La nature, texte primitif, est donc l'être-chaos qui se manifeste comme procés signifiant dont les figures dessinent, non un systéme ou un chaos, mais justement un voile" (ibid., 135).

94. "Das Gleiche—l'être qui revient dans l'éternel retour, rendent co-présents la nature et la voile dans l'unité equivoque du texte" (ibid.).

95. "le Même, qui désigne l'être du phénomène, ajointe la nature et le griffonage des interprétations de telle manière que le texte est, pour tout interprète, énigmatique" (ibid.).

96. Ibid., p. 137.

97. "Le réel est un chaos sans aucune commune mesure avec les exigences humaines" (ibid., p. 143).

98. Heidegger, *Nietzsche II* (trans. Krell), p. 129. "Der Gedanke

der ewigen Wiederkunft des Gleichen *macht fest, wie* das Weltwesen als Chaos des Nothwendigkeit ständigen Werden ist" (Heidegger, *Nietzsche I*, p. 392).

99. Heidegger, *Nietzsche II*, p. 147. "ist je nach seiner Art immer der Wiederklang aus dem Ganzen und der Einklang in das Ganze" (Heidegger, *Nietzsche I*, p. 408).

100. Heidegger, *Nietzsche II* (trans. Krell), p. 147. "in Wahrheit der Gedanke der Gedanken [Eternal Return] in das Seiende für jeden Augenblick die höchsten Schärfe und Entscheidungskraft [bringt]" (Heidegger, *Nietzsche I* pp. 408–9).

101. Cf. KSA XI, p. 553; XIII, pp. 34–35.

102. "Il est bien entendu que l'éternel retour ne peut pas être l'éternel retour des contenus" (A. Philonenko, "Mélancholie et consolation chez Nietzsche," p. 95). In this same connection, Granier writes, "le Même, et non l'identité logique de l'ontologie" (p. 135). "La pensée Nietzschienne de chaos."

103. Kahn, *Anaximander and the Origins of Greek Cosmology*, 220.

104. Ibid., 220. The metaphysical cast of this "cosmetic" vision is obvious: "all extant examples of *kosmos* or *diakosmos* in the early philosophical fragments illustrate the idea of an all-embracing 'arrangement' or ordering of parts: the natural world is conceived as a structural whole in which every component has its place" (ibid., 229).

105. "La logique, en dernier instance, n'est q'une façon de s'approprier le réel en le déformant à partir de besoins déterminés, qui fixent le sens même de notre perspective, à nous hommes." Granier, "La pensée nietzschienne du chaos," p. 146.

106. "La Duplicité de l'Etre: le jeu de la vérité mortelle et de l'illusion-valeur, selon l'infinité des interprétations,—le jeu du chaos et de la vie" (ibid., p. 147).

107. Cf. Kirchhoff, "Zum Problem der Erkenntnis bei Nietzsche."

108. "Non pas une probabilité répartie sur plusieurs fois, mais tout le hasard en une fois; non pas une combinaison finale désirée , voulue, souhaitée, mais la combinaison fatale, fatale et aimée, *l'amor fati*" (Deleuze, *Nietzsche,* p. 31).

109. Cf. JGB 5.

110. "Car le coup de dés est l'affirmation multiple, l'affirmation du multiple. Mais tous les membres, tous les fragments sont lancés en un coup; tout le hasard en une fois. Cette puissance, non pas de supprimer le multiple, mais de l'affirmer en une fois, est comme le feu" (Deleuze, *Nietzsche,* p. 34).

111. "Die formlos-unformulirbare Welt des Sensationen-Chaos— also *eine andere Art* Phänomenal-Welt, eine für uns 'unerkennbare'" (KSA XII, p. 396). There are any number of inspirational sources that

could be cited with respect to Nietzsche's view of nature as interpretationally "chaotic": i.e., as formless but in-formable. For example, in addition to the Dionysian worldview of the culture Nietzsche admired most, we may count the critical philosophy of Kant and Schopenhauer outlining the constitutional, representational character of the human understanding, which for Schopenhauer could serve as access to the essence of the world. Although there are parallels between Nietzsche and Schopenhauer in terms of the analogistic power of self-reflection, what Schopenhauer takes to be a trustworthy track (if one eliciting a pessimist's account of the sorrows of existence) is for Nietzsche the way of illusion: our illusion to be sure, but nature's illusion too. If art is an antidote to the privileged Schopenhaurian analogistic vision, it is the essence of the Nietzschean reflective analogy. Ontogeny recapitulates phylogeny for Nietzsche too, yet not without changing the whole point of view. We forget this metaphysical difference between Schopenhauer and Nietzsche at our peril.

112. "Versuchen wir den Spiegel an sich zu betrachten, so entdecken wir endlich Nichts als die Dinge auf ihm. Wollen wir die Dinge fassen, so kommen wir zuletzt wieder auf Nichts, als auf den Speigel.— Diess ist die allgemeinste Geschichte der Erkenntniss" (M 243, KSA III, pp. 202–3).

113. "War es nicht etwas sonderbar zu verlangen, dass ein Werkzeug seine eigene Trefflichkeit und Tauglichkeit kritisiren solle? dass der Intellekt selbst seinen Werthe, seine Kraft, seine Grenzen 'erkennen' solle? war es nicht sogar ein wenig widersinnig?" (M iii, KSA III, p. 13).

114. "Der Intellekt kann sich nicht selbst kritisiren, . . . um den Intellekt zu kritisiren, wir ein höheres Wesen mit 'absoluter Erkenntniß' sein müßten" (KSA XII, p. 188).

115. "Das Problem von Werthe der Wahrheit" (JGB 1, KSA V, p. 15).

116. "Die Welt ist *nicht* so und so: und die lebenden Wesen sehen sie, wie sie ihnen erscheint. Sondern die Welt besteht aus solchen lebenden Wesen, und für jedes derselben giebt es einen kleinen Winkel, von dem aus es mißt, gewahr wird, sieht und nicht sieht. Das 'Wesen' fehlt: Das 'Werdende,' 'Phänomenale' ist die einzig Art Sein" (KSA XII, p. 249).

117. "Die Menschheit hat zu allen Zeiten das Activum und Passivum verwechselt, es ist ihr ewiger grammatikalischer Schnitzer" (M 121, KSA III, p. 115).

118. "Das Lebende ist nur eine Art des Todten" (FW 109, KSA III, p. 468). Cf. Nietzsche's chemically inspired observation in KSA XI, p. 537: "Der Übergang aus der Welt des Anorganischen in die des Organischen ist . . . unsicheren, unbestimmten."

119. "Ob das Perspektivistische zum Wesen gehört? Dies wäre möglich, wenn alles Sein essentiell etwas Wahrnehmendes wäre" (KSA XII, p. 188).

120. "Gesetzt, aber, wir legen in die Dinge gewisse Werthe hinein, so wirken diese Werthe dann auf uns zurück, nachdem wir vergessen haben, daß wir die Geber waren" (ibid., p. 192).

121. "Jedes Kraftcentrum hat für den ganzen Rest seine Perspektive" (KSA XIII, p. 371), WP 567.

122. "Die 'Welt' ist nur ein Wort für das Gesammtspiel dieser Aktionen" (KSA XIII, p. 371).

123. JGB 36.

124. "Ein Machtquantum ist durch die Wirkung, die es übt und der es widersteht, bezeichnet. Es fehlt die Adiaphorie: die an sich denkbar wäre. Es ist essentiell ein Wille zur Vergewaltigung und sich gegen Vergewältigungen zu wehren. Nicht Selbsterhaltung: jedes Atom wirkt in das ganze Sein hinaus" (KSA XIII, p. 258).

125. "Die Mechanik als eine Lehre der *Bewegung* ist bereits eine Übersetzung in die Sinnensprache des Menschen" (KSA XIII, p. 258).

126. "Es liegt ein tiefes und gründliches Glück darin, dass die Wissenschaft Dinge ermittelt, die Stand halten und die immer wieder den grund zu neuen Ermittlungen abgeben:—es könnte ja anders sein!" (FW 96; ibid. III, p. 411.

127. "Der Gesammt-Charakter der Welt . . . ist in alle Ewigkeit Chaos" (FW 109, KSA III, p. 468).

128. "Die Griechen nannten diess Reich des Unberechenbaren und der erhabenen ewigen Bornirtheit Moira" (KSA III, p. 120).

129. The range of this egoism is expressed by what Nietzsche names "Zuschauer-Göttlichkeit" (M 130, KSA III, p. 120).

130. "Wenn ihr wisst, dass es keine Zwecke giebt, so wisst ihr auch, dass es keinen Zufall giebt" (FW 109, KSA III, p. 468).

131. Cf. Aristotle, *Physics*, 2:6.

132. "Jene eisernen Hände der Nothwendigkeit, welche den Würfelbecher des Zufalls schütteln, spielen ihr Spiel unendliche Zeit: da müssen Würfe vorkommen, die der Zweckmässigkeit und Vernünftigeit jedes Grades vollkommen ähnlich sehen" (M 130, KSA III, p. 122), D 130.

133. "Von unserer Vernunft aus geurtheilt, sind die verunglückten Würfe weitaus die Regel, die Ausnahmen sind nicht das geheime Ziel, und das ganze Spielwerk wiederholt ewig seine Weise, die nie eine Melodie heissen darf,—und zuletzt ist selbst das Wort 'verunglückter Wurf' schon eine Vermenschlichung, die einen Tadel in sich schliesst" (FW 109, KSA III, p. 468).

134. "Die Natur . . . ist kein Model. Sie Ubertreibt, sie verzerrt, sie lässt Lücken. Die Natur ist der Zufall" (G-D, KSA VI, p. 115).

135. "Wie weit der perspektivistische Charakter des Daseins reicht oder gar ob es irgend einen andren Charakter noch hat, ob nicht ein Dasein ohne Auslegung, ohne 'Sinne' eben zum 'Unsinn' wird, ob, andrerseits, nicht alles Dasein essentiell ein auslegendes Dasein [ist]" (FW 374, KSA III, p. 626).

136. "eine contemplative und übersieht dabei, dass er selber auch der eigentliche Dichter und Fortdichter des Lebens ist" (FW 301, KSA III, p. 540).

137. This distinctive context is reminiscent of Hannah Arendt. Nietzsche's own expression is more daring: "Ihm, als dem Dichter, ist gewiss vis contemplativa, und der Rückblick auf sein Werk zu eigen, aber zugleich und vorerst die vis creativa"(FW 301, KSA III, p. 540) (UB II:6, KSA I, p. 285).

138. "Der Mensch ist das Zeugniß, welche ungeheure Kräfte in Bewegung gesetzt werden können, durch ein kleines Wesen vielfachen Inhalts . . . *Wesen, die mit Gestirnen spielen*" (KSA XII, p. 40).

139. "*Im Menschen ist* Geschöpf und *Schöpfer* vereint: im Menschen ist Stoff, Bruchstück, Überfluss, Lehm, Koth, Unsinn, Chaos; aber im Menschen ist auch Schöpfer, Bildner, Hammer-Härte, Zuschauer-Göttlichkeit und siebenter Tag" (JGB, KSA V, p. 161).

140. "Wir erst haben die Welt, *die den Menschen Etwas angeht*, geschaffen! Gerade dieses Wissen aber fehlt uns, und wenn wir es einen Augenblick einmal erhaschen, so haben wir es im nächsten wieder vergessen: wir verkennen unsere beste Kraft und schätzen uns, die Contemplativen, um einen Grad zu gering,—wir sind *weder so stolz, noch so glücklich*, als wir sein könnten" (FW 301, KSA III, p. 540), GS, p. 242.

CHAPTER 5

Nietzsche's Genealogy of Science: Morality and the Values of Modernity

THE GENEALOGY OF MORALS
AND THE VALUE OF SCIENCE

Nietzsche's genealogical hermeneutic of the cultivation and enculturation of human life, concerning morality as the basis of culture and society (the latter being only in part "religion"), leads to an analysis of science in terms of moral value. Morality for Nietzsche is not a simple social code but e-valuation. E-valuation or setting values is, of course, to be understood as interpretation, that is, Will to Power. The moral scheme manifests in a covert but essential way, the same "lust to rule" that Nietzsche regards as characteriztic of every drive or Will to Power: "each one has its perspective that it would like to compel all the other drives to accept as a norm."[1]

How does a genealogy of morals or of truth or, indeed, of science, work? What is involved? For Nietzsche at the outset, one has to question the hithertofore ignored origins of things, not in terms of intentions or other routine causes or reasons, but (and thus the word *genealogy* with its biological overtones, the study of the kind in question of the nature of morals or truth of science, as the sort of thing that it is) in terms of the organic necessity at the root of our being apart from intentions and desires.[2] As Nietzsche expresses it, the necessities that move us, bringing forth our ideas, our values, assents, and dissents, respond to the same ecological course that consummates autumn in fire and death.[3] But if the origin of morality is to be put in question, in particular if morality is shown to be a product of organic necessity (and not a triumph over such necessity), the very idea of freedom is at stake. If Nietzsche can claim that freedom and self-reflective distance are mere illusions, does he not claim this very insight into the essence of the human being for his own philosophical project? We have seen this protest before as the recurrent or standard Nietzschean counterfactual, but we have also seen that such a criticism overlooks the self-consciously complex sense of Nietzsche's thinking.[4] To understand this complexity in the current context and apart from the popularity of recent readings rendering genealogical analysis in terms of Foucault's very different project, Tracy B. Strong notes that Nietzsche's method of genealogical analysis shares the descriptive orientation of Husserl's, likewise differently developed, phenomenological method. Strong writes that,

> Husserl's litany, *"zu den Sachen selbst"* is an exhortation to bracket all that is casuistically human and interpretive in order to arrive at the "things themselves," a world of direct and non-mediated experience. Genealogy on the other hand, does not seek out and describe the "things" that phenomenology holds to be the world but rather delineates the manner in which the "things are made into facts." Nietzsche tries to bring out precisely how a particular world is put together and made a world.[5]

Nietzsche's own words anticipating the now classic perspective of Husserl's phenomenological method: "No retrograde hypotheses! Rather a standpoint of ἐποχή instead!"[6] are poised against anthropomorphic or even nonanthropomorphic presumptions, explaining "The task: to *see* things *as they are!* The means: to see them out of a hundred eyes, out of *many* persons!"[7] Thus Nietzsche recommends a "Cosmological Perspective" as transformative, as salutary in a *Nachlaß* projection of the history of European nihilism and its overcoming.[8] The cosmological sensitivity recommended here cannot be an impersonal, objective perspective. Because there can be no escape from the individual perspective, Nietzsche advocates multiple personal perspectives.

Against such a pluralistic—today we might say: postmodern—ideal, the objective ideal of science demands the impossible, "that we should think of an eye that is completely unthinkable, an eye turned in no particular direction, in which the active and interpreting forces, through which alone seeing becomes seeing *something*, are supposed to be lacking."[9] In contrast, Nietzsche distinguishes and defines "objectivity," "not as 'contemplation without interest' (which is a nonsensical absurdity), but the ability to *control* one's Pro and Con and to dispose of them, so that one knows how to employ a *variety* of perspectives and affective interpretations in the service of knowledge."[10] Such "objectivity" is a perspectival objectivity, incorporating in its vision the "active and interpreting forces, through which alone seeing becomes seeing *something*." By advocating a cosmological perspective as the background augmenting this genealogical scheme, Nietzsche repeats his fundamental perspectivalism.

Nietzsche's genealogical approach is neither metaphysical nor epistemological—genealogy does not uncover first causes or truths; rather, and this makes it so poular among today's radical hermeneutic and postmodern thinkers, a genealogy develops and exposes *suspicions* at the root of things—and it is not axiological. Nietzsche neither analyses morality in Schopenhauer's spirit as an abstract concept or ideal nor considers the nature or utility of moral actions as such. For the philologist Nietzsche, following his earlier call for "*critique* of moral valuations,"[11] one best questions morality by analyzing those who at any time are called "good" rather than what good is done. This recalls the archaic Greek personal ascription of value where the being (position, endowment) of the actor, rather than an "objective" judgment of the subject's action itself, determined the quality of the act. For the Greek, there is no subject who acts, the subject is the action, or, better, the action is the subject. So Nietzsche writes, "It is immediately obvious that designations of moral value were everywhere first applied to *human beings*, and only later and derivatively to *actions*."[12] The "Good" were the powerful:

> The noble type of man feels *himself* to be the determiner of values, he does not need to be approved of, he judges "what harms me is harmful in itself," he knows himself to be that which in general first accords honor to things, he *creates values*. Everything he know to be a part of himself he honours: such a morality is self-glorification.[13]

Because people fear and thus misjudge the effective political and social implications of this description of the noble or "higher" human being, we do need to add the corrective qualification that such a noble morality is also terribly innocent.[14] For Nietzsche, this naïveté counts as a good will.[15] The values of the noble human type are the values of overflowing power: "In the foreground stands the feeling of plenitude, of power which seeks to overflow, the happiness of high tension, the consciousness of a wealth which would like to give away and bestow."[16] So that we do not miss the point here, although it will be necessary to come back to it again and again, the mark of, the *proof* of overflowing power is evidenced in what is given out. For Nietzsche, the greatness of a great human being "lies in the fact that he *expends* himself."[17] This valorization of self-expenditure is a description not only of Aristotle's "great-souled man" but also of Nietzsche's appropriation of Lucian's term ὑπεράνθρωπος reflected as the *Übermensch*. From this height of power and the innocent inclination to self-expression that springs from greatness, such a human being will consequently aid others, not out of pity "but more from an urge begotten by the superfluity of power."[18] For Nietzsche and for the Greeks, generosity and kindness are benedictions bestowed on others not in response to the imperatives of duty or as gifts in a structural economy of exchange. Rather, like a star's shining exuberance, like the sun dripping gold, the gold of light, sparkling reflection, of *unreal* shining glory, the greatness of will or benevolence cannot be withheld. The noble, great-souled individual is compelled to and is capable of an excess that makes its own measure. Thus Nietzsche defines good in terms of the noble, monumental human type who may be characterized further by the desire and capacity for difficult tasks and dangerous undertakings. The Greek ideal of moderation, for we are echoing the Greeks in this celebration of nobility, is thus preserved, for the essence of moderation is a relative matter, and power has its own measure.[19]

On the other hand, the values possible for those not blessed with such potent natures, the values that can be set by such weaker human types will reflect their specific inadequacy as the values of impotence. Note here such values are exactly not impotent values. Moral valuation from the side of weakness pronounces weakness a good and, in terms of its own interests and inherent neediness, expresses that very same weakness as an ideal, deserving promotion and support. In this same way, injury may be interpreted as culpable (originally implicating the agent, but, Nietzsche shows, permitting an aetiological application to the patient's desserts).

Note that the perspective from the side of impotence is non-naive (in contrast to the naïveté of the noble) and marks the beginning of social and cultural consciousness as such.[20] What is important in this extension of the morality of impotence is its position vis-à-vis life. The perspective of weakness results in a resentment against life. This is a resentment against everything that hinders the inadequate longing for an easy (less demanding) existence. What Nietzsche names *Ressentiment* opposes all expressive life-values because they appear to endanger and even worse to waste life rather than preserve it as such. This view of life-expression, erroneous when conceived from the side of life, is not erroneous from the side of weakness. For the weaker perspective is not in error when it estimates those life-expressive values articulating the power of an individual perspective as values potentially injurious both to others and to the originator. Noble or life-affirmative values *are* actually extravagant—Nietzsche's word is *verschwenderisch*—a squandering, wasting, frivolous (expression) expenditure of life.

THE ASCETIC IDEAL: THE INVERSION OF VALUES

The life-preservative ideals characteristic of the aforementioned weak or slave morality are not the values of some early Christian or Jewish past. They are today's living, ruling ideals. If noble and slave values are thought to be aboriginally opposed, the resultant of such an opposition is invariably the ascendance of slave values. It would be misguided to interpret Nietzsche's distinction between noble and slave morality as setting a possible program for future state revolution, that is, to read the slogan "Grand Politics" in an ordinary (reactive) light.[21] Nietzsche's programmatic goal, insofar as the ideal of an elite culture does interest him, is qualified by his recognition of the prevalence of reactive morality and so its correspondingly necessary societal dominance. This refinement would thus account for the difference between my own necessarily philosophical perspective on Nietzsche's political reflections on the workings of social *Ressentiment* and other theoretical accounts, such as, for one example, those of the political theorist William Connolly in his recent *Identity/Difference*.

Qualifying his own invocation of Nietzsche in a context discussing the problem of "Democratic Negotiations of Political Paradox," Connolly rightly observes, "A democrat who draws upon Nietzsche to think about contemporary issues of identity, ethics, and politics will soon be compelled to take several steps away from him."[22] If Nietzsche is fundamentally and perhaps best seen as antipolitical, he is also and obviously no democrat. Connolly underlines a theoretical point of interpretation that is essential for any understanding of Nietzsche as Connolly advances his own political appropriation of Nietzsche. Connolly reads Nietzsche as offering "a tragic conception of life with non-theistic reverence." For Connolly, taken "together these provide a human basis for agonistic care and self-limitation."[23] Connolly thus articulates his own political scheme as one conveying the "paradox of ethicality . . . an *agonistic* ethic of care that ambiguates assumptions it iself is often compelled to make about the truth of the identity it endorses."[24] For the purposes of the present context, it should be noted that Connolly adverts to the singular problem of the "overman" in a postmodern, that is, not merely machine- but media- and ecology-conscious, culturally pluralist age. For my reading, without the aid of what may be called "vertical" thought, expressing Nietzsche's concept of higher and lower, eso- and exoteric, the concept of the "overman" is an empty metaphor: a foresworn possibility. Unless the human being is something that can be overcome in its humanity, unless a post-human is possible, the concept of the overman must be regarded as an aspect of Zarathustra's arch irony, where the author of Zarathustra, a book for all and none, must be said to have known that the projected realization of the overman reflects the paucity of opportunity that heightens its romantic appeal. In a brilliant dialectic account of the postmodern era and Zarathustra's program for the overman, Connolly finds Nietzsche's themes and perspectives imbricated in a contradictory interplay that would be deadly if this same imbrication were not also the effective (concinnous) expression of Nietzsche's own (romanticizing) point:

> the metaphors of wildness, hermits, eagles, snakes, caves, silence, deep wells, high mountains, solitude, mob, flight, and earth that populate Nietzsche's invocations of the overman no longer do *double* duty today. The "hermit" has become an anonymous member of a regulated multitude who are homeless; the "eagle" has become a protected species; the "mob" has become a criminal network entangled with official intelligence agencies; the "deep well" accumulates pollutants from road maintenance, toxic wastes, and fertilizer runoffs; urban "caves" have become nightly residences for homeless outcasts who restlessly haunt the streets by day; the "earth" has become a deposit of finite resources for late-modern production.[25]

Connolly concludes that "even the metaphors have become infiltrated by the significations they would rise above." And this point is surely well taken. But the catalogue given above shows only the preternatural self-absorption endemic to any age. More critically, we may note that there are actually only two themes in Connolly's reductive list. These are, not surprisingly for a political theorist, social problems and the environment—and the latter reduces to the first. As guileless in our day as Hesiod was in his own, we tend to overestimate the golden difference between our era nearing the high end of the twentieth century and the time that was Nietzsche's own near the end of the nineteenth. Nietzsche would not have had to be reminded of the decline of "nature" in its purity, and in using such glorious metaphors for the spirit, given the text in question, it is likely that Nietzsche may have had this very contradiction in mind.

In the next section we shall see why Connolly's exciting promise that as a result of "the late-modern possibility" of human self-extinction "preservation and overcoming are now drawn closer together, so that each becomes a term in the other: the latter cannot succeed unless it touches the former,"[26] must remain an optimistic syncretism that fails to think the overwhelming, difference-collapsing force of self-preservation as a value. This is all the more regrettable where Connolly otherwise recognizes the problem so well in his discussion of technological millenarianism that he addresses its breadth and command by naming it the "ontological narcissism"[27] of our day. Connolly details the imperializing tactics of this ontological narcissism in the pseudo-alternative between "conquest and conversion"[28] and (here reflecting Todorov) in the impossibility of truly dissident or revolutionary thought as the interior side of conversion where "identity and difference are bound together."[29]

It is Deleuze, in spite of all his Kojévian-Hegelian limitations, who is able to detail with more than French-Hegelian finesse that in the conflict between active and reactive values the reactive mode always triumphs—this is the price the noble pays for innocence![30] Thus although an elite culture may be the fruit of a broader, reactive culture, an elite culture cannot serve a ruling function.[31] Such legislative goals are inevitably limited to mob interest.[32] Nietzsche offers this description of the conflict between mass culture and the so-called rare individual: "The more similar, more ordinary human beings have had and still have the advantage, the more select, subtle, rare and harder to understand are liable to remain alone, succumb to accidents in their isolation and seldom propagate themselves."[33] Nietzsche makes his point more explicitly in a late note echoing the same point in the third part of the *Genealogy of Morals*: "as odd as it sounds: one must always protect the strong from the weak."[34] The Nietzschean call to protect the strong from the weak (the noble from the slave)[35] derives from the explicitation of this apparently counterintuitive notion in a reactively polarized dynamic of opposition and typological conflict, "the strongest and happiest are weak when they have the organized herd-instinct, the fearsomeness of the weak, the majority against them."[36] Driving the logical and linguistic schemes of human and social cultural life, it is this "fearsomeness" that enables *Ressentiment* to become active.[37]

It appears to be contradictory for Nietzsche to describe reactive values as creative (as a rule, he associates creativity with activity). But he explicitly claims that "The slave revolt in morality begins when *Ressentiment* itself becomes creative and gives birth to values."[38] The same creative, defining power of active noble life, namely, "the lordly right of giving names,"[39] is now employed in the service of reactive morality. The distinguishing mark of reactive conceptual representation is the articulation of difference as such. Rather than being named from the wealth of a naive oversufficiency, or the artistic, playful, egotistical coining of names, what is named by the slave is named negatively, what is conceived is nihilistically evaluated. Accordingly, Nietzsche defines the reactive valuation as syllogistic, or dialectical. Because all that is not slave or weak must be named evil, and because this that is evil receives its definition in contrast with the slave, the slave becomes what is good. As this naming action compensates for weakness by defending that weakness, which is to say "preserving" it, such a nomination is reactive.

As an evaluating scheme, slave morality comes into being as a social reaction against the strong (self-sufficient individual). But it does not stop there: this reactive valuation works against—that is, it denies *and* preserves—weakness as such, and in this way it works against life and ultimately against the nature of existence, insofar as it cannot be conserved, because existence is nothing but change and becoming. Existence itself is wasteful and destructive, full of threatening danger and full of heedless (from the vulnerable perspective of the slave) external sufficiency. Valued from a slave perspective, the designation "good" is to apply to the inoffensive, the mild—to all values that succor and sustain life at even its lowest level (so conserving life "at any price"). Such a definition is unable to comprehend the perspective of the powerful because, favoring the impotent, it can exalt only whatever assists or sustains weakness.

Because all that is not slave or weak must be named evil, and this that is "evil" receives its definition in contrast with the slave, the figure of the slave becomes what is good. As this naming action compensates for weakness by defending that weakness, which is to say by preserving weakness, this nomination is essentially reactive. The dynamic of this reactive-inversion of positive valuation is articulated in the subtlety that brings it to birth: it is not merely because we live in the inherited tradition of this successful revolt of life-preservative values that we find its working difficult to perceive.[40] The roots of this subtlety go deeper than simple circumstantial immediacy. This complication demands clarification. The impulse of the inversion of values from a pristine self-glorifying, life-glorifying, "great-souled" (noble morality of good and bad: of power and its expression) to a squinting, self-diminishing, conservation of diminished life (slave morality of good and evil or of pity and forgiveness), lies in their sensitive differences, that is, the difference of sensibilities that characterizes the two moral schemes.

From the noble side, "the pathos of distance" expresses self-sufficiency: the non-comprehending distance separating the powerful from the impotent. On the slave side, "the pathos of distance" becomes a "pathos of *difference*": expressing the impotent misunderstanding of (but desire for) power. Thus it may be said that the interpretational expression articulating an original impotence in the desire for power—"the tyrant-madness of impotence (*der Tyrannen-Wahnsinn der Ohnmacht*)"—manifests its Will to Power.[41] The fictional force from the perspective of a slave morality is derived from its own "pathos of distance," articulating what is good in a reactive, negative expression. But, because the Will to Power is represented from the side of a lack of power (that is, plainly, the side of impotence),[42] it becomes what it never is when seen from the perspective of power: a drive not to *express* power (which is the essence of organic nature) but to *acquire* power. Accordingly, in the movement from bad conscience to asceticism, *Ressentiment* is employed as a means to insure control over oneself and, eventually, over the impulses and instincts of others, so, ultimately to insure control over one's life-circumstance, over nature and life itself. Thus we see that Nietzsche's philosophy of Will to Power, so far from advocating "a ruthless philosophy in which a few exercise mastery over other humans and nature,"[43] highlights the importunate aggression against the strong hiding in this imputation of danger from the side of the politically righteous and the mild.

Asceticism, as Nietzsche sees it, is more than a compensation for the sick constitution. Asceticism transfigures the power economy of weakness and represents it as restraint. Thus impotence literally becomes the signifier of power: self-control as manifest strength. In this way, the ascetic ideal is not merely the therapeutic project of a weak or impotent culture; the ascetic ideal (as it is employed by the ascetic priest) becomes the means for preserving life, not for healing, but for merely preserving the sick as sick (thereby, needless to say, preserving the role of the priest). Such preservation requires that sickness or impotence not be regarded as "sickness" or inadequacy but reconstructed in the light of subjective-intentionality, as the virtue of restraint. This evaluation is based on the potential for freely making and acting upon elections requiring power beyond one's nature. The working of this evaluative scheme is the achievement of a negative finesse—the *representation*, the rehabilitation of weakness as strength.

With this projection of freedom, Nietzsche means to explain the recriminative glance from the perspective of the weak, which charges that the powerful individual could do (have done) otherwise than (however thoughtlessly or needfully) injure. The implication is that, if he or she could only be made to respect the "rights" of others even where those "rights" cannot compel their own recognition, a powerful person could conduct him or herself more carefully. This reactive perspective projects a (nihilistic) notion of freedom which assumes "the right to make the bird of prey *accountable* for being a bird of prey."[44] Here, the bird of prey is expected to counter its ownmost possibilities, as if "*the strong man is* free to be weak and the bird of prey to be a lamb."[45] The efficacy of *Ressentiment* is nothing other than the source of its evaluative inconsistency, here the equation between weakness and strength. It is this illusion that is operative in the moral invention of negative culpability, the notion that one *could* do or be otherwise than one actually does or is: the grammatical separation sundering the actor from the action. This is the foundational exigence of slave morality become creative: the values to which it gives birth are those of the illusion of self-(subject!) denial or asceticism. But this illusion is not unreal. Following Nietzsche's account, the weak *really* do take over the values of the strong: the bad of noble valuation thereupon *become* the good, while the erstwhile good become the evil ones of slave or reactive valuation.

The conceptualization of this conviction in the ideal of equality before the law means that the constitution of the particular individual is irrelevant. For the strong, overabundant power can be defined as a weakness: representing a lack of (self-)control. For the weak, failures are represented as strengths: testifying to (self-) control. Thus when we read in a later text, "The saint in whom God takes pleasure is the ideal castrate . . . Life is at an end where the 'kingdom of God' *begins*,"[46] we see concealed in the the image of ascetic control what Nietzsche would regard as a lack of potency. By means of the ideal of the ascetic, that is, through the ascetic ideal, one who is lacking in power may yet be praised for putative restraint (constraint). Virtue is to be had, then, not in terms of what one is or what one does, but much more through what one is not and by means of what one does not do. In this, passivity is judged an action, inaction itself becomes a deed. A revaluation of this kind is possible only where the value of the individual as an individual has been eclipsed and the subject (so the social order) made the arbiter and source of value. Responsibility can either be between equals in a noble society or a mass (the subject's consciousness or conscience: Nietzsche's *"höheren Egoismus"*) responding equally to everything. In the latter case, the supreme good must be the social good rather than an individual good. This "socialization," the subjectification (subjection) of the individual to society is not a difficult achievement. A "mass" supremacy corresponds to a natural species constructive order, moreover, it is reflected in language.[47] Only the monumental society of a heroic (individual) order briefly overcomes it. But, in the mass rule of the social herd the value of an action is determined by its harmlessness and the range of its benefits. The significance of inaction is similarly evaluated. Here the standard of valuation, the interpretive perspective is an external, reactive measure: categorical and not individual, dialectical and not monological. For Nietzsche, of course, all external determination is essentially reactive. Thus, Nietzsche defines the distinction between kinds of artworks in terms of the conditions involved in their production,

"All thought, poetry, painting, compositions, even buildings and sculptures, belong either to monological art or to art before witnesses."[48]

Communication and not expression is prized in the social perspective on art. Not surprisingly, Nietzsche favors monological art because it is actively generated. Such an artless (i.e., unmannered, unsocialized) self-expression[49] must be contrasted with the image and effect essential for reactive social perspective. While one man's pacifist comportment may express innate gentleness, a coward's or a weakling's presumption can stake a claim to the same temperance without risking refutation. Expression as such loses its original importance in favour of its communicative effect. In the hiatus of expression essential to the conceptual constancy (monotony) of language and inactive action, the example of celibacy would prove a potent cover for impotence, or what may well be the same in homophobic societies: homosexuality. In the old fable illustrating the dynamic between actual effect and communicative image, wherever there are words and terms together with a comprehending audience, the fox failing to reach the grapes of his desire need not also lose face. Because all of the grapes could well be sour, all of the grapes can be *said* to be sour and the fox's miss thus interpreted or communicated (hermeneutics and communicative theory are the same here) as prudent restraint. The representation of an inadequacy as an excellence is a human (or a humanized fox's) trick.[50] Seen as a representation of strength, an inversion of values, the ascetic perspective reveals its motivation as an expression of the Will to Power: "The will of the weak to represent *some* form of superiority, their instinct for devious paths to tyranny over the healthy—where can it not be discovered, this will to power of the weakest!"[51]

In the end, asceticism emerges in many forms. There is an asceticism proper to the strong, to the philosopher, and to the free spirit. Asceticism in these contexts is a means of self-overcoming: it is an instrument of self-creation, representing the capabilities and discipline of an artist. But, the same ascetic ideal in the hands of those unable to create yet still seeking to dominate—for example, the weak, and especially the priests of the weak—serves as a means to the preservation of their inadequacy, because what characterizes the weak is their fearfulness, and what they fear most is to be overpowered, overcome. Simply put, using the words of Nietzsche's Zarathustra: these are those who do not want to go under. In the herd society of mediocre, democratic, modern life, the artistic self-overcoming that might be possible with the aid of the ascetic ideal is prevented. Only the self-overcoming necessary for self-preservation or useful for mediocre and safe satisfaction is permitted or, indeed, desired.

The difference between self-overcoming and self-preservative structures repeats the difference between the two economies of value: good/bad and good/evil. For Nietzsche, the "bad" regarded from the noble perspective can be no more than a "side issue," an irrelevant, dismissing assessment while, conversely, the notion of "evil" bubbling from "the cauldron of unsatisfied hatred" is not incidental but rather "its creative *deed*."[52] The latter inversion of life values is possible if the original value of life as the expression of power has been re-valued and what cannot express itself, what is impotent or without burgeoning, exceeding, blossoming life is declared worthwhile and becomes what is good.

The inversion of values is not a "faulty" perception of "true" values, a fall from an original, universal state of noble grace, or an odd quirk—a passing error—of the fashions of the day. Nietzsche's claim is broader than that: the supreme product of this inversion of values is nothing less than culture. The ascetic control needed for an individual ruled by *Ressentiment* is presented as the precondition for the accomplishment of culture: humanity had to be tamed—mastered and herded together—before it could tame the world.[53] In this, Nietzsche's characterization of Western culture as the product of *Ressentiment* or the ascetic ideal may be applied to the scientific project of culture.

RESSENTIMENT: SCIENCE AND CULTURE

The reciprocal involvement of science and culture is now expressed as a manifestation of *Ressentiment*. On the one hand, science itself is promoted by the ascetic values operative in *Ressentiment*. Such values include all the virtues essential to scientific research on the everyday level and for any further revolutionary possibility. The goal of the scientific ascetic ideal is not the internal and self-directed focus of inverted values working for the sake of the individual;[54] these values are reversed and expressed in an externally measured achievement. On the other hand, the purposes of the inverted values operating on the level of the ascetic ideal are themselves supported by a transformation in scientific practice. The scientific project provides an effective compensation for weakness. It is a perfect instrument of revenge conceived from the perspective of the inadequacy of weakness, particularly as the scientific project finds its expression and application in technology. This has less to do with tracing the roots of science in technology to the original Greek meaning of τέχνη, than recognizing science as the reactive development of βάναυσία. The significant sense of βάναυσία is preserved in the English word banausic, derived from the Greek βάναυσικός. Modern technology reflects the positive definition of βάναυσία in mechanical practice, but it deflects its negative (vulgar) connotations as it evokes τέχνη in its own name. With the repetitive, vulgar (freely: banal) focus of the reactive orientation, technology may have lost its once inherent "saving power." Because technology serves brutal, mechanical interests, its orientation toward the earth is characterized by the violence that for Heidegger attends the definition of science as unreflective representation.[55]

The institutional, practical, social, and epistemic success of science as a logico-mathematic simplification of the chaotic complexity of the world and the practical advantages of its technological procedures are nothing less than support systems for weakness. Science employs the ascetic ideal in furthering its own progress (it is dependent upon such an ideal for that progress), but the value of that same ideal (from its priestly, therapeutic heritage) employs science in its turn as a means to successfully preserve, sustain, and thus excuse weakness. What the religious manifestation of the ascetic ideal in the Judeo-Christian heritage of the West accomplished in (interior to) the subject (the original inversion of values) is now externally directed and expressed.

For Nietzsche, the direction of scientific asceticism is highlighted in the prior achievement of the religious ascetic ideal.[56] In an earlier chapter I drew attention to the need for an interpretive transposition redirecting Nietzsche's comments contra Christianity to the scientific culture in the Western tradition, and I must now add that this is so because science and religion are hardly diametrically opposed projects of the human understanding but instead represent different manifestations of *Ressentiment* along the same ideal ascetic continuum in Western culture. The Judeo-Christian priestly ascetic ideal effects nothing less than the preservation of weakness. In this the sick resentment of suffering is redirected against itself. The dialectical equation, if suffering is bad, then suffering is an injury; forms the first stage in the logic of negative valuation. But this resentment is useless and life-discouraging unless one can (manage or) interpret it. From the slave (original Judeo-Christian) perspective, once in existence with a "right,"[57] the individual subject can feel the responsibility for his or her own suffering in his or her own created being. But, thinks the subject in this context, if I am responsible for my suffering, suffering is a punishment: this is the second stage (fulfillment of the Judeo-Christian perspective). Note here that in order to attain this second stage one must presuppose a culture, that is, an already-cultured culture. For Nietzsche, punishment is employed in the life of the species to breed, to cultivate, and so to civilize a people. But, once personal guilt begins, once the reactive individual subject is held personally responsible for the pain of existence, once that pain is nonnecessary but free, optional pain, what more need be done to generate the conscious, self-conscious subject of modern conscience? Guilt increases suffering, or, better, a subjective responsibility holds the reactive subject ready for a sensitive reception of pain. For the individual in a dominant slave culture (eventually: the subject), a consciousness of suffering creates new pain where there was none before, orients one toward the past, inculcating regret, longing, and repudiation and, conceiving the failed freedom potential in the past, hurls one ineluctably, nihilistically toward the future.[58] What is thus synthesized is not an individual who might be capable of keeping his or her word according to his or her own law. Instead we see the evolution of subjection, of the subject before the law: a subject whose life is structured in fear of "witnesses," of convention and opinion. A good deal is achieved with this, if one may so speak: synthesized and regulated subject. This subject understands pain in terms of his or her own failing. More importantly, the subject knows how to avoid such failure or else to reduce suffering by regimented, addictive distraction. It is for this very reason that words like *pain* and *suffering* are inadequate to describe the psychological climate of modern life without hyperbole. Throughout the third essay of *On the Genealogy of Morals*, Nietzsche writes that pain can be obliterated in wild socially regulated orgies of feeling—in war, by means of seasonal festivals, with ritualistic self-examinations, penances, and so on. But the ultimate therapy is available through the technological and practical business affairs of the modern era:

here, one is really numbed or anaesthetised by repetition and the regularity in work and leisure. Thus it often seems labored to talk of suffering as *necessary* today. Life is no longer, and one no longer understands how it could be, *tragically* problematic: modern difficulties seem less indices of suffering than poor practical management.

This last shows the fundamental advantage of the ascetic ideal in its scientific manifestation. Where one has—in the Western metaphysical tradition and its Judeo-Christian articulation—a right to life, suffering is (at least on the religious, moral level) a result or trial of sin. In the tradition of Platonic metaphysics, suffering is a consequence of error or ignorance. The unification of the two concepts, linking sin and error or indigence is historical, and the implications are world building. Suffering has been made corrigible. This is secular millenarianism.

A technologico-scientific culture of *Ressentiment* is sustained by the same physician who created it: the ascetic priest in the technician's culture of a scientific priestly caste. The drive to uncover nature's "truths" may lead to contradictions, even to a loss of certainties. But that is only problematic from a critical philosophic perspective, for life otherwise becomes more and more problem-free. If the threat of annihilation continues to press on the wordly horizon, from the perspective of the everyday, so long as the relevant economy continues to work as it should, life has never been easier. Thus generally regarded, the success of technologically mediated science is proof that the scientific drive is an inexhaustible therapy for the preservation and expression of average desire.

For Nietzsche, a mediocre culture is a sick culture. A culture of sickness, we recall, requires only that it be able to live its sickness, and in this sense: to *preserve* its weakness.[59] Accordingly, the "sick herd" or society requires nothing from its advocate but a means for self-preservation. Science, as the latest expression of the ascetic ideal, provides nothing less. The ascetic priest acting in the service of a degenerating life is its conservative force: "You will see my point: this ascetic priest, this apparent enemy of life, this *denier*—precisely he is among the greatest *conserving* and yes-creating forces of life."[60]

A life that is degenerating may only be sustained through a regimen of restrictions. Against the suffering and sickness of the spirit, and against the fullness of life as well, the ascetic priest employs "means that reduce the feeling of life in general to its lowest point. If possible, will and desire are abolished altogether; all that produces affects and 'blood' is avoided . . . no love; no hate; indifference; no revenge; no wealth; no work; one begs; if possible, no women, or as little as possible."[61] Ultimately then, for Nietzsche, "the ascetic ideal is an artifice for the *preservation* of life."[62] But the slave or impotent individual needs the kind of therapy that can deal with the consequences of a reactive response to life—a response that reads every event from its own petty position and takes that limited position to be supreme. The redirection of *Ressentiment* is as effective as it is because of this prior reactive propensity to banal self-importance. An internal direction of feeling collimates its range and at the same time reduces available perspectives.[63] What is important in this consideration of a reactive focus is the connection revealed between the drive of the ascetic ideal and the office of mechanical activity.[64] In the heightening of any one feeling all other sensibilities are ablated, *deadened*: "by means of a more violent emotion of any kind, a tormenting secret pain [becomes] unendurable, and [drives] it out of consciousness at least for the moment."[65]

Nietzsche explains the "instinctive hatred of reality" associated with the religious manifestation of the ascetic ideal as the "consequence of an extreme capacity for suffering and irritation which no longer wants to be 'touched' at all because it feels every contact too deeply."[66] This is not opposed to stoicism[67] or "saintly tolerance" or (on another level) buddhistic equanimity. A heightened sensibility is the precondition for such therapeutic countermeasures.

To explain the rise of ascetic ideals, Nietzsche hypothesizes a feeling of phsyiological inhibition, that is, fearfulness vis-à-vis the exertions and consequences of life ("a *feeling of physiological inhibition*") that inspires "a grand *struggle against the feeling of displeasure*."[68] He delineates stages in the procedure (though not the goal): "This dominating sense of displeasure is combatted, *first*, by means that reduce the feeling of life in general to its lowest point," just to the point "at which life will still subsist without really entering consciousness."[69] Yet this spiritual achievement is far from common. Hence, failing this achievement, Nietzsche, with particular prescience describes today's culture in proto-postmodern terms, writing: "Much more common than this hypnotic muting of all sensitivity, of the capacity to feel pain . . . is a different *training* against states of depression which is at any rate easier: *mechanical activity*. It is beyond doubt that this regimen alleviates an existence of suffering to a not inconsiderable degree."[70] That is, the sense-deadening goal of ascetic spirituality remains the automatic—and in hindsight more than chilling—consequence of a mechanical, apparently spiritless regimen, "and what goes with it—such as absolute regularity, punctillious and unthinking obedience, a mode of life fixed once and for all, fully occupied time, a certain permission, indeed training for 'impersonality,' for self-forgetfulness."[71] In this ordinary way—for Nietzsche the Nazi excuse of duty or "unthinking obedience" is on a par with the compulsion/desire to watch the "news" or read the daily paper with breakfast and, I might as well add here, the masculine-bonding enthusiasm for sports—the everyday human being is brought to the saint's equanimity. Nietzsche explains this democratic triumph: "The alleviation consists in this, that the interest of the sufferer is directed entirely away from his suffering—that activity, and nothing but activity enters consciousness, and there is apparently little room left in it for suffering: for the chamber of human consciousness is *small*!"[72] However quickly the religious image of asceticism comes to mind as the ideal manner of achieving the goal of a diversion of feeling, no religious regimen has ever been as effective (not to mention as generally available) as today's bureaucratically scientific/technological life-style. The vulgar, bureaucratic life-style is more generally effective just because, for Nietzsche, spiritual achievement is rare[73] and requires more than belief but is a lived attitude or practice: "The word 'Christianity' is already a misunderstanding—in reality there has been only one Christian, and he died on the Cross. . . . only Christian *practice*, a life such as he who died on the Cross *lived*, is Christian . . . Not a belief but a doing, above all a *not-*doing of many things, a different *being*."[74] The reactive expression of the ascetic ideal dissolves (or absolves) the distinction between a Christian life and Christian nonlife.

To summarize this surface-level connection between science and religion, we may speak of duplicity as a kind of self-collusion. Modern science holds itself unmistakeably distinct from religion and most scholarly accounts concede that science is unique. For science has been proposed as a champion of free thought specifically against the tyranny of religious introversion and its ascetic ideal. Nietzsche plainly asserts that science is the "latest and noblest form" of that ideal. If he traces a continuum between religious morality and science, today's science repudiates just that connection. Science represents an advance beyond religion. Science, it is thought, is not a further step along the same ascetic trajectory. We have seen that this deliberate opposition reveals nothing but the priestly machinations of a lifestyle yet characterized by *Ressentiment*. The scientific interpretation of science as I have described it in current philosophy of science perpetuates this difference according to the same inspiration that effected "*the slave revolt in morality.*"[75] The mendacious mechanic of the dissembling representation, the representation of the vulgar (the bad, for the noble vision of life) as the "good," casting the noble good as "evil," recedes behind a fundamental subtlety. Thus the distinction currently emphasized between science and religion is in every sense deceptive. But such underground machinations (pre-)serve the (diminished) interests of life at any price. For Nietzsche, "all *protracted* things are hard to see, to see whole."[76]

The difference claimed between science, morals, and religion works to obscure their commonality as expressions of the ascetic ideal, as means for maintaining a diminished life in the interests of the general life of humanity, which Nietzsche names the herd, the (incurably) mediocre: "Handcraft, trade, agriculture, science, a greater part of art—all of this can only exist on a broad foundation: on a strong and healthy consolidated mediocrity. Science and even art works in its service and is served by it in turn."[77] On the level of the herd, Nietzsche describes the technological/scientific culmination in our day as a "hypnotic muting of all sensitivity" by means of a regimen of "mechanical activity." The social pattern of regularization and impersonality characterizing our work and business world is the explicit criterion not merely of capitalist but also of scientific efficiency. Science is more effective than any other ascetic technique in attaining the ideals of the smaller man, or slave type, or as Nietzsche's Zarathustra expresses this accomplished ideal: the "last" man. This psychological ideal repeats the Christian prescription, "love thy neighbor"—represented for modern tastes, as a Lockean civilization of the instincts both personal and national. According to this ideal we can predict the utility of economic apologetics for an industrialist, capitalist system. The progressive ideal allows a reciprocal exchange of selfish goals for the transcendence of particular interests, or, as Nietzsche continues, "The happiness of 'slight superiority,' involved in all doing good."[78]

WITHOUT PRICE:
THE WILL TO TRUTH AS THE WILL TO LIFE

As we have seen, the democratic or (for Nietzsche, decadent) drive of scientific culture is expressed in the Will to Truth. The Will to Truth reigns in science because the decadent moral ideal of its culture proclaims not merely a will to life, but a will to life at any price. The motif of self-preservation, or survival is, by its own definition, an insistent, desperate one. What is desired in the will to life at any price is not at all life or living, per se. What is willed is much rather simply the preservation of life, perhaps as little as possible, perhaps so that one may have it for as long as possible, perhaps as painlessly as possible.[79] What is essential is merely that one "have" and not that one "live" life.

Our words betray our values. Thus we tend to say, "Life involves risk" as if it were possible to exclude risk with a little care. This possibility is impossible. Life is fundamentally risk. In what Anaximander expresses as the supreme principle of the cosmos (which Schopenhauer understood so well), the contradictory heart of the living thing sounds the promise of its evanescence.[80] The longing for life at any rate, at any price denies physiological finitude. And life is nothing but physiological finitude. The desire for immortality manifests the nihilism of a longing for life at any price, *sans aucun risque*. Because life cannot be held in stock nor ultimately preserved, the will to life is fundamentally opposed to the essence of life.

It has been argued that for those seeking to preserve life at all costs, it is not merely death that threatens. The expenditure of life (as it can be found in those who endanger life by celebrating its joys or spending its resources recklessly) is to be countered as well. The will to (preserve) life is plainly aligned with morality. This opposition reveals two life-orientations. The first orientation—that of saving or preserving life—is consumptive and acquisitive, expressing its own need and indigence, while the second—the literal spending of life—is an expression of inherent power. The artist's desire to create, to give, and to enjoy which also means to suffer life, stands in contrast to the fearful, acquisitive longing for life. And here it should be said that the "artist" in question is not the creative individual of paints and brushes nor a dancer nor even a musician nor any kind of cultural architect. The meaning of art must be understood in terms of the grand style, a matter more of character than *métier*.

All of life manifests desire or Will to Power. That is, all life expresses Will to Power, even in its most decadent expression. Yet, a decadent expression of the Will to Power is radically different from the strong pessimism of the Will to Power that is, as Nietzsche expresses it in 1886, "prompted by well-being, by overflowing health, by the *fullness* of existence."[81]

The material acquisitiveness of a decadent Will to Power battles emptiness on the unidimensional plain of fantasy in its drive to possess. Nietzsche, describing the correspondence between the constant search and constant boredom of modern culture, writes, "it is a tragic spectacle to see how the dance of its thought rushes longingly toward ever-new forms, to embrace them, and then, shuddering, lets them go suddenly as Mephistopheles does the seductive Lamiae."[82]

This is more than an account of modern culture between the agitations of the mode, but it describes the broad compass of science in its endless project, the achievement of a "unified vision of the world."[83] Driving this blithe commitment to truth "at any price," what is then operative in this life-preservative orientation—the need to preserve life "at any price"—is a powerful thrust toward world appropriation.[84] This drive can be understood as Nietzsche understands psychological and organic drives: "The course of logical ideas and inferences in our brain today corresponds to a process and a struggle among impulses that are, taken singly, very logical and unjust. We generally experience only the result of this struggle because this primeval mechanism now runs its course so quickly and is so well concealed."[85] We have seen that this primordial mechanism is the basis of perceptual and conceptual (that is logical) knowledge. The working of this mechanism corresponds to Nietzsche's expression of the Will to Power.

This mechanism can now be defined. Elucidating the direction of the moral structure of science, desire alone should be understood simply as will. Power (*Macht*) is correlative to desire: it is its articulation. But the Will to Power as a concept goes beyond desire; it is equivalent neither to an unconditioned or indeterminate *Wille*,[86] nor to any kind of conatus. Thus, for Nietzsche, apart from a negative definition, the idea of power cannot be given an eudaimonistic expression, because the positive significance of power is to be found only in the activity or expression of power. The end of the active expression of power is power itself. Where the end of expressed power, where what is risked, is no more than a bid for *more* power, an original (and reactive) lack is confirmed. The expression of power, then, has a double sense corresponding to the degree and type of power (that is, active or reactive). We may explicate this duality as the desire for power (impotence) and the desire of power (abundance). The first reactive expression of the Will to Power is from the side of a lack of power and a need for power; this is neediness of desire: articulated want. The second (active) expression of the Will to Power is from a superabundance of power and a need for creative expression; this is the plenitude of desire: articulated affirmation.

SCIENCE AND INADEQUACY

The vision of science popularized is the teleological project to preserve/exploit the world for the benefit of human interests and needs. Thus, insofar as science is an achievement of salvific culture, as the millenarian idol of Connolly's "ontological narcissism," the popular viewpoint of science is vindicated by the factual efficiency of science (in scientific, i.e., technological goods). However, while the average subject is thereby at least materially redeemed, this redemption is effected at the expense of life in its fullest expression. For Nietzsche, "it is a sign of degeneration when eudaimonistic valuations begin to prevail as the highest."[87] The noble, creative expression is, as we have already noted, made superfluous. And we have also had occasion to note that by noble in this context, by the elevated ideal of the *Übermensch*, Nietzsche means the individual directed against easy pleasure and "peace of mind": "The higher human being may be distinguished from the lower by his fearlessness toward and his readiness to challenge misfortune."[88]

Nietzsche tells us that concealed in the goal of universal happiness the culture of science seeks no less a goal than the transcendent revelation of truth. This measure contains the coda that returns the popularized scientific view to its specialized, relatively sophisticated purity. Thus for a culture of ontological narcissism, science is a salvific cultural achievement in complete alignment with the Judeo-Christian tradition and its Enlightenment tradition in the West. As Deleuze precises the process: "Morality is the continuation of religion but by other means; knowledge is the continuation of morality and religion, but by other means. The ascetic ideal is everywhere."[89] What Deleuze emphasizes in this universal scheme is fundamental to Nietzsche's analysis of the reactive type and the means by which reactive values triumph over active values. Nietzsche explains the wish behind the duplicitous project in religion and science: "God created man happy, idle, innocent, and immortal: our actual life is a false, decayed, sinful existence, an existence of punishment—Suffering, struggle, work, death are considered as objections and question marks against life, as something that ought not to last; for which one requires a cure—and *has* a cure!"[90] It is in light of this imaginary conviction that religion and science may be seen to share the same foundation, "The determinism of science and the belief in the fact of salvation stand on the same ground: they are alike in that, granting humanity a right to happiness, they judge contemporary life according to this measure."[91] The fundamental duplicity of this common goal is suggested by the technological wonders that past muster for ordinary expectation as demonstrations of the scientific achievement. Let me quickly acknowledge that reference to applied science in terms of technological achievements is often judged irrelevant to (pure) science, while observing that the genealogist or the perspectivalist philosopher is distinguished by attending precisely to what is thought irrelevant.

If scientific knowledge admits its limitations, confronting infinitely extendable limits, this sophistication proves the resourcefulness of the scientific manifestation of the ascetic ideal. Thus such sophistication may not be thought to herald the inception of "tragic culture," as Nietzsche expresses this vision in his first book, because it sidesteps the incorporation of the tragic wisdom that is the necessary prelude to such an aesthetic.[92]

It will not do to follow Kaufmann by reading Nietzsche as if he were Hegel when Nietzsche writes, "All great things bring about their own destruction through an act of self-overcoming." Even if Kaufmann had not suggested it, the association still seems to be illicit, just because Nietzsche repudiates Hegel at least as an aesthetic taste. Nietzsche nevertheless goes on to illuminate this self-destruction with a typically Hegelian expression as "*Selbstaufhebung*."[93] The one who can overcome himself, who can go under, is the noble, creative individual living beyond himself and so at the height of life. Opposed to this prospect of self-overcoming, Nietzsche writes that the ideal of ultimate mediocrity has assumed the upper hand: "The *herd animal ideal*: now prominent as 'society's' *highest attribution of value.*"[94] Thus Nietzsche's note that "strictly speaking, there is no such thing as science 'without presuppositions,'"[95] presumes the slave-interest of decadent life.

It is the impoverished life that impels scientific interest, that needs the healing and prophylactic measure of scientific asceticism. Nietzsche suggests that what is revealed at this weakened limit of life, through the Socratic ideal of and search for truth, leads to the bright night of nihilism. But what is found in this search is less the full vision of world-being in the affirmative midday of its wild chaos, than an image betraying its own origin in its fragmentation and its limits. Nihilism is not (yet) tragic wisdom.

The motor, because the reward of the scientific ascetic ideal, is the happiness of the "good life." For Nietzsche, this is a decadent life: "*Decadence* betrays itself in this preoccupation with 'happiness.'"[96] Nietzsche continues, explaining the provenance of this preoccupation: "The fanaticism of their interest in 'happiness' reveals the pathological at its foundation: it was a life-interest. To be reasonable *or* to go to ground, these were the alternatives they faced."[97] The last thing a reactive type desires is to "go to ground," that is, to go under. Instead such a human type seeks to remain at whatever level it can, using whatever means it can.

We know what the ascetic ideal preserves and signifies through the consideration of its current manifestation in science. The very notion of progress allows an ersatz production.[98] Rather than affirming the aesthetic dimension of life and the creativity of illusion, with all the good conscience that characterizes art, what we have instead is the lame consciousness of a science that takes its vision for truth or near-enough. Thus, we seek the illusion of truth, the permanence of the absolute, in some future life where we are not vulnerable, not mortal, or insignificant in unsure protest against the manifold variety of reality in the breadth and depth of Becoming/Being: we fear the abyss.

Axiomatic in this future-faith is our belief in the saving power of science in an utopian redemption and the convicted centrality of this technologico-scientific ideal in our lives. It is this belief that permits us to remain as we are. Where religion and enlightenment once held sway as societal arbiters of value, it is now science that prevails (in the same sphere). Thus disposed, our lives are oriented toward the perpetuation of weakness because devoted to developing the support systems of such weaknesses. Ontological narcissism is not, for all its involuted concern, an ontological hedonism. We are still talking about ascetism. But ascetic conservation of weakness as such does not mean that as a culture we are any "*less desirous* of a transcendent solution to the riddle of [human] existence."[99] Salvation is so needed by a culture of *Ressentiment* that one failure to attain it is met by the construction of another approach, another tack. And, here in fact, we see that "salvation" is an indirect expression of scientific technology. It is important to resist the conclusion that lacking the aim of another spiritual world, because science has "replaced" religion, it is manifestly obvious that "Since Copernicus, man seems to have got himself on an inclined plane—now he is slipping faster and faster away from the center into—what? into nothingness? into a *'penetrating'* sense of his nothingness?"[100] Nietzsche's resolute counterpoint is, "Very well! hasn't this been the straightest route to—the *old* ideal?"[101] Ontological narcissism betrays the same transcendent ideal as its antecedent: theological narcissism.

Today's science plays on the limits of knowledge, it exploits what it knows and uses it as a template for what it does not know. The claim to absolute truth made by contemporary science is prudently approximative, asymptotic, a peripheral movement: one does not claim to have the truth, and this is one's claim to truth. In this way science need not acknowledge its dissembling tactics.[102] What is known and what is not known are thus continuously connected. In discussing Nietzsche's Copernican revolution in epistemology, we have already seen that the claim not to know and the claim to know alike express epistemic excesses. Thus Nietzsche admonishes the champions of the current ideal that presents the evolution of truth as the value of the limit. The pride of science that represents "its own austere form of stoical ataraxy, consists in sustaining this hard-won *self*-contempt of man as his ultimate and most serious claim to self-respect (and quite rightly, indeed: for he that despises is always one who 'has not forgotten how to respect' . . .)."[103] It will do to recall that, for Nietzsche, the ascetic ideal common to the aesthetic projects of both religious and scientific morality is a monolithic ideal: "it permits no other interpretation, no other goal; it rejects, denies, affirms, and sanctions solely from the point of view of *its* interpretation . . . it believes that no power exists on earth that does not first have to receive from it a meaning, a right to exist, a value."[104] As the latest (and the "best"—indeed, the "noblest") instance of the ascetic ideal, the worldview of science removes us from the center of the theological universe, placing us on the sidelines of the objective universe, as spectators, spectacular knowers. We thus take over the vantage point reserved for the Olympian gods by the Homeric Greek, for the Jewish God by the Talmudists, and for the Christian God by the medieval scholastics. For Nietzsche, we claim in this move to provide our own justification, and the one-time fantasies of Baron Münchhausen become our own current and not-so-farfetched accomplishments. Nietzsche recalls the theological motif as evidence to speak of "knowledge as a means to power, to Godlikeness."[105] In today's scientific knowledge, in the evaluation of knowledge as the highest ideal, we are progressing toward what Deleuze speaks of as "the replacement of God by man": "But what is this replacement if not the reactive life in place of the will to nothingness, the reactive life now producing its own values?"[106] The ultimate salvation, as Sartre surmised, may well be the deification of the manipulative knower.

DUPLICITY: SCIENCE AND THE ASCETIC IDEAL

We are now in a position to consider the deeper association of science and religion through Nietzsche's account of the evolution of the ascetic ideal. Although the determining opposition of the Judeo-Christian worldview opposes life and nature,[107] and to this degree is opposed to science as well,[108] an ultimate rapprochement, on the analogy of the same "Jewish" mechanism that began "the slave revolt in morality," may be anticipated. Let us see how this common reductive mechanism works. For Nietzsche,

> from the trunk of that tree of vengefulness and hatred, Jewish hatred—the profoundest and sublimest kind of hatred, capable of creating ideals and reversing values, the like of which has never existed on earth before—there grew something equally incomparable, a *new love*, the profoundest and sublimest kind of love—and from what other trunk could it have grown?[109]

From Jewish hatred and impotent desire for revenge blossomed the flower of Christian love. But this did not vanquish Jewish *Ressentiment*; Christian love is merely its "threefold" fulfillment:

> One should not imagine it grew up as the denial of that thirst for revenge as the opposite of Jewish hatred! No, the reverse is true! That love grew out of it as its crown, as its triumphant

crown spreading itself farther and farther into the purest bright-
ness and sunlight, driven as it were into the domain of light and
the heights in pursuit of the goals of that hatred—victory, spoil,
and seduction—by the same impulse that drove the roots of that
hatred deeper and deeper and more and more covetously into
all that was profound and evil.[110]

As such a subversion, the "Jewish" seduction of Rome was Jesus
himself, styled by Nietzsche (Jesus was the *evangel* opposed to
Paul, the *dysangel* in the *Antichrist*). The Jewish rejection of the
incarnate salvation of *"eben jenen jüdischen Werthes und Neuer-
ungen des Ideals:"*

> Did Israel not attain the ultimate goal of its sublime vengeful-
> ness precisely through the bypath of this "Redeemer," . . . that
> Israel must itself deny the real instrument of its revenge before
> all the world as a mortal enemy and nail it to the cross, so that
> "all the world," namely all the opponents of Israel, could
> unhesitatingly swallow just this bait? . . . What is certain, at
> least is that *sub hoc signo* Israel, with its vengefulness and reval-
> uation of all values, has hitherto triumphed again and again
> over all other ideals, over all *nobler* ideals.[111]

Nietzsche traces the anti-Semitism that characterizes the historical
opposition between Christianity and Judaism to this shared root.
So Nietzsche draws attention to "the chosen people" in terms of
their understanding of every other people:

> once the chasm between Jews and Jewish Christians opened up,
> the latter were left with no alternative but to employ *against* the
> Jews the very self-preservative procedures counselled by the
> Jewish instinct, while the Jews had previously employed them
> only against everything *non*-Jewish.[112]

The non-Jew does not share the same relationship with the God
of Israel that the Jew enjoys. This frees the non-Jew from the
responsibilities of the "covenant" or traditional conventions that
structure this relationship. He need not conform to halakic ritual;
to do so would hardly profit him. Because chosen, the Jews can
define themselves as the "holy people." This definition is essentially exclusive. The non-Jew is nonsanctified and, in this context,
excluded from sanctification. Yet with Paul's account of the
Christian evolution/transformation of Judaism (it is important to
note that Nietzsche qualifies Paul's reading as nonpriestly, but

interpretive, hermeneutic, or "rabbinic") into the oppositional equation of Jews and non-Jews, slaves and Romans, men and women alike, are to be the same, as Christians, all rendered as *equal* before God. Thus Nietzsche can conclude "The Christian is only a Jew of a 'freer' confession."[113]

In much the same way, the impulse of the ascetic ideal to deny nature and the body is converted to the interests that mark the development of science. The original opposition is retained: the scientist is in the business of mastering or subduing nature. The drive to dominate nature is not an accidental one: it too grows out of *Ressentiment* and fear. The original vulnerability to feelings of *Ressentiment* is preserved (and this means that *Ressentiment* is conserved rather than surpassed or overcome) through science. And how could this not be the case? For Nietzsche no matter how rigorously science appears to be practiced in opposition to religious, metaphysical ideals,

> where it is not the latest expression of the ascetic ideal—and the exceptions are too rare, noble, and atypical to refute the general proposition—science today is a *hiding place* for every kind of discontent, disbelief, gnawing worm, *despectio sui*, bad conscience,—it is the unrest of the *lack* of ideals, the suffering from the *lack* of any great love, the discontent in the face of involuntary contentment.
>
> Oh, what does science not conceal today! how much, at any rate, is it *meant* to conceal![114]

Scientists, as practitioners of a technique Nietzsche names "self-narcosis," in the face of this lack of passion, must be conceived despite their surface character as contented "modest and worthy laborers . . . happy in their little nooks" in the full awareness of their own contorted nature. What Sartre would later call "bad faith" is the prime characteristic of these cheerful ascetics. For Nietzsche, when one has to do with scientists, one typically has to do with "*sufferers* who refuse to admit to themselves what they are, with drugged and heedless men who fear only one thing: *regaining consciousness* . . . "[115] As the mechanism of contentment in the face of nothing, in the wake of nihilism, science is the "latest and noblest form" of the ascetic ideal. It does not merely champion (revalue) the defects or deflect the poisons of inadequacy (impotence) but effectively, actually enables the perpetuation of weakness and impotence through technological advance and prevents one from suffering more spiritual pains through mechanization of activities and the satisfactions of providing an account of the world that is more and more "correct," or true. And it must be said, confirming Ricoeur's convention of matching Nietzsche with both Marx and Freud: science outdoes religion.

THE ASCETIC IDEAL: THE COST OF PERPETUATION

The cultivation of the subject (the Western cultural ego) has a price, and, in its own economy of values, it works to depreciate the value of life. This must be the case where culture is a static achievement, that is, where culture (and the individual self or subject) is something that wants to stay the same. This conservative interest however is antithetical to the expression of life as Will to Power or possibility. Nietzsche writes, "Against the (self)*preservational drive* as radical drive: the living being wants far more to *release* its power (it *'wants'* to and it *'must'*)."[116] The demand for the beneficial illusion of homeostasis involves the trickery and the illusion of sufficient permanence. The organic body, itself a being in constant flux, presents itself as a static thing. More: the body is manifest as if it interacted with other static things. This (unconscious and above all nondeliberated—hence nontelic) play of illusion is the essence of what Nietzsche refers to as the life-giving lie of life. This expression manifests its Will to Power.

To understand the life-giving power of illusion or the lie, as Nietzsche understands these terms, we can recollect that the ill-constituted (the herd) are the cultural goal, the social end of Western culture. As the so-called herd, the slaves of Nietzsche's genealogical critique of morality and culture, compose the bulk of society, the values of society are their values, and society is constituted to preserve them. This is a reversal of the valuations that support the noble and the rare. As a consequence, the noble are today thoroughly rarified: indeed, they have effectively vanished. This relational sequence is capital. The fundamental, willed evanescence (self-expenditure) and, more crucially in this regard, the tacit and effective impossibility of the "noble" means that every fascistic interpretation of Nietzsche's work and thought is impossible. The (aboriginal) noble is not the end of culture: such "higher" beings cannot be maintained in society, the esoteric order cannot be supported by the exoteric.

To Nietzsche's mind, and qualifying his still romantic conception of genius, there are no (longer any) noble or higher individuals as such: this lack is the point of entry for, because the point of the significance of, the concept of the *'Übermensch'*. Nietzsche's further project of overcoming present-day humanity is dyadic: to regard present-day possibilities in light of the decadence of everyday ideals (i.e., as inadequacies) and, from this perspective, so to overcome what is in us toward what is possible for each of us at the extreme of our being. Yet, the second projective aspect shows that even such a resolve toward our most authentic possibility is not yet noble. This resolve is merely underway; there is still too much that is human. Even for the best among us—and that will also be to say, for the best among us even on our best days—our greatest possibility may well remain that of a bridge beyond ourselves. Furthermore, Nietzsche's melancholic voice here contends that all this teetering rhapsodizing, all the evening-reflection that marks philosophic promise remains irrelevant, even for philosophers. Rather than the affirmation of the greatest and rarest possibilities as the heart of the human potential, as the song of sufficient reason faced with the tragic challenge of existence and nothingness, an equalization, that is, a *tacit* equality, a *levelling* of differences, of rights and potential has gained cultural favor. In this estimation of human quality, the genius of the herd, that is, the mediocre talent accessible, useful to all is celebrated. This focus lends scientific interests a special currency.

We recall that, for Nietzsche, a decadent culture or weakened life is symptomatic of a diminished capacity for life. A culture lacking vitality is decadent. But religion and morality as forms of the ascetic ideal are the means by which a reduced life can be maintained and promoted. Science, as the embodiment of the Will to Truth at all costs is itself an expression of the Will to Life at any price. As such it opposes what is changeable in life in its search for the ultimate truth of things. Insofar as science embodies this metaphysical drive, it denies life, which is: change and passing away and, indeed, illusion. But Nietzsche's counterclaim is that there is no absolute or foundational truth. Thus to the extent to which science embodies the Will to Truth, it cannot avoid a collision with the untenability of its own metaphysical ideal. Such a collision is so subtle that it can occur a number of times without report. Science, in its natural and social research expressions, perpetuates its own loyalty to its absolute by the same expedient

Nietzsche shows always to have been employed in the service of the ascetic ideal: that is, one renounces the appearance of one's ideal. Thus, science renounces metaphysics as it renounces any claim of its own to an absolute status. But in this, it searches for a knowledge that cannot be called "knowledge" in the traditional (metaphysical) sense, and it delights in the exploration of the limits of this knowledge—this is its metaphysical sense—so transgressing its own boundaries. But all this is symptomatic of a reactive Will to Power (and Nietzsche thinks, a decided lack of taste). In the train of this reactive Will to Power, in the culture that it defends and perpetuates, a certain type—the clerical, scholarly, or scientific type, that is today the technician or the managerial type: the businessman—is thereby advanced as a supreme social type. This type, needless to say, is detrimental to the noble possibilities suggested above. At the same time, the conditions for this expression of nihilism are based on the limits of knowledge.

Here, of course, science's own artistic potential is revealed. Because it is the most efficient expression of the ascetic ideal, science is its most extreme (its "noblest") form. Seeing this, Nietzsche looks to a new possibility for science. The conditions for a tragic knowledge are already in play, their repetition is found in the orbit of practical benefit. The possibility of a tragic culture, a Dionysian pessimism, and that is to say, a Dionysian aesthetic, may yet again figure on the human horizon.

Still, as we have seen before, Nietzsche finds science to be, like religion but unlike art, the dissembling practice of an illusion. For Granier, science is a "consoling illusion and nothing else."[117] If illusion is not affirmed, and affirmed with a good, artistic will, the result cannot be truth but only the illusion of truth or delusion. Only a person of sufficient artistic power can affirm illusion as such. Through a full expression of power, one may overcome the style of impotence. But an important question remains: can such a revision ever take place in an emancipated, democratic culture? Nietzsche feared not, and this was the tragedy of his personal life and his philosophy.[118] But deferring the extent of this question, it can be observed that if it *is* possible, it can be attained only by transforming the person of the scientist, that is, by making the technician into an artist. Here, a crucial observation: the artist can never be a scientist—whenever that happens, what results is no more than art with a bad conscience. Yet the scientist is already an (self-deceiving) artist—albeit unconsciously! By transforming the scientist, refashioning the technical worker into an artist, one can go beyond the fearful nihilism that is the heritage of impotence. To do this requires no further technology, no new mythology: what is necessary is not at all the re-presentation of impotence as power (which consecrates impotence) but (what is more difficult) impotence as the failure of power is to be overcome. Rather than employing scientific illusion as a negative, absolute ideal against the ambiguity of being, the redemptive thought—the thought with taste and style, and so, above all, feeling for existence in all its rich ambiguity—affirms life. This affirmation once again sets τέχνη apart from mere βαναυσία as art. The goal of this properly aesthetic effort is (like science's own popular goal) a redemptive one, if the redemption here is not material but stylistic. Yet, for this the inconvenience, the pain, the inconstancy, and the ambiguity of existence are not to be redeemed (or denied): "One must reach out and try to grasp this astonishing *finesse, that the value of life cannot be estimated.* Not by a living man, because he is party to the dispute, indeed its object, and not the judge of it; not by a dead one, for another reason."[119] This "astonishing finesse" involves what he names the "innocence of becoming."[120] The subject is not to be promised a redemption here: this is no substitute for religion, no new version of the ascetic ideal. What is to be redeemed is grander than the ambiguity and pain of life: it is the whole of existence itself.

SCIENCE AS AN AESTHETIC ACHIEVEMENT: *MÉCONNAISSANCE*

Ultimately, for Nietzsche, science is to be conceived aesthetically. While science is not articulated as an artistic or deliberate aesthetic creation it is nevertheless effective (as interpretation) on an artistic level. Nietzsche's critique of science, and by extension its philosophy, is directed to the common failure of science and scientific philosophy to realize an adequately aesthetic—for Nietzsche, an artistic—self-conception. Instead of a conscious artistic presentation, science deludes itself about its own project, taking it to represent an absolute career. Thus Nietzsche compares the laws proclaimed by science to the magician's sleight of hand, and criticizes its fanfare: "If someone hides an object behind a bush, then seeks and finds it there, that seeking and finding is not very laudable: but that is the way it is with the seeking and finding of truth in the rational sphere."[121]

Nietzsche finds the thought of the artistic power of creation and its implications inspiring.[122] We represent the world according to a scientific image: reciprocally, our interest in the facts of our experience is determined by a scientific perspective.[123] For this reason, we should not be surprised to find that "in fact" the world presents this same appearance to us.[124] Instead, Nietzsche counsels,

> Everything marvelous that we admire in the laws of nature and that promotes our explanation and could mislead us into distrusting idealism, consists exclusively of the mathematical stringency and inviolability of time- and space-perceptions. But we produce these perceptions within ourselves and out of ourselves with the same necessity as a spider spins its web. If we are compelled to grasp all things only under these forms, then it is not surprising that in all things we really grasp only these forms: for they all must carry the laws of number in themselves and number is the very thing that is most astonishing about things.[125]

The experimental direction of Nietzsche's own style can once again be recognized in this passage. The metaphoric image of a spider (with variations) and the vision of organic necessity together with its creative/illusionary/delusionary emphasis (by way of metonymy to be heard in the German, *spinnen*) is an image that recurs throughout Nietzsche's work (representing human nature, human perception, existence, God, singularities in time, and so on).

How should we hear this early passage and its recurrent resonance throughout Nietzsche's late work? In a concinnous horizon, the passage works across two tonal registers. The first tone level expresses scientific loyalties and its telos in Western understanding; the second tone level operates as, is troped in accord with, an aposematic appeal to a selective, select reader. When the reader is all too general, this tone is apotropaic. Hence, readers are set "apart" according to their affinities: rare readers are challenged to think the thought beyond the writing itself, while others are seduced by content and style to either reject or idolise (misconceive) the text. Jean Granier expresses this two-edged expressive character in terms that are relevant to the topic of the present chapter, in his own didactic textual projection:

We must first of all highlight all the elements of science on which Nietzsche relies in order to demolish the edifice of Metaphysics, then show how it is, nevertheless that "objective" science remains subject to Idealism. That done, we will retrace at the same time, the evolution of those conceptions of Nietzsche concerning the rapport between metaphysics and science, by illuminating the necessity of the *inversion* by which, after having celebrated the antimetaphysical virtues of science, he was led to recognize the ultimate expression of modern nihilism in science.[126]

In this way, Granier uncovers Nietzsche's stylistic ruse in his "positivistic" project. He explains its working further:

certain aspects of science . . . may be easily employed as arguments against idealism, but considered with regard to its essential project, science is not thereby hindered from accomplishing the destiny of metaphysics on its own.[127]

In the double play Granier describes, he observes that "it might seem paradoxical"[128] that this style, on the one hand, could encourage readers with an approbative view of science, while, on the other hand, this same style could single out those rare readers who do not share in the cultural adulation of science, where Nietzsche's expressed interest is not science as such, but the larger moral phenomenon of Western culture.[129]

From a genealogical standpoint, Nietzsche questions the cultural project in which science (as morality) is meant to be efficacious by asking, what human type needs science? needs safety? needs a blunt and bland happiness of this kind? Again: what human type needs "Ruhe, Sicherheit und Consequenz," that is, "rest, security, and predictability?" Plainly this typology describes the human being marked at the very least—that is, if also a weak, sick, and decadent kind of human being—by a certain sensitivity.[130] It is this human type that needs to be protected and needs the moral precept of self-preservation. This typology would characterize most of us today. The eudaimonistic goal[131] (as a happiness of contentment and ease) accepted in both religion and the practice of science (and technology) in the West is an index of the generality of this social focus. In Granier's words, "the scientific will to know is hardly more than a subtle disguise for the ancient moral will, the expression of a need which drives man to create the fable of an *intelligible world*' excluding change, suffering, contradiction, movement, ambiguity, and becoming."[132] Although Nietzsche affirms that the intellect is a tool for simplification, oriented toward the biological exigencies of security in a world of change and becoming, such a bio-pragmatic goal is neither valid (true) nor worthwhile in itself. This is Nietzsche's hardest point; it shows too that his "selectivity" or his "elitism" can be directed to no political aim. For Nietzsche, the ideals of Western science serve the interests of life-preservation at the expense of creative possibilities for the full expression of life.

VESUVIUS: *"GEFÄHRDETE MENSCHEN, FRUCHTBARER MENSCHEN"*[133]

Nietzsche suggests that a certain level of individual preservation is requisite for life-expression. Arguing that life- and self-preservation have been hitherto overvalued, he emphasizes full life-expression (style) instead. Yet, beyond the vulgar celebration of life-value, such a plenary achievement of life (in the grand style) is compatible with a failure to sustain the expression of that life. Hence, two interests proper to life, the preservation of life and full life-expression are neither identical nor ultimately compatible. In "On the Uses and Disadvantages of History for Life," Nietzsche explains, "he lives best who has no respect for existence."[134] For Nietzsche, the life-preservative style of living is nonartistic in a decadent, noncreative sense. Valuing the conservation of life as if it were the highest value in life (so opposing change or becoming) does not affirm but denies life.[135] For Nietzsche, if the organic impulse of life is an expression of power, *Selbsterhaltung* as such can be no more than a kind of *Nebenwirkung* of the life-expressive (power-expressive) Will to Power. Yet with science, the indirect expression of the Will to Power, that is, negative or reactive Will to Power, the Will to Power of inadequacy, is rendered consistently effective.

Science directly transforms the project of self and life-preservation (as the religious ascetic ideal had indirectly transformed it) into what it never is within the raw physical dynamic of the Will to Power. Life-preservation, incidental to the expression of life, becomes a viable, transcendent goal. Science (together with its banal technology) renders the goal of life- and self-preservation an achievable, already-realized goal. Vulgar (self-) interests, the

worldviews and world-historical expectations of ontological narcissism, then, are successfully advanced at the expense of creative,
artistic interests.

The orientation of Nietzsche's devastating refinement of the
Darwinian principle of the survival of the fittest reveals this essential common advantage. Nietzsche assumes that the struggle to
survive not only is among species individuals but also includes
cultures as well as the human spirit (the mind). For Nietzsche, the
variety of cultures and different human types prohibit the characterization of the result of this struggle as the "brute" survival of
the fittest. Writing in a section of *Twilight of the Idols* entitled
"Anti-Darwin," this nuanced reading of "fitness" is made
explicit: "the weaker dominate the strong again and again—the
reason being they are the great majority, and they are also clev-
erer."[136] In this, Nietzsche affirms the importance of shared interests and the ability to communicate those interests toward the end
of survival in a human world. As communal, these interests cannot be select, special, or esoteric. Nietzsche explains:

> on the whole the easy *communicability* of need, that is to say
> ultimately the experiencing of only average and *common* expe-
> riences, must have been the most powerful of all the powerful
> forces which have disposed of mankind hitherto. The more sim-
> ilar, more ordinary human beings have had and still have the
> advantage.[137]

Natural social interests are not neutral; they decide the value of
values (*die gesammte Rangordnung ihrer Werthe*) and reveal the
makeup (*Aufbau*) of the human spirit. Yet, once again, this is not
to say that decadence is a simple sign of weakness, because it is far
from that. Nietzsche explains that "strong" or "noble" individuals
"are subsequently weaker, more devoid of will, more absurd than
the weak average."[138] Because Nietzsche's idea of a powerful or
creative individual typology affirms that "they are *squandering*
races,"[139] the less powerful, herd-human beings as he calls them
must be conceived "fitter" in all regards—if fitness is defined in
terms of survival.[140] Thus "Tremendous counter-forces have to be
called upon to cross this natural, all too natural progressus in simile, the continuing development of mankind into the similar, ordinary, average, herdlike—into the *common*!"[141] For Nietzsche,
"The mediocre alone have the prospect of continuing on and prop-
agating themselves—they are the men of the future, the sole sur-

vivors."[142] This focuses the point of Nietzsche's later declaration: "Darwin forgot the mind . . . the weak possess more mind . . . To acquire mind one must need mind."[143] As he continues to explain, "One will see that under mind I include foresight, patience, dissimulation, great self-control, and all that is mimicry."[144] In these terms, then, the common "mimic" knows best how to take advantage of every cover, and he can use his tricks under all circumstances because he is, in the common jargon, a survivor. Conversely, for Nietzsche, the "genius"

> in his work, in his deeds—is necessarily a prodigal: his greatness lies in the fact that *he expends himself* . . . The instinct of self-preservation is as it were suspended; the overwhelming pressure of the energies which emanate from him forbids him any such care and prudence . . . He flows out, he overflows, he uses himself up, he does not spare himself—with inevitability, fatefully, involuntarily, as a river's bursting its banks its involuntary.[145]

In the contrast between self-preservation and self-expression, we can find the significance of the well-known discriminatory standard Nietzsche proposes for creative quality in terms of the Will to Power, the canonic question determining artistic creativity: "is it hunger or super-abundance that has here become creative?"[146]

In an economy of power and its interpretive expression, the ideal of self- and life-preservation is successfully employed by those fundamentally lacking in power (taking power as the expression of life). But Nietzsche writes, "The wish to preserve oneself is the symptom of a condition of distress, of a limitation of the really fundamental instinct of life which aims at *the expansion of power*."[147] This conservative orientation characterizes science, and this means that the typology of science expresses a lack of power. In the effective ethos of life-preservative morality, the absolute character of science and the drive of knowledge knows its practical limits and the complexities that compromise its desire: it affirms just these limits. The effective or practical recognition of limits to knowledge combined with the absolute ideal of truth (progress toward truth) is an expression ("the latest and noblest") of (ascetic) nihilism in Nietzsche's view. In this nihilism, we have the duplicity of a science abhoring contradictions, but, in all innocence, absolving its own project from this stricture. Strong carries this point further: "what then is the epistemology of nihilism[?] . . . Men arrive at it when they find both that there is no truth and that they should continue to seek."[148] An "epistemology of nihililism" may be an ironic description of Nietzsche's own project, but in a more trenchant fashion it describes the self-conscious, "sophisticated" (fallibilistic excellence) of the scientific knowledge project. The difference between Nietzsche's critical nihilism and the blind nihilism of productive science—its *"souveräne Unwissenheit"*[149] —is a matter of the "courage (*Tapferkeit*) of conscience."[150] The commonality is life-redemptive knowledge and the shared focus is the human orientation of truth.[151]

While the original Copernican (or Galilean) rendition of the research ethos of modern science overcomes the nihilism implicit in the Judeo-Christian metaphysical tradition, it involves another (more durable) kind of nihilism as it represents the very same ascetic ideal, for the same banal culture of impotence and fear. Thus we can read Nietzsche's note: "Since Copernicus, the human being has rolled from the center into x."[152] The unknown, or "x," named here is Nowhere or Nothing, not affirmed to be dangerous, but instead courted as an advantage. To be nowhere is to attain the best vantage point for the knowledge project of science conceived as the ascetic ideal of mediocre life-preservation. What better camouflage? The danger of seeking an advantage beyond one's right or power seems ever to lead to a desire to disappear into thin air (consider the temptation of invisibility for Gyges). But this advantage is fundamentally nihilistic. If one escapes life's threats by hiding in the margins, by becoming invisible, does one ultimately escape life as well? The mature Nietzsche challenges the conservative scientific perspective with increasing vigor: "Death, change, age, as well as procreation and growth, are for them [scientists] objections—refutations even."[153] Death, change, age, procreation, and growth are features of life conceived in time, life as a whole, as a historical, organic finite affair. If these are objections for science, the scientific conception of life rejects the essence of life as a whole: that is, life as $\phi\acute{v}\sigma\iota\varsigma$, as a dynamic, and indeed chaotic process of becoming, as the manifestation of the Will to Power that is life.

To recapitulate here: from a ecophysiological viewpoint, the intellect is the tool of the sensitive organism serving the interests of self-preservation. As affirmed above, we are, each of us, beings of the self-preservative kind. Beyond this common pettiness— "those who think only of narrow utility"[154] —there is a selective or higher—esoteric in Nietzsche's sense—orientation to overcome the drive for security, which is the desire to have life, affirming the value of living life, that is, risking endangering life. The esoteric, artistic, or poetic (in Heidegger's poetizing sense) perspective does not shrink before the affirmation of such danger: to live life is to expend it. This unflinching distinction, defining the essence of life as the Will to Power, fulfills Nietzsche's criterion for life-expressive action.[155] Instead of an ordinary self-preservation, what is needed and still not attained is a rare self-glorification, not the (still-all-too) average *egotist's* vanity observed by Schopenhauer and certain romantic poets but an authentically Dionysian, thoroughly *generous* self-celebration "in the truth" of art.[156]

NOTES

1. "Jeder hat seine Perspektive, welche er als Norm allen übrigen Trieben aufzwingen möchte" (KSA XII, p. 315), WP 481.

2. For Nietzsche these intentions and desires necessitate a genealogy. "Warum sieht der Mensch die Dinge nicht? Er steht selber im Wege: er verdeckt die Dinge" (M 438, KSA III, p. 268).

3. GM ii, KSA V, pp. 248–49.

4. I have addressed the Nietzschean counterfactual in chapter 4 above, but in another discipline, the political theorist William E. Connolly, adverts to this same problematic accusation, "Don't you presuppose truth (or reason, subjectivity, a transcendental ethic) in repudiating it?" and points to the fact that "The simple answer is yes on all counts. Yes, yes, yes, yes , yes. Although many modernists seem to skip over these passages, Foucault, Derrida, and Nietzsche constantly give that answer after posing such questions to themselves. But because they think within a different problematic, they also find that answer insufficient" (*Identity/Difference*, p. 60). It is worth noting as Connolly does here that if Nietzsche would find the "answer insufficient" the aforementioned modernist treats the problem constellation as inherently, triumphantly unanswerable and "as a discovery accepted by every healthy, rational, red-blooded academic" (ibid.) .

5. T. Strong, *Friedrich Nietzsche and the Politics of Transfiguration*, p. 52.

6. "Keine rückläufigen Hypothesen! Lieber ein Zustand der $\epsilon\pi o\chi\acute{\eta}$!" (KSA XI, p. 170).

7. "Aufgabe: die Dinge *sehen, wie sie sind! Mittel*: aus hundert Augen auf sie sehen können, aus *vielen* Personen!" (KSA IX, p. 466; cf. GM III:12).

8. See KSA XII, p. 244.

9. "Hier wird immer ein Auge zu denken verlangt, das gar nicht gedacht werden kann, ein Auge, das durchaus keine Richtung haben soll" (GM III:12, KSA V, p. 365), G, p. 119.

10. "Nicht als 'interesselose Anschauung' verstanden (als welche Unbegriff und Widersinn ist), sondern als das Vermögen, sein Für und Wider in der Gewalt zu haben und aus-und ein-zuhängen: so dass man sich gerade die Verschiedenheit der Perspektiven und Affekt-Interpretationen für die Erkenntnis nutzbar zu machen weiss" (GM III:12, KSA V, pp. 364–65), G, p. 119.

11. Nietzsche writes, "I see nobody who ventured a critique of moral valuations" and thereby finds the project of his own philosophic question vindicated: "Nobody up to now has examined the *value* of that most famous of all medicines which is called morality . . . the first step would be—for once to *question* it" (GS 345, pp. 284–85), KSA III, p. 578 .

12. "Es liegt auf der Hand, dass die moralischen Werthbezeichnungen überall zuerst auf Menschen und erst abgeleitet und spät auf Handlungen gelegt worden sind" (JGB 260, KSA V, p. 209), BGE, p. 176.

13. "Die vornehm Art Mensch . . . weiss sich als Das, was Überhaupt erst Ehre den Dingen verleiht, sie ist wertheschaffend. Alles, was sie an sich kennt, ehrt sie: eine solche Moral ist Selbstverherrlichung" (KSA V, p. 209), BGE, p. 176.

14. This fear stems in part from the tendency to conceptual solecism that equates Nietzsche's thought with that of Ayn Rand or other more classic proponents of aristo-vulgar esoteric doctrines. Thus Connolly, despite his perceptiveness in other regards, characterizes this prejudice by noting that in addition to Nietzsche's patent "disdain of democracy" it seems evident that "the coiner of the phrase 'will to power' must endorse a ruthless philosophy in which a few exercise mastery over other humans and nature" (*Identity/Difference*, p. 185).

15. Consider the concluding (and even naively romantic) reflections of "Über Wahrheit und Lüge im aussermoralischen Sinne" (KSA I, pp. 888–90).

16. "Im Vordergrunde steht das Gefühl der Fülle, der Macht, die überströmen will, das Glück der hohen Spannung, das Bewusstsein eines Reichtums, der schenken und abgeben möchte" (JGB 260, KSA V, p. 209).

17. "Dass as sich ausgiebt" (G-D, KSA VI, p. 146), TI, p. 108.

18. "Sondern mehr aus einem Drang, den der überfluss von Macht erzeugt" (JGB 260, KSA V, p. 210).

19. Nietzsche notes that the Greek adage of measure , "nothing in excess," is not meant for mediocre types: "Die Lehr μηδέν ἄγαν wendet sich an Menschen mit überströmender-Kraft—nicht an die Mittelmäßigen" (KSA XI, p. 105), WP 940.

20. This suggests that the noble (naive) individual lacks the modern social consciousness. The noble society appears to be a society of individuals—precisely as MacIntyre challenges. But the circumstances of this society are not those of a liberal society, as MacIntyre suggests in *After Virtue*. Nietzsche opposes the democratic fundament of liberal society: equality.

21. On the matter of Nietzsche's political relevance (a thematic of burgeoning theoretical interest that even this long note can do no justice to) see (again) Connolly's *Identity/Difference* as well as his earlier *Politics and Ambiguity*. See too Henning Ottman, *Philosophie und Politik bei Nietzsche*, and for an overview of Nietzsche's politics in connection with Foucault, Keith Ansell-Pearson, "The Significance of Michel Foucault's reading of Nietzsche." In addition see Mark Warren's excellent *Nietzsche and Political Thought*, Robert Eden, *Political Leadership and Nihilism*, Ofelia Schute, *Beyond Nihilism* and "Nietzsche's Politics," and a recent collection edited by Michael Allen Gillespie and Tracy B. Strong, *Nietzsche's New Seas*. In an earlier, continental generation, including Karl Löwith at least in spirit, Georg Lukàcs is known for assuming a notoriously reactive and damning stance against Nietzsche, along with Mann and many post-Nazi authors, since he possessed the Hegelian conception of history that permits aligning Nietzsche as a direct progenitor of the fascist vision of National Socialism (see Warren in English for a clear statement of this position, its implications and a critical, postmodern political alternative. For a recent polemic in German on these revivified connections, see Bernard Taureck, *Nietzsche und der Faschismus*. For a clear biographical account, Peter Bergmann, *Nietzsche: "the Last Antipolitical German"*). Finally, Strong, *Friedrich Nietzsche and the Politics of Transfiguration* remains unusually nuanced and important.

22. Connolly, *Identity/Difference*, pp. 184–85.

23. Ibid.

24. Ibid., 14; emphasis added. Connolly's is a masterful book, "post" any number of positions—"post" that is, in the sense of having thought about and through these positions. Connolly offers a reflection on (rather than the simple advocacy of) theory respecting the tension between identities and differences and besetting the topic of his book on identity/difference.

25. Ibid., p. 189.

26. Ibid., p. 187.

27. Ibid., p. 30.

28. Ibid., p. 43.

29. Ibid., p. 44.

30. Cf. Deleuze, *Nietzsche and Philosophy*, especially chapter 2. Above, I observe that a society of nobles would be not a society in the

true sense but merely an association of individuals. A corollary suggests that society as such can only be the expression of a dominant slave culture.

31. Those who like to find examples of Nietzsche's celebrated inversion of Platonism may find an opportunity here. If Plato reserves the instrumentality of the "philosopher (king)" for the larger interests of the society, Nietzsche affirms society as a means to his idea of the "philosopher."

32. So Nietzsche's Zarathustra rails against the democratic ideal of social equality: "You preachers of equality, thus from you the tyrant-madness of impotence cries for 'equality'; thus your most secret tyrant appetite disguises itself in words of virtue" (Z, "Of the Tarantulas," p. 123). "Ihr Prediger der Gleichheit, der Tyrannen-Wahnsinn der Ohnmacht schreit also aus euch nach 'Gleichheit': eure heimlichsten Tyrannen-Gelüste vermummen sich also in Tugend-Worte!" (KSA IV, p. 129). The dominant culture in a society will always be characterized by a reactive, or slave, mentality. The reactive value system, after all, has the great mass on its side. See G-D, KSA VI, pp. 120–21.

33. "Die ähnlicheren, die gewöhnlicheren Menschen waren und sind immer im Vortheile, die Ausgesuchteren, Feineren, Seltsameren, schwerer Verständlichen bleiben leicht allein, unterliegen, bei ihrer Vereinzelung, den Unfällen und pflanzen sich selten fort" (JGB 268, KSA V, p. 222), BGE 268, p. 187.

34. "So seltsam es klingt: man hat die Starken immer zu bewaffen gegen die Schwachen" (KSA XIII, p. 304).

35. Or less controversially, and therefore contextually illuminating: one must protect the healthy from the sick.

36. "die Starksten und Glücklichsten sind schwach, wenn sie organisirte Heerdeninstinkte, wenn sie die Furchtsamkeit der Schwachen, der Überzahl gegen sich haben" (KSA XIII, p. 303).

37. But, we may ask what it would take to overthrow the (noble) values Nietzsche saw as originally active in the opposition between noble and slave moralities? The answer is a revolution that is at the same time an inversion. Such a revolution is nonevolutionary. Slave perspectives were not originally noble ones. Only a creative power (or a reactive power become creative) can accomplish the revolt in slave values, vanquishing noble values and leaving the meek, utilitarian concepts of social duty and responsibility behind.

38. "Der Sklavenaufstand in der Moral beginnt damit, dass das *Ressentiment* selbst schöpferisch wird und Werthe gebiert" (GM 10, KSA V, p. 270), G. p. 36.

39. "Das Herrenrecht, Namen zu geben" (GM 2, KSA V, p. 260).

40. Cf. M 438.

41. "Von den Taranteln" (Z, KSA IV, p. 129).

42. GM I:15.

43. Connolly, *Identity/Difference,* p. 185. This is, of course, the view of the loyal, eternal opposition.

44. "Damit gewinnen sie ja bei sich das Recht, dem Raubvögel es zuzurechnen; Raubvögel zu sein" (GM I:13, KSA V, p. 280).

45. "*Es stehe dem Starken frei,* schwach und dem Raubvögel, Lamm zu sein" (GM, KSA V, p. 280), G, p. 45).

46. "Der Heilige, an dem Gott sein Wohlgefallen hat, ist der ideale Castrat . . . Das Leben ist zu Ende, wo das 'Reich Gottes' anfängt" (G-D, KSA VI, p. 85), TI, p. 45.

47. Care should be taken not to reflect a Hobbesian account of primordial (or blond) bestiality into Nietzsche's genealogy.

48. "Alles, was gedacht, gedichtet, gemalt, componirt, selbst gebaut und gebildet wird, gehört entweder zur monologischen Kunst oder zur Kunst vor Zeugen" (FW 367, KSA III, p. 616), GS, p. 324.

49. For Nietzsche, creative art cannot be calculated to elude a popular response: it is not a statement but depends upon creative absorption or world-forgetfulness "all monological art . . . based *on forgetting,* it is the music of forgetting" (GS 367, p. 324).

50. Cf. A 44; and GM III, 23.

51. "Der Wille der Kranken irgend eine Form der Überlegenheit darzustellen, ihr Instinkt für Schleichwege, die zu einer Tyrannei über die Gesunden führen,—wo fände er sich nicht, dieser Wille gerade der Schwächsten zur Macht!" (GM III:14, KSA V, p. 370), G, p. 132.

52. "Die eigentliche *That* in der Conception einer Sklaven-Moral" (GM I:II, KSA V, p. 274), G, p. 36.

53. This is the point of GM II.

54. Nietzsche lists four interests guiding the promotion of culture for the sake of popular benefit and thirteen motives for scholarship proposing the possibility of an active self-overcoming (UB III:6).

55. Heidegger, "Wissenschaft und Besinnung," in *Vorträge und Aufsätze,* pp. 41–66.

56. Cf. KSA VII, p. 429.

57. Nietzsche's charge: "that you have a *right* to existence seems irrefutable to you" (FW 335, p. 264); "dass du aber ein Recht auf Existenz habest, dünkt dich unwiderleglich!" (KSA III, p. 561). But as noted in the previous chapter: "man hat kein Recht, weder auf Dasein, noch auf Arbeit, noch gar auf 'Glück': es steht es mit dem einzelnen Menschen nicht anders als mit dem niedersten Wurm" (KSA XIII, p. 98), and this means, as we will see below, that "insgleichen darin daß sie dem Menschen ein Recht auf Glück zugestehen; daß sie mit diesem Maßstab das Maßstabe des gegenwärtige Leben verurtheilen" (KSA XIII, p. 98).

58. GM III:11.

59. So Nietzsche writes, "the ascetic ideal springs from the protec-

tive instinct of a degenerating life which tries by all means to sustain itself and to fight for its existence" (G III:13, p. 120). "das asketische Ideal entspringt dem Schutz- und Heil-Instinkte eines degenerirenden Lebens, welches sich mit allen Mitteln zu halten sucht und um sein Dasein Kämpft" (KSA V, p. 366).

60. "Man versteht mich bereits: dieser asketische Priester, dieser anscheinende Feind des Lebens, dieser Verneinende,—er gerade gehört zu den ganz grossen conservirenden und Ja-schaffenden Gewalten des Lebens" (GM III:13, KSA V, p. 366), G, p. 120.

61. "Mittel, welche das Lebensgefühl überhaupt auf den niedrigsten Punkt herabsetzen. Womöglich überhaupt kein Wollen, kein Wunsch mehr; Allem was Affekt macht, was 'Blut' macht, ausweichen . . . nicht lieben; nicht hassen; Gleichmuth; nicht sich rächen; nicht sich bereichern; nicht arbeiten; betteln; womöglich kein Weib, oder so wenig Weib als möglich" (KSA V, p. 379), G, p. 131.

62. "Das asketische Ideal ist ein Kunstgriff in der Erhaltung des Lebens" (GM III: 13, KSA V, p. 366), G, p. 120.

63. Cf. GM II:7.

64. Cf. FW 21.

65. "[Man] will . . . einen quälenden, heimlichen, unerträglich-werdenden Schmerz durch eine heftiger Emotion irgend welcher Art betäuben und für den Augenblick wenigstens aus den Bewusstsein schaffen" (374) (GM III:15; G, p. 127).

66. "Folge einer extremen Leid- und Reizfähigkeit, welche überhaupt nicht mehr 'berührt' werden will, weil sie jede Berührung zu tief empfindet" (AC 30, KSA VI, p. 200), A, p. 152.

67. Nietzsche makes a reference to the cynical outcome of hedonism to illustrate this point in this same context (AC 30, KSA VI, p. 200).

68. "Im grössten Stil ein Kampf mit dem Unlustgefühl" (GM III: 17, KSA V, p. 378), G, p. 131.

69. "Man bekämpft erstens jene dominirende Unlust durch Mittel, welche das Lebensgefühl überhaupt auf den niedrigsten Punkt herabsetzen" and "das Leben gerade noch besteht, ohne eigentlich noch in's Bewusstsein zu treten" (KSA V, p. 379).

70. "Viel häufiger als eine solche hypnotistische Gesammtdämpfung der Sensibilität, der Schmerzfähigeit . . . wird gegen Depressions-Zustände ein anderes training versucht, welches jedenfalls leichter ist: die machinale Thätigkeit. Dass mit ihr ein leidendes Dasein in einem nicht unbeträchtlichen Grade erleichtert wird, steht ausser allem Zweifel" (GM III: 18, KSA V, p. 382), G, p. 134.

71. "Und was zu ihr gehört—wie die absolute Regularität, der pünktliche besinnungslose Gehorsam, das Ein-für-alle-Mal der Lebensweise, die Ausfüllung der Zeit, eine gewisse Erlaubnis, ja eine

Zucht zur 'Unpersönlichkeit', zum Sich-selbst-Vergessen" (GM III: 18, KSA V, p. 382), G, p. 134.

72. "Die Erliechterung besteht darin, dass das Interesse des Leidenden gründsätzlich vom Leiden abgelenk wird,—dass beständig ein Thun und wieder nur ein Thun in's Bewusstsein tritt und folglich wenig platz darin für Leiden bleibt; denn sie ist eng, diese Kammer des menschlichen Bewusstseins!" (KSA V, p. 382).

73. "Solche . . . schon seltenere Kräfte, vor Allem Muth, Verachtung der Meinung, 'intellektuellen Stoicismus' voraussetzt" (GM III: 18, KSA V, p. 382).

74. "Das Wort schon 'Christenthum' ist ein Missverständnis—im Grunde gab es nur Einen Christen, und der starb am Kruez. . . . bloss die christliche *Praktik*, ein Leben so wie der, der am Kreuz starb, es *lebte* ist christlich . . . *Nicht* ein Glauben, sondern ein Thun, ein Vieles-*nicht-thun* vor Allem, ein andres *Sein*" (AC 39, KSA VI, p. 211; A, p. 151).

75. GM 1:7.

76. "Alle *langen* Dinge sind schwer zu sehn, zu übersehn" (GM 1:8, p. 34, KSA V, p. 268), G, p. 37.

77. "Das Handwerk, der Handel, der Ackerbau, die Wissenschaft, ein großer Theil der Kunst—das Alles kann nur stehen auf einem breiten Boden, auf einem stark und gesund consolidirten Mittelmäßigkeit. In ihrem Dienste und von ihr bedient arbeitet die Wissenschaft—und selbst die Kunst" (KSA XIII, p. 368).

78. "Die kleine Freude, die des gegenseitigen Wholthuns gepflegt wurde" (GM III: 18, KSA V, p. 383), G, p. 135.

79. The meaning of this non-Nietzschean "perhaps" is evident in its diffidence, its decadent fearfulness—it is hardly a "dangerous perhaps"!

80. Anaximander offers the first (tragic, non-ascetic) expression of the pessimistic moral claim. A life deserving its existence should continue to exist—as it would have a right to its existence. But life as such does not continue. So perishing must serve as a general expiation for, because proof of, the general crime or presumption of existence itself. As is well known, this grand vision takes the justice of time to requite the offence of being and life in the passing away of the existent. Becoming itself works as the justification or rectification of the offence of being. But Anaximander's pessimism does not touch the everyday banality of pain and suffering. And, the question of (subjective) suffering is not so grandly regarded by a human culture that counts every injury, suffering as such, as a potential contradiction, i.e., as a threat to insured existence and reacts to this threat with fear. Anaximander assumes the guilt of existent being against all that thereby is hidden from manifestation by its own presencing. All revelation of life is possible only at the expense of other possibilities. Guilt is not a qualification of the existent being as such, rather for Anaximander it belongs to the essence of individuation.

This perspective is radically opposed to the Judeo-Christian ideal of a divine blessing of the particular created being, just because the grand, pagan perspective fails in the latter conviction of divine right and creaturely indigence.

81. "Aus wohlsein, aus überströmender Gesundheit, aus *Fülle* des Daseins" (GT i, KSA I, p. 12).

82. "So ist es ein trauriges Schauspiel, wie sich der Tanz ihres Denkens sehnsüchtig immer auf neue Gestalten stürzt, um sie zu umarmen, und sie dann plötzlich wieder, wie Mephistopheles die verführerischen Lamien, schaudernd fahren lässt" (GT 18, KSA I, p. 119), BT, p. 115.

83. Sellars, "Theoretical Explanations," 335.

84. GT 15.

85. "Der Verlauf logischer Gedanken und Schlüsse in unserem jetzigen Gehirne entspricht einem Processe und Kampfe von Trieben, die an sich einzeln alle sehr unlogisch und ungerecht sind; wir erfahren gewöhnlich nur das Resultat des Kampfes: so schnell und so versteckt spielt sich jetzt dieser uralte Mechanismus in uns ab" (FW iii, KSA III, p. 472), GS, p. 172.

86. This, after all has been said, can never be equated with a Schopenhauerian metaphysical World-Will.

87. "Es ist ein Zeichen von Rückgang, wenn eudaemonistische Werthmaaße als oberste zu gelten anfangen" (KSA XIII, p. 53), WP 222.

88. "Der *höhere* Mensch unterscheidet sich von dem *niederen* in Hinsicht auf die Furchtlosigkeit und die Herausförderung des Unglücks" (KSA XIII, p. 53), WP 222. Cf. Nietzsche's definition: "'Der Glückliche': Heerdenideal" (KSA XIII, p. 38).

89. Deleuze, *Nietzsche & Philosophy*, p. 98.

90. "Gott schuf den Menschen glücklich, müssig, unschuldig und unsterblich unser wirkliches Leben ist ein falsches, abgefallenenes, sündhaftes Dasein, eine Straf-existenz . . . Das Leiden, der Kampf, die Arbeit, der Tod werden als Einwände und Fragezeichnen gegen das Leben abgeschätz, als etwas Unnatürliches, etwas, das nicht dauern soll; gegen das man Heilmittel braucht—und *hat!*" (KSA XIII, p. 98), WP 224.

91. "Der Determinism der Wissenschaft und der Glaube an die That der Erlösung stehen darin auf gleichem Boden. Insgleichen darin, daß sie dem Menschen ein *Recht auf Glück* zugestehn; daß sie mit diesem Maßstabe das gegenwärtige Leben verurtheilen" (KSA XIII, pp. 99–100).

92. GT 17, 18.

93. "Alle grossen Dinge gehen durch sich selbst zu Grunde durch einen Akt der Selbstaufhebung" (GM III: 27, KSA V, p. 410), G, p. 161.

94. "Die *Heerdenthier Ideale*—jetzt gipfelnd als *höchste Wertheansetzung* der 'Societät'" (KSA XIII, p. 65).

95. "Es giebt, streng geurtheilt, gar keine 'voraussetzungslose' Wissenschaft" (GM III: 24, KSA V, p. 400), G, p. 151.

96. "Die décadence verräth sich in dieser Präoccupation des 'Glücks'" (KSA XIII, p. 270.

97. "Ihr Fanatatismus des Interesses für 'Glück' zeigt die Pathologie des Untergrundes: es war ein Lebensinteresse. Vernunftig sein oder zu Grunde gehen war die Alternative vor der sie alle standen" (KSA XIII, p. 270).

98. The real benefits of technology are glossed with illusion: the life saved by medical technologies is often only generously called "life," and modern flight may be more like Jonah's confinement within the bulk of his transport than Icarus' dangerous freedom.

99. "*Weniger bedürftig* nach einer Jenseitigkeits-Lösung seines Räthsels von Dasein geworden" (GM III: 25, KSA V, p. 404).

100. "Seit Kopernicus scheint der Mensch auf eine schiefe Ebene gerathen,—er rollt immer schneller nunmehr aus dem Mittelpunkte weg—wohin? in's Nichts? in's 'durchböhrende Gefühl seines Nichts'?" (KSA V, p. 404).

101. "Wohlan! dies eben wäre der gerade Weg—in's *alte* Ideal?" (ibid.).

102. Unlike the more nearly Nietzschean possibility envisioned by Prigogine and Stengers in the book of the same title, the "new alliance" of the sophisticated humanist pursues what can be called the limits of a desperate knowledge, but without ever succumbing to the creative possibility of a tragic culture. The tragic knower (nihilistic epistemology in its positivist sense) need not yield to tragic art: the illusion need never be willed because the topology of delusion is circular and it likes peripheral movement.

103. "Ihre eigene herbe Form von stoischer Ataraxie darin, diese mühsam errungene Selbstverrachtung des Menschen als dessen letzten, ernsten Anspruch auf Achtung bei sich selbst aufrecht zu erhalten (mit Recht, in der That: denn der Verachtende ist immer noch Einer, der 'das Achten nicht verlernt hat')" (GM III: 25, KSA V, pp. 404–5), G, p. 156.

104. GM III: 23, p. 146; translation altered. "Lässt keine andere Auslegung, kein andres Ziel gelten, es verwirft, verneint, bejaht, bestätigt allein im Sinne seiner Interpretation . . . es glaubte daran, dass Nichts auf Erden von Macht da ist, das nicht von ihm aus erst einen Sinn, ein Daseins-Recht, einen Werthe zu empfangen habe" (KSA V, p. 396).

105. "Erkenntniß als Mittel zur Macht, zur Gottgleichheit" (KSA XII, p. 373).

106. Deleuze, *Nietzsche and Philosophy*, p. 159.

107. "Der Priester entwerthet, entheiligt die Natur: um diesen Preis besteht er überhaupt." *The Antichrist* (AC 26, KSA VI, p. 196).

108. Using a distinction indebted to Kant, Nietzsche observes "Der Priester kennt nur Eine grosse Gefahr: das ist die Wissenschaft . . . Der Schuld- und Strafbegriff, die ganze 'sittliche Weltordnung' ist erfunden gegen die Wissenschaft,—gegen die Ablösung des Menschen vom Priester" (AC 49, KSA VI, p. 228).

109. "Aus dem Stamme jenes Baums der Rache und des Hasses, des jüdisches Hasses—des tiefsten und sublimsten, nämlich Ideale schaffenden, Werthe umschaffenden Hasses, dessen Gleichen nie auf Erden dagewesen ist—wuchs etwas ebenso Unvergleichliches heraus, eine neue Liebe, die tiefste und sublimste aller Arten Liebe" (GM I: 8, KSA V, p. 268), G, p. 34.

110. "Dass man aber ja nicht vermeine, sie sei etwa als die eigentliche Verneinung jenes Durstes nach Rache, als der Gegensatz des jüdischen Hasses emporgewachsen! Nein, das Umgekehrte ist die Wahrheit! Diese Liebe wuchs aus ihm heraus, als seine Krone, als die triumphirende, in der reinsten Helle und Sonnenfülle sich breit und breiter entfaltende Krone, welche mit demselben Drange gleichsam im Reiche des Lichts und der Höhe auf die Ziele jenes Hasses, auf Sieg, auf Beute, auf Verführung aus war, mit dem die Wurzeln jenes Hasses sich immer gründlicher und begehrlicher in Alles, was Tiefe hatte und böse war, hinunter senkten" (GM I: 8, p. 35, KSA V, p. 268), G, 35.

111. "Hat Israel nicht gerade auf dem Umwege dieses 'Erlösers,' dieses scheinbaren Widersachers und Auflösers Israel's, das letzte Ziel seiner sublimen Rachsucht erreicht? . . . dass Israel selber das eigentliche Werkzeug seiner Rache vor aller Welt wie etwas Todfeindliches verleugnen und an's Kreuz schlagen musste, damit "alle Welt," nämlich alle Gegner Israel's unbedenklich gerade an diesem Köder anbeissen konnten . . . Gewiss ist wenigstens, dass sub hoc signo Israel mit seiner Rache und Umwerthung aller Werthe bisher über alle anderen Ideale, über aller *vornehmeren* Ideale wieder triumphirt hat" (KSA V, p. 269), G, p. 35.

112. "Sobald einmal die Kluft zwischen Juden und Juden-Christen sich aufriss, blieb letzteren gar keine Wahl, als dieselben Prozeduren der Selbsterhaltung, die der jüdische Instinkt anrieth, gegen die Juden selbst anzuwendedn, während die Juden sie bisher bloss gegen alles Nicht-Jüdische angewendet hatten" (AC 44, KSA VI, pp. 220–21), A, p. 159.

113. "Der Christ ist nur ein Jude 'freieren' Bekenntnisses" (AC 44).

114. "Wo sie nicht die jüngste Erscheinungsform des asketischen Ideals ist . . . ist die Wissenschaft heute ein *Versteck* für alle Art Missmuth, Unglauben, Nagewurm, despectio sui, schlechtes Gewissen,—sie ist die *Unruhe* der Ideallosigkeit selbst, das leiden am *Mangel* der grossen Liebe, das Ungenügen an einer *unfreiwilligen* Genügsamkeit. Oh was verbirgt heute nicht Alles Wissenschaft! wie viel *soll* sie mindestens verbergen!" (GM III: 23, KSA V, p. 397).

115. "*Leidenden*, die es sich selbst nicht eingestehn wollen, was sie

sind, mit Betäubten und Besinnungslosen, die nur Eins fürchten: *zum Bewusstsein zu kommen*" (GM III: 23, KSA V, pp. 397–98), G, p. 148.

116. "Gegen den *Erhaltungs-Trieb* als radikalen Trieb: vielmehr will das lebendige seine Kraft *auslassen* (es '*will*' und es '*muß*')" (KSA XI, p. 222).

117. "Illusion consolant, et rien d'autre" (Granier, *Probléme de la verité*, p. 80).

118. Philonenko argues this point in his intriguing article, "Mèlancholie et Consolation chez Nietzsche."

119. "Man muss durchaus seine Finger darnach ausstrecken und den Versuch machen, diese erstaunliche Finesse zu fassen, dass der Werth des Lebens nicht abgeschätzt werden kann" (G-D, KSA VI, p. 68), TI, p. 30.

120. KSA XI, p. 533, u.a.

121. "Wenn Jemand ein Ding hinter einem Busche versteckt, es eben dort wieder sucht und auch findet, so ist an diesem Suchen und Finden nicht viel zu rühmen: so aber steht es mit dem Suchen und Finden der 'Wahrheit' innerhalb der Vernunft-Bezirkes" (UWL, KSA I, p. 883). Cf. JGB 11; TL, p. 251.

122. "Der Mensch ward wieder einmal Herr über den 'Stoff'—Herr über die Wahrheit! . . . Und wann immer der Mensch sich freut, er ist immer der Gleiche in seiner Freude: er freut sich als Künstler" (KSA XIII, p. 194) should be understood in light of Nietzsche's reverent attitude (the attitude of an artist) toward the very fact of knowledge. Cf. FW 110.

123. "Es giebt keinen 'Thatbestand an sich', sondern ein Sinn muß immer erst hineingelegt werden, damit es einen Thatbestand geben könne" (KSA XII, p. 140).

124. "Gesetzt aber wir legen in die Dinge gewisse Werthe hinein, so wirken diese Werthe dann auf uns zurück" (KSA XII, p. 192).

125. "Alles Wunderbare aber, das wir gerade an den Naturgesetzen anstaunen, das unsere Erklärung fordert und uns zum Misstrauen gegen den Idealismus verführen könnte liegt gerade und ganz allein nur in der mathematischen Strenge und Unverbrüchlichkeit der Zeit- und Raum-Vorstellungen. Diese aber produciren wir in uns und aus uns mit jener Nothwendigkeit, mit der die Spinne spinnt; wenn wir gezwungen sind, alle Dinge nur unter diese Formen zu begreifen, so ist es nicht mehr wunderbar, dass wir an allen Dingen eigentlich nur eben diese Formen begreifen" (UWL, KSA I, pp. 885–86), TL, p. 253.

126. "Nous allons donc d'abord mettre en relief tous les éléments de la science sur lesquels Nietzsche a pu s'appuyer pour démolir l'édifice de la Métaphysique, puis montrer en quoi, néanmoins, la science 'objective' demeure asservie à l'Idéalisme. Ce faisant, nous retracerons du même coup l'évolution des conceptions de Nietzsche concernant les rapports de la Métaphysique et de la science, en éclairant la nécessité du *ren-*

versement par lequel, apres avoir célebré les vertus antimétaphysiques de la science, il a été amené à reconnaître dans la science l'éxpression ultime du Nihilisme modern" (Granier, *Probléme de la vérité*, p. 75). In this same deconstructive spirit, Granier advises, "Bref, Nietzsche a exploité les ressources *polémiques* que lui offrait la science dans le but de ruiner la Métaphysique traditionelle, quitte à se retourner ensuite contre la science elle-même" (Granier, *Probléme de la vérité*, pp. 74–75).

127. "Certains aspects de la science . . . se laissent bien effective-ment utiliser comme arguments contre l'Idéalisme, mais cela n'empêche pas que la science, considérée dans son projet essentiel, ne soit l'accom-plissement du destin de la Métaphysique elle-même" (ibid., 75).

128. "Il peut sembler parodoxal" (ibid.).

129. Cf. Kaulbach, "Nietzsches Kritik an der Wissensmoral."

130. Cf. KSA XII, p. 365.

131. Cf. AC 15.

132. "La volunté de connaître scientifique n'est donc qu'un déguisement subtil de l'ancienne volunté morale, elle est l'expression du besoin qui pousse l'homme à créer la fable d'un *'monde intelligible'* d'où seraient exclus le changement, la douleur, la contradiction, la lutte, l'ambiguïté, le devenir" (Granier, *Probléme de la vérité*, p. 84).

133. FW 283.

134. "Der am schönsten lebt, der das Dasein nicht achtet" (UB II: 2, KSA I, p. 260), U, p. 69.

135. Nietzsche will ask why human interest should be prejudiced against the the transitory, the disappearing, the metamorphosing: why not delight in becoming? (cf. KSA XII p. 365). He finds the answer in a contradictory desire for immortality, for permanence which loses itself still clinging to life "at any price" (UB II:2).

136. "Die Schwachen werden immer wieder über die Starken Herr,—das macht, sie sind die grosse Zahl, sie sind auch klüger" (G-D, KSA VI, p. 120), TI, p. 75.

137. "Die leichte *Mittheilbarkeit* der Noth, dass heisst im letzten Grunde das Erleben von nur durchschnittlichen und *gemeinen* Erlebnis-sen, unter allen Gewalten, welche über den Menschen bisher verfügt haben, die gewältigste gewesen sein muss. Die ähnlicheren, die gewöhn-licheren Menschen waren und sind immer in Vorurtheile" (JGB, KSA V, p. 222), BGE 264, p. 187.

138. "Sind hinterdrein schwächer, willenloser, absurder als die durchschnittlich-Schwachen" (KSA XIII, p. 370).

139. "Es sind *verschwenderische* Rassen" (KSA XIII, p. 370), WP 364.

140. Opposing this standard of sheer survival Nietzsche adds, "Die 'Dauer' an sich hätte ja keinen Werth" (KSA XIII, p. 370).

141. "Man muss ungeheure Gegenkräfte anrufen, um diesen natür-lichen, allzunatürlichen progressus in simile, die Fortbildung des Men-

schen in's Ähnliche, Gewöhnliche, Durchschnittliche, Heerdenhafte—in's *Gemeine*! zu kreuzen" (GM III: 268, p. 187, KSA V, p. 222).

142. "Die Mittelmässigen allein haben Aussicht, sich fortzusetzen, sich fortzupflanzen—sie sind die Menschen der Zukunft, die einzig überlebenden" (JGB 262, KSA V p. 217), BGE 262, p. 182.

143. "Darwin hat den Geist vergessen . . . die Schwachen haben mehr Geist . . . man muss Geist nöthig haben, um Geist zu bekommen" (G-D, KSA VI, p. 121), TI, p. 76.

144. "Ich verstehe unter Geist . . . die Vorsicht, die Geduld, die List, die Verstellung, die gross Selbstbeherrschung und Alles was mimicry ist" (G-D, KSA VI, p. 121).

145. "In Werk, in That . . . nothwendig ein Verschwender: dass es sich ausgiebt, ist seine Grösse . . . Der Instinkt der Selbsterhaltung ist gleichsam ausgehängt; der übergewaltige Druck der ausströmenden Kräfte verbietet ihm jede solche Obhut und Vorsicht" (G-D, KSA VI, p. 146), TI, p. 98.

146. "Ist hier der Hunger oder der Überfluss schöpferisch geworden?" (FW 370, KSA III, p. 621), GS, p. 329.

147. "Sich selbst erhalten wollen ist der Ausdruck einer Nothlage, einer Einschränkung des eigentlichen Lebens-Grundtriebes, der auf *Machterweiterung* hinausgeht" (FW 349, KSA III, p. 585), GS, p. 241.

148. Strong, *The Politics of Transfiguration*, p. 71.

149. KSA XII, p. 189.

150. JBB 5, KSA V, p. 19.

151. "Naturwissenschaft ist zwar als Kritik am Princip der anthropozentrischen Zweckmässigkeit der Natur entstanden aber das schließt nicht aus, daß sie hinsichtlich der von ihr vorausgesetzten und ihre Geltung begründen den humanen Relevanz der Wahrheit an der teleologischen Prämmisse festgehalten hat" (Blumenberg, *Säkularisierung und Selbstbehauptung*, p. 162).

152. "Seit Copernikus rollt der Mensch aus dem Centrum ins x" (KSA XII p. 127). Cf. GM 25.

153. "Der Tod, der Wandel, das Alter ebensogut als Zeugung und Wachstum sind für sie Einwände—Widerlegungen sogar" (G-D, KSA VI, p. 74), TI, p. 35.

154. "Der an die Nützlichkeit Denkende" (JGB 260, KSA V, p. 209).

155. The connection with the thought of the eternal return is the result of thinking the thought of pessimism. "Das Ideal des übermüthigsten, lebendigsten und weltbejahendsten Menschen, der sich nicht nur mit dem, was war und ist, abgefunden und vertragen gelernt hat, sondern es, *so wie es war und ist*, wieder haben will, in alle Ewigkeit hinaus"(JGB 56, KSA V, p. 75).

156. Cf. "Funf: Vorreden. Über das Pathos der Wahrheit" (KSA I, p. 759).

CHAPTER 6

Toward a Perspectival Aesthetics of Truth

A PERSPECTIVALIST PHILOSOPHY OF SCIENCE

We have seen that Nietzsche represents science and scientific truths as world-interpretations by underlining the constructive, interactive character of such interpretational positions. The human knowing process is an expression of the Will to Power in a world collectively characterized as Will to Power, participating in the chaos of individual world-construing or world-interpretational forces. As world-constructive, such interpretation is not to be thought as the expression of a merely or arbitrarily subjective viewpoint, just because, for Nietzsche, apart from relational positions, apart from interpretations, there is nothing, there is no *other* objectivity.[1] Perspectivalism de-privileges the position of the human knower-spectator and so sheds critical light on the way one comes to know what is known.

The epistemic value of the known (and that is also to say, for Nietzsche, the knowable as such) corresponds to its perspectival "truth." What Nietzsche regards as the perspectival "lie" then is not opposed to this truth as such, for as perspectival the "lie" is not to be calumniated, but is rather a particular viewpoint among others. But denying its perspectival character, the perspectival *illusion* of truth crowns one value (morality, practical knowledge) supreme among others, fixing reality as an absolute object of knowledge, as being, as (approximate) truth. Against this static being, against the only possible stasis, the stasis of ascetic nihilism or denial, the truth of perspectivalism regards life and change without denial, as alteration, nature taken on its worst terms, that is, *becoming* in its most dreadful appearance: "my truth is *dreadful*: for hitherto the *lie* has been called truth."[2] This dreadful truth is resisted by a decadent culture, that is, as we have seen the culture of religion and science, of ascetic nihilism. This culture is grounded upon and grows out of "the *desire* not to see at any price what is the fundamental consitution of reality."[3]

The perspectivalist context of interpretation defined by a world of chaotic and necessary Will to Power reveals the creative interactional character of cultural and so scientific truths. The interactive interpretive process is a precondition for the nonorganic and organic knowing systems. In other words, interpretation is not merely nonsubjective but is literally an empirical, physical—indeed, in Heideggerian terms—*poietic* event.[4] Because the universe as a whole resists and exceeds the grasp of the individual perspective, we may call Nietzsche an "active" or "creative" empiricist. This creative empiricism confounds—it does not mitigate—Nietzsche's hyperrealism, and it is the keystone of his perspectivalism. Creative empiricism not merely construes scientific entities and theories as constructive and instrumental but conceives the empirical, measureable objects of its concern, together with experimental procedures and entire institutional programs, as constructively instrumental. This view can be softened and expresssed in terms amenable to traditional philosophy of science and nature. Thus B. van Fraassen expresses a much tamer and more sophisticated "constructive empiricism," noting that "scientific activity is of construction rather than discovery: construction of models that must be adequate to the phenomenon and not discovery of truth concerning the unobservable."[5]

In addition to van Fraassen, we have seen that different thinkers in traditional philosophy of science offer accounts supporting a constructive instrumental, that is historical and interpretive account of scientific reality (some of these thinkers even go so far as to express a *creative* realism—which would almost accord with a perspectivalist hyperrealism!). Already cited are Duhem, Poincaré, Hanson, Feyerabend, Hacking, Heelan, and, just above, van Fraassen. Although these are not likely to support the extreme interpretive position Nietzsche adopts (much less his analysis of the motivational identity between science and religion), the proposition that interpretation involves a direct interaction with a counteractive—that is, reactional—reality suggests, although exploring that issue cannot be our project here, that such accounts may be given an articulation in terms of a Nietzschean perspectivalism. To the degree to which Nietzsche is not an idealist, his efforts to shed light on the nature of reality apart from our ideals—that is, apart from the prejudices of current (including current scientific) culture—are properly *realist*. To the degree to which he lends no special status to objective reality, transcending the traditional scientific distinctions that allow science to take itself so seriously and to take its results to be approximative truth, he is *antirealist*.

We have seen that Nietzsche explains the scientific reading of facts and laws of nature as deriving from "bad arts of interpretation and bad 'philology'—it is not a fact, not a 'text.'"[6] His own interpretive program demands that one ask why we cannot consider the world of our concern to be the result of the interpretive projection of our own interests, a fiction generated by our own Will to Power? For Nietzsche, this possibility is not to be read as a subjectivism or relativism:

> And he who then objects: "but to the fiction there belongs an author?"—could he not be met with the round retort: *why?* Does this "belongs" perhaps not also belong to the fiction? Are we not permitted to be a little ironical now about the subject as we are about the predicate and object? Ought the philosopher not to rise above the belief in grammar?[7]

It is in this fictional or creative sense that world-interpretation is an expression of the Will to Power. An interpretation is a "fiction" only if it is thought to be a universally applicable truth, otherwise it is a projection of a particular will. The universal (objective, scientific) domain remains viable because interpretation is nothing less than world-interpretation. But in this universal domain, a Will to Power is a will that works against the universe, which universe is likewise to be understood as a constituted-constituting collective Will to Power. This is *"Die Welt von innen gesehen"*—"the world seen from within, the world described and defined according to its 'intelligible character.'" For this insight, Nietzsche explains, "one must venture the hypothesis that wherever 'effects' are recognized, will is operating upon will—and that all mechanical occurences, in so far as a force is active in them, are force of will, effects of will."[8] On this account, the "fictions" or "truths" of science and art are nothing more than artifacts of mutually interpretational Wills to Power. The essence of "truth," so understood, manifests the backwards and forwards balancing dynamic of the interpretive process.

A PERSPECTIVAL AESTHETICS OF TRUTH

For Nietzsche, there is no truth, there are no "facts," no metaphysical or absolute world, no world of appearances as such. Instead there is only interpretation. In an earlier chapter it was noted that this is no harmless relativism; nor does Nietzsche pretend to a superempirical insight. It is not Nietzsche's claim that science or politics or a pot of boiling water can be interpreted according to any perspective whatever, without at the same time entering into a delusionary vision from the point of view of interpreting science or politics or the boiling water. Nietzsche's understanding of interpretation means that to engage in a science, say physics, as it is practiced or has been practiced is to enter into an interpretational project. And interpretation is, for Nietzsche, a supremely relational matter. It is by denying this relational quality (all the while, as we have seen above, employing it nonetheless for its own efficacy) that science so far has proposed itself as (approximative) knowledge of the objective world.

The notion of perspectival interpretation invokes physical as well as psychological (intellectual) attitudes. For Nietzsche who describes the personality as the outcome of warring drives, no fast distinction can be made between the two. Within the range of perspectives, however, interpretational efforts or structures can claim an absolute truth. It is useful to see things then, so Nietzsche writes, as atoms or monads because this affirms the "inner" orientation of an individual perspective, an orientation involving the entire universe. The individual perspective projects its own absolute. Expressing or extending its power as far as it can, the individual Will to Power presents its perspective as an absolute. The limited range of interactional expressions manifests the opposition of the rest of the world, testifying to the total character of the world as Will to Power itself.

For Nietzsche, the world seen from within and as a whole is Will to Power *and* nothing else. Yet he also claims that the total character of this world is chaos to all eternity. In addition, that is, conversely, as Heidegger has made notoriously clear for his part, Nietzsche also claims that the total character of the universe, the whole seen in time, is closed upon itself. This closed process entails the Eternal Return of the Same. In the present chapter we cannot discuss the significance of the last contradictory position, which must be left for the final chapter. Here we sill consider the notion of Will to Power, conceived as "chaos to all eternity" and reprise Nietzsche's understanding of chaos considered earlier.

The two definitions of the world with which we have to do here, that (seen from within) as will to power (reflected in the Zarathustran gnomon, *Es giebt kein Aussen!*), and that definition of the world as chaos to all eternity, permit neither simple translations nor mutual elaborations with one another. Will to Power, understood as thoroughly interpretive, as a perspectival rendering of the world according to an individual perspective, is not chaos but, as Heidegger notes, ordering as such. And yet how is the world as Will to Power (and nothing else) to be understood to be—*to* all eternity—chaos? Let us take some care to be exact. We have already considered the complexities of the idea of originative, generative chaos. The totalized character of the world as original chaos is not the same as chaos from and to all eternity. But for Nietzsche, the total character of the world, taken in all its eternity, is chaos.

If the Will to Power expressed from the viewpoint of an individual perspective is expressed as the will to order the world according to that original perspective (interpretation), then the chaos of the world in its collective, eternal character need not be the same as the individual Will to Power (or even the same as the individual being, itself a consummation of mutiple Wills to Power). Thus clarified, the world as Will to Power is not the world as the manifestation of the Spirit, nor is it the world as Will. The world is composed of a *multiplicity* of Wills to Power.[9] In the collective expression of these power constellations at any given moment, the world has its phenomenal character determined by the expression of the forces composing it. There is no "law" to this expression, save that of the Will to Power. Where every power perspective seeks to express its perspective in opposition to all others, this dynamic can well be described as "chaos." In this sense, the image of a chaos expresses constantly shifting power perspectives; chaos describes the world-character as perpetual becoming. We are talking less of the Germanic concept/term *Fluß* here than of flux. Thus we understand more of the Heraclitean notion of river by thinking the Nietzschean metaphor of the sea. This dual character of the world as a multifarious flux (chaos) of interpretive power perspectives (orderings) is to be understood (interpreted) through Nietzsche's perspectivalism. Each perspective provides a world-interpretive position; each perspective constructs or interprets the world.

But this requires that the passive connotations of the term *interpret* be left to one side, discarding as well any special emphasis upon the original individual perspective (that is, denying that what Nietzsche has in mind is some purely subjective point of view). Interpretation is the fundamental constructive or world-making act. For Nietzsche, "Life is not the adaptation of inner circumstances to outer ones, but will to power, which, working from within, incorporates and subdues more and more of that which is 'outside.' "[10] Thus the problem of saying that there is no outside is a problem of perspective once again.

Seen in the light of *organic* life, that is, altogether, seen in the light of life as such, interpretation is not a passive reaction or response to the world. To explain this point further we can extend an image Nietzsche supplies.[11] The amoeba engulfing a bacterium may be said, in Nietzsche's usage of the term here, to *interpret* the bacterium as the amoeba absorbs or incorporates it. This incorporation is the fundamental expression of the Will to Power as the active imposition of one perspective (the amoeba) over another (the bacterium). The bacterium is interpreted by and from the perspective of the amoeba. This incorporation is an (active) expression of interpretive power, because the amoeba overreaches its bounds and thereby engulfs and incorporates what it can. In similar fashion, it can be added, when the amoeba is overfull it responds to the weight of its own being in the world and divides. For Nietzsche, the reader of the biological theorist Wilhem Roux, the example of the amoeba, of protoplasm in general, is particularly fortuitous: "A protoplasm divides in two when its power is no longer adequate to control what it has appropriated: procreation is the consequence of an impotency."[12] The interactive world-environment is likewise to be interpreted. The amoeba may encounter an electric shock or a noxious chemical gradient. In this case, by shrinking from such contact it also, like the earthworm discussed in the text above, effectively interprets its environment. This interpretation manifests a reactive Will to Power. For the amoeba, reaction is as interpretive as the action of engulfing the bacterium, and both are usually protoplasmically mediated, Nietzsche notices, through the pseudopodia. Yet this last point is not simply an observation of cytological structure—astute as that is, for the pseudopodia are not *there* as such but are constantly changing articulations of the presence of the amoeba in and to its "world."[13] Just in this sense, constructive or creative interpretation must be understood as world-interaction: a direct, manipulative, effective assertion of a perspective. For Nietzsche, the functioning of logic reflects this same structure and genesis:

> The fundamental inclination to *posit as equal*, to *see things as equal*, is modified ... by considerations of *success*. ... This whole process corresponds exactly to that external, mechanical process (which is its symbol) by which *protoplasm* makes what it appropriates equal to itself and fits it into its own forms and orders.[14]

Thus conceived, Nietzsche's perspectivalism—and his corresponding assertion that there are no facts but only interpretations—expresses a directly world-shaping activity rather than simply an affective point of view. Does this make Nietzsche's perspectivalism rather more Kantian and more pragmatic (in a Greek sense) than it is often conceived to be? Perhaps, but that is not my problem in this context. The representation of understanding interpretation (and so perception) as primarily an action (construction) is problematic for an alignment of Nietzsche's perspectivalism with philosophy of science. That interpretive action is determined from its original position means that Nietzsche speaks of it as a matter of perspective. Accordingly, as interpretation retains the relational connotations of this original (context) situating perspective, the term interpretation is to be preferred over terms such as *sense-perception* or *intuitional knowledge*.

In the example given above, the difference between the amoeba's interpretive response to an encountered bacterium and an encountered electric shock is its corresponding physical comportment. The signficant interpretive difference is that between absorbing or acting and recoiling or reacting. It is a simple yet critically important amendment to this observation to note that a larger amoeba (or indeed a Plasmodium or a coelenterate on another level) does not react to the shock (or the bacteria or the same internal bulk membrane distension) at all.[15] With its elevated threshold of response, there simply "is" no shock for the larger amoeba. In common parlance, such a perceptual difference is explained as "a matter of perspective." But for Nietzsche, perspectival differences represent differences in the interpretive expression of the Will to Power.

Interior to its given perspective, a particular vision reigns as truth. The particular perspective is nothing (as Will to Power) but the effort to impose that vision or point of view upon the world. What is "true" is then what is "fitting" for this perspective. In line with this construction but in a more elaborate direction, Paul Churchland offers a structural, neurological illustration of the nature of this "fitting" regarded as Nietzsche intends it (although, to be sure, Churchland is attempting to accomodate van Fraassen, not Nietzsche, in this context). For Churchland, if one considers

> the great variety of cognitively active creatures on this planet—sea slugs and octopi, bats, dolphins, and humans; and when we consider the ceaseless reconfiguration in which their brains or central ganglia engage—adjustments in the response potentials of single neurons made in the microsecond range, changes in the response characteristics of large systems of neurons made in the seconds to hours range—then van Fraassen's term "construction" begins to seem highly appropriate. There is endless construction and reconstruction, both functional and structural. Further, it is far from obvious that truth is either the primary aim or the principal product of this activity. Rather its function would appear to be the ever more finely tuned administration of the organism's behaviour.[16]

Indeed, the organism does not aim for an ideal truth, and Nietzsche suggests that such an aim is impossible. But, if one thinks of "truth" as a construal of the world experienced from a given perspective, this quality varies with the scheme of interpretation and between perspectives. In this way one can comprehend how, for certain schemes, what was originally no more than a "fitting truth" becomes *the* truth. Indeed, such a perspective can become the truth that is valid for the world even where the power of the originating perspectival orientation is insufficient to underwrite this truth in itself.

In the previous chapter we saw that this is the very effective force of the reactive perspective. In an interpretive gambit to manipulate the project of truth in the religious-ascetic or metaphysical perspective, the existing world is declared an illusion, and a better world (the true world) is then proposed beyond it. The most recent development of this (world) nihilistic process is the scientific ascetic perspective, which summons a calculating and mechanical force to interpret the world after its own image, and, succeeding, regards what escapes its interpretive grasp as irrelevant and inconclusive. The scientific perspective, like the religious, is a denial of the world that exists. Inasmuch as science is an effective interpretational effort, its nihilistic extension is a negativity technically defined and obscured.

Although rooted in τέχνη, the idea and practice of science itself is not τέχνη (properly speaking). Instead, as noted in chapter 5, science owes its great success in making practical sense of (effectively interpreting) the world to its affinity for and practical expression of the goals of mechanics (not as a theory in physics but rather as a practical mechanic's, better: a *technician's* world-comportment). Thus I suggested above that science has more to do with βᾱναυσία in practice and connotations than τέχνη. The calculation of βᾱναυσία never frames the work of art in truth and such calculation is all too often to be found behind what counts as art in practical political and business dealings. The practical efficiency of βᾱναυσία is not relevant to the project of looking at science in the light of art because, in a Nietzschean context, the problem of science cannot be conceived on the ground of science. It is from the perspective of a *philosophy of science*, that is, from the perspective of philosophy or art construed in the light of life that science is to be conceived (interpreted) as itself a perspectival seeing. In this way, the progress of science can be seen according to the exigence of an exclusive world-orientation (a world-interpretation that regards that world as its object and nothing more).

TRUTH AS ILLUSION

Insofar as truth may be regarded as a fitting expression of an interpretive construction, it is an aesthetic event. This aesthetic event is construed from the vantage point of the interpretive expression. Consequently, everything has its "own" truth and truths, as aesthetic representations are necessarily multiple.

We have seen that the ideal projection of Western culture in science, as in religion, appeals to an absolute vision of truth. In this vision, the nonreliable emerges as a sign of error (this reflects the banal, mechanical emphasis noted above). But since the ideal of constancy, that is, an unwavering, unambiguous absolute, or ultimate ground of being guarantees the demarcation of error as its reciprocal correlate, the nonreliable can be no more than a conditional or limited or relative index of error. Because the ideal of an absolute is (theoretically) invoked to explain what is perceived, the relationship between the constancy of the absolute (which is the theoretical, approximative truth of things) and the inconstancy of the non-reliable (mere appearances) is circular. Even on its own terms, scientific knowledge is not truth but merely an image or approximation of truth.

By this assessment, Nietzsche does not mean to diminish or ridicule the scientific project in its essence but in its image of itself—that is, if it takes its project (and it does) to offer an account of the objective world by means of its approximative truths.[17] If science were honorably, that is, for Nietzsche, candidly expressed, one would expect science to do more than affirm its short-sightedness (after all, it does this: this is scientific modesty) or its predisposition to error (this recognition is part of scientific sophistication). Science is well able to represent its limitations as achievements. Characteristic of the deceptive play of the ascetic ideal, a weakness is represented (either deliberately or innocently) as a strength. One affirms weakness in this, to be sure, but such an affirmation is merely a ruse. In fact, in precise fact, weakness is conceived under another aspect, that is, as if it manifested a contrary (and invisible) power, weakness according to the mechanic of the ascetic ideal can be represented as a strength.

This misrepresentation works through the interpretive ambiguity of any appearance in a fundamentally perspectival, that is, diverse world. Let us consider a few variations upon the fundamental ambiguity of appearances. We all know that a surface can conceal a secret—being mere appearance—or that a surface can be (no more than) what it seems to be. In terms of social discourse, the original distinction between a lie and the truth can only be the liar's concern (seen from the liar's perspective as a liar). The liar himself usually ignores such distinctions. Conversely, it is well known that when one expects prevarication, the (so-called true, present or not) contents need never come to light in order to be counted as hidden, which in this case is also to say: as present. Once the trick is up, once the liar has been found out, all his "truths" are seen to be, are now renamed as "lies." This discovery tends to be as categorical as its antecedent credulity. Even where there are no "real" lies, suspicion will now uncover or invent one, as the boy who too often cried "wolf" learned to his sorrow. Truth and lie are mattters of interpretation, and they have the same insistence. Because putting over a lie or standing in the truth are correlevant actions, the result is that the ideal of truth remains. Expressing this same interpretive dynamic on another, more broadly cultural level: the invisible God has the same face as the illusion of a nonexistent God. Again, here more ordinarily, the confidence artist shuffles empty walnut shells with the same attention to his cover as he would if a pea were actually there to be found if once they were all turned over. The ideal of (invisible) virtue determines the ascetic ideal.

If the only truth is the truth of illusion, and the measure of the success of an illusion (value or truth) is found in the durability of the values it serves, and, further, where the art of illusion works in the service of life, truth may be understood as such an aesthetic value. Moreover, the significance of truth is its function in the illusions that support existence. The highest illusion in all this is the one that knows what it is doing. This is (creative) art. In the foregoing chapters, I have underlined Nietzsche's distinction between merely preserving life and exalting or expressing life. And in this we had occasion to observe that life-expression frequently opposes the goals of simple self-preservation: in the former, life's aesthetic expression or celebration is at stake. The expression of life as such never aims to preserve existence, it seeks no more than an aesthetic justification.

In the above context, absolute truth as it is conceived in accordance with the ascetic ideal, that is, metaphysical (and scientific) truth lacks a creative aesthetic quality. Its aesthetic appeals to a passive spectator: it offers the promise of comfort, not celebration. Absolute truth cannot be a fully artistic or abundantly creative truth, just because it may not disregard the danger of full life-expression. Metaphysical truth pretends to the truth of its illusion: it believes in its own truth and that is its strength. An artistic truth, however, offers only the illusion of truth as illusion, it believes only in the illusion of its "truth" and that would be, at last, the good will to illusion.

To conceive the thought that places truth and illusion as the sides of a single coin, as markers good only for certain equivalent exchanges, is to challenge the possibility of any knowledge at all. Only the select appeal of a claim to knowledge and to truth is relevant in this economy. The "charm" that is the appeal of a knowledge claim is the occasion of the truth value of any claim. The centrality of charm and interest means that an aesthetics of truth does not follow Kant's definition of aesthetic value or evaluation. A perspectival aesthetic of truth is determined by its style, whether the style of impotence or the grand style of a pessimimistic insight overcome through Dionysian affirmation. An aesthetics of truth makes no claim to an absolute criterion for certainty in the illusion of truth; it is an approach to thinking truth in all its possibilities—or, as Nietzsche would have it, in "all its realities." The prudence of reserving judgment (what I have all along dubbed the "sophistication" of contemporary scientific culture) must give way to the joyful affirmation of the impossibility of making good potential truth claims. There is no truth! This affirmation is joyful in the same way in which a collector can delight in carnival scrip. Its curiosity value is the focus of appreciation. This originally esoteric appeal generates what cash value it may come to have because in itself it has none. The collector knows this but can go so far as to forget the value of the object of his delight as a pure favoring, as merely a pleased response to a charm. This forgetfulness, the threat of taking a light favouring as a serious value, as genuine, or true, is the only danger.

THE ILLUSION OF TRUTH AND
THE QUESTION OF THE ETERNAL FEMININE

It must be said that an illusion, even the illusion of truth, can never succeed in masking the entire register of the Real. But this is not to say that the Real obtrudes in truth; what we have are multiple illusions. What is significant here is the symptomatology of a culture expressed through its "truths" or illusions. And in such terms, the strength of an illusion corresponds to the complexity of the problematic of suffering a life and what that can mean for a reactive culture. In this context D. Breazeale writes

> It seems to me quite impossible to appreciate Nietzsche's struggle with nihilism so long as one does not see how closely connected this is to the problems of giving to suffering a positive value. Schopenhauer was right, "as deeply as one looks into life, just so deeply does he look into suffering" . . . Nietzsche's project is to accept this without succumbing to a world-denying nihilism.[18]

Breazeale observes that the problem of what Lacan in his turn would name the Real is a persistent one: the ideal of truth works as a veil in place of the raw reality of existence. The problematic character of the Real and the ideal of (rational) truth opposing it recalls the philosopher from the fact of existence (for Lacan, again, the impossible) to the critical project of questioning. As Breazeale has it:

> How are we to understand the "rationality of the actual"? As long as one could confidently appeal to an "other world" as a source of values and criteria, these questions were of secondary importance. But if such a recourse is no longer a possibility, then the question of the relationship between critical reason and concrete understanding becomes the question: the question which questions the possibility of questioning itself.[19]

This provocative invocation of the Heideggerian (that is, also Hegelian on Breazeale's reading) project of questioning concerns the interrelation of life, truth, and illusion. But for neither Heidegger nor Nietzsche is the drive to life the drive to the truth of illusion. Nietzsche's denomination of traditional truths as "false truths" retains the name of "truth." The illusion of truth—the ideal of truth—expresses the will to truth even as illusion. This connection does not favor the illusion of truth, which latter illusion lies only in the perspective of truth. For Nietzsche, as for Heidegger, truth is most like itself when it hides.[20] While for Heidegger, because truth is ἀλήθεια, truth and untruth are manifestly codependent, for Nietzsche truth and the lie are rather to be thought coequal. So Nietzsche writes, "Indeed, what compels us to assume there exists any essential antithesis between 'true' and 'false'? is it not enough to suppose grades of apparentness and as it were lighter and darker shades and tones of appearance—different *valeurs*, to speak the language of the painters?"[21] It is then that Nietzsche expresses this insight when he compares truth with a woman. The connection is to be seen in the disguises and seductiveness emminent in the cultural illusion of both truth and woman. Note here that both terms play on the imaginary and the symbolic level, the level of functioning of metaphor and metonymy, that is for Lacan the whole of human society and culture. As Derrida could observe of woman and as he might as well describe the allure of the vision of truth: "it is impossible to resist looking for her."[22]

But what about woman and truth? For my part, I find it ever essential to recall that Nietzsche, writing metaphorically of truth as a woman is also speaking in metaphors when he speaks of woman as such (*Das weib-an-sich*). Nietzsche's discussion of woman is not a piece of natural history or biographical revelation.

For Lacan, (in)famously and (here no doubt unintentionally) echoing Nietzsche: "Woman does not exist."[23] Yet although there *is* nothing in the place of woman, although there is no woman (just as there is no truth), Nietzsche's proto-Heideggerian insight is that we are not equipped to find truth without untruth. The (cultural) image of woman affords metaphoric exemplification to the ideal of truth. Woman, that is, the fiction that is woman, in her concern and regard for appearance above all else, is apparently, emminently dissimulative. Nietzsche's favorite image for this is appearance regarded in its nature as illusory, as seduction, as a lie. Thus woman, for Nietzsche, is "a little dressed-up lie" or a bit worse "Young: a cavern decked about. Old: a dragon sallies out."[24] Here it is clear that the ideal of illusion works in the same way in this (patently and celebratedly misogynistic) context for Nietzsche as for ordinary social conventions concerning beauty (and truth). Both philosophical context and social convention presuppose the conviction that to be *true* beauty/truth must be eternal and profound. Completing the Platonic picture here, as in the so-named fairy-tale character, Beauty is not only beautiful but good. Beauty's vain (and ugly) sisters are as evil as their beauty is an artifice (claimed as but not real or natural) purchased through long primping and costly cosmetics and refinement. (True) Beauty is effortless: it reflects the soul and, if genuine, endures forever. But true beauty affirms that it is surface and nothing more. It is the Greek delight in the fold, especially where it comes to the surface of the (ever eternally, ever thoroughly, however revealingly draped) female body. It is for this reason that feminist and other Nietzsche interpreters puzzling over these comments find it necessary to suspend their usual antagonism to the complexities of Nietzsche's irony, adverting to the significance of Nietzsche's scare-quoted "convictions,"[25] which Nietzsche himself explains:

> One sometimes comes upon certain solutions to problems which inspire strong beliefs is *us*; perhaps one thenceforth calls them one's "convictions." Later—one sees them only as footsteps to self-knowledge, signposts to the problem which we *are*—more correctly to the great stupidity which we are, to our spiritual fate, to the *unteachable* "right down deep."[26]

Although Nietzsche as generously as ever saves his commentators the labor of interpretation the problem recurs precisely because of the nature of what he proceeeds to call *his* truths.

The problem of woman is not (only) a matter of misogyny; rather it is linked to the fate of Western objectivity and truth. In his account of this destiny Derrida, as much a lover of (true) beauty—that is to say, truth—as any man, argues by shifting terms that at this moment Nietzsche transforms "the truth of woman, the truth of truth . . . Woman (truth) will not be pinned down. In truth, woman, truth will not be pinned down. That which will not be pinned down by truth is, in truth—*feminine*."[27] Nietzsche, susceptible as he himself is to the allure of the Platonic metaphor, the metaphor of beauty and truth, is nevertheless too much of a stylist to forget the play of metaphor as such, Platonic or not. Hence the drift of this Platonic metaphor must not be thought to be a merely superfluous baiting on Nietzsche's part of the pretensions (and the position) of metaphysicians. Much more, and much rather the depth of the Platonic metaphor as such, must be thought to affect everything Nietzsche writes about woman. Rather than mere psychological digs and constructions, rather than a simple expression of his own misogyny,[28] Nietzsche's philosophic expression of the nature of woman reflects and repeats the possibilities of the affirmation or denial of illusion. This is Nietzsche's understanding of truth, and to this extent Nietzsche was able to exploit his own misogyny, in style, tracing the Platonic metaphor as such.

Woman, precisely as her appearance, exemplifies the connection between truth and untruth, truth and illusion.[29] So Derrida writes, "It is impossible to dissociate the questions of art, style and truth from the question of the woman."[30] Since the most apparent things about woman are disguises, ornaments, adornments—since these are her lures, that is, her allure, Nietzsche can explain vis-à-vis men, the creators and victims of this ideal, this image, the myth/metaphor *woman*: "What is truth to a woman! . . . her great art is the lie, her supreme concern is appearance and beauty. Let us confess it, we men: it is precisely *this* art and *this* instinct in woman which we love and honor."[31] The truth about woman, the unmasking of the myth, woman is her untruth: what is valuable, honorable, desirable in the true woman is this untruth. That this is a *vision* of woman or of the illusion of woman is clear in Nietzsche. It does not touch the issue of being female as such: "That which Dante and Goethe believed of woman the former when he sang *'ella guardava suso, ed io in lei'*, the latter when he translated it 'the eternal womanly draws us *upward'*—I do not doubt that every nobler woman will resist this belief, for *that* is precisely what she believes of the eternal manly."[32] We have again nothing less than the perspectival position that expresses the truth of illusion, in opposition to absolute truth—which last we have seen to be only the illusion of truth. In Derrida's analysis of Nietzsche's style, he expresses this movement of truth as illusion in a feminine metaphor. It is the metaphorical synonomy of truth and woman in Nietzsche that allows the concept of truth to function against itself. Thus it may be said that if the truth about truth (woman) is untruth, the metamorphosis in the idea of truth as untruth corresponds to that of the "becoming female of the idea."[33]

Here we can elaborate the dramatic sense in which Nietzsche hypothesizes, "Supposing truth to be a woman—."[34] If woman's art—decoration, or as Nietzsche says "adornment" where he writes "self-adornment pertains to the eternal-womanly, does it not?"[35]—is an illusion, if the cultural ideal of woman is this art, truth as illusion (i.e., as a true woman) is a woman in every sense. Yet it will only be as far as woman bodily appropriates a conscious, artful illusion that one may say that a woman (truth) is a woman in truth. Hence within the proper domain of this metaphor, the truth of truth (the truth of illusion) is never a *false woman* but always, as woman, essentially false. This is because, as we have seen before in our consideration of the illusion of truth, a genuinely false woman would claim to be in truth what she appears to be, that is, to be true without recourse to illusion. Such a false woman presents everything about her as true, claims to be truly as she appears. Appearance is not then affirmed as such but rather denied as such, that is, appearance is denied in terms of *illusion*, and re-presented as truth. In the case of woman herself (in this cultural image), that is, looking at the history of woman in the West, one smiles at this whole effort of dissimulative presentation. Thus the English version of the Latin rhyme: "The golden hair Fabula wears, is hers who can deny it? / 'Tis hers she swears, and true she swears, for I did see her buy it." The misogynist *lies* when he claims to have been deceived. After all, to use a Nietzschean phrase in this Nietzschean context, one knows better. Seeing through the veil of illusion, truth is not the insistent claim of illusion as truth. Instead, truth (the genuine truth of truth as mere illusion) can only be the truth of a genuine woman. A true woman is nothing—making no claim to be anything—but appearance. Here the romantic man knows that her seeming, her seemliness is by design; a design calculating everything, but nothing beyond its own expression.

There is of course no such woman; but there is then again no such thing as woman either. And in any case, we have been talking about truth. It is not for nothing that Derrida offers his avowal of the metaphor at the start of his essay on style: "The title for this discussion will be the question of style. But my subject shall be woman."[36] Derrida therefore took the license to unveil his subject—even where as he again with reason, would admit, at the end, the project of such an unveiling would amount "to destroying a fetish"[37] —on the issue of style and woman, of woman and truth.

CONTRA-MORALITY—AGAIN

To write against nihilism, as Nietzsche does, seems to serve the metaphysics and ideals of our time. The contemporary attitude opposing nihilism is motivated by the vainest of reasons, indeed, by vanity itself raised to a social level. On the basis of democratic (social) self-interest,[38] one employs relational prophylactic measures to limit the destructive tendencies of science. We have seen that, for Nietzsche, these destructive tendencies correspond to the decadence of a fearful life (a life lived in fear of the cost of life—the connection between living and dying and of pain, difficulty, annoyance, suffering, etc.) and the fictionalizing nature of all knowledge as interpretation. This recalls Nietzsche's analysis of Socratism, where Nietzsche interprets the exigence of rationality as a response to decadence and its ascetic project of logical optimism.

Did the necessary antinihilism, as antidestructive threat, of his antiscientific writings escape Nietzsche? Many of those who write on the cultural-therapeutic connection between Nietzsche and Socrates, for example, seem to think so.[39] But I have been at pains to emphasize that this association misses the esoteric connection, mistaking the importance of Nietzsche's stylistic dictum that one understands only as much as one is capable of understanding, that is, that character and power determine one's capacity for thought. It is one thing to recognize the vulgar interest in the Will to Life as the will to life at any price. Yet it would be another thing to define the threat of nihilism exclusively in terms of life-endangerment. If Nietzsche details the necessary self-violation of logical and scientific analysis, if he observes the denial of life required in the service of the ideal of life for all and at all costs, the self-cancelling and self-diminishing project is to be opposed for reasons beyond petty vanity.

In fact, the vulgar threat of nihilism is furthest from Nietzsche's concern.[40] A prudential concern with the threat of individual or cultural death (e.g., the normal interest we have in news reports that certain popular activities, such as smoking and drinking, may have mortal consequences) is not directly at issue. Nietzsche's philosophy is first and foremost a philosophy of possibility. And what is possible in the Nietzschean sense can be limitless only once the threat of the ruling culture (with its attendant nihilism but, most of all, with its supreme societal power) has been successfully overcome. Nietzsche did not believe he had succeeded in this project. This reservation underlines the difference between Nietzsche's opposition to nihilism and the broader basis for our normal antipathy to the thought and image of cultural decline.

Here, it is probably necessary to repeat a clarification that is as tired as it is ineffective. Nietzsche really was not championing the need for a new German *Führer*, nor was he proclaiming the Germans to be a master race. And, beyond the Germans, he was not proposing a transformative human ideal for all.[41] He was, instead, seeking a possibility that he himself, despite himself, that is despite his dearest hopes, did not believe could exist. Crudely put, he was looking for an overarching, self-overcoming individual of genius. Beyond his own (failed) search for a friend, he was looking not simply for a single exemplar; perhaps better said, he was actually, in the traditional philosophical fashion, looking for the *possibility* of the friend as such.[42] The social exchange and life of what Heidegger calls *"Mit-Sein"* is more essential for Nietzsche's conception of the *Übermensch* insofar as he or she would be said to have a correspondingly exceptional need for comparable companions in life and growth. So Nietzsche explains in a declaration that matches the spirit of the final books of Aristotle's *Nicomachean Ethics*, "NB: There must be many *Übermenschen*: everything good develops only among its own kind."[43] The hero's agonistic need for an interactive challenge is obviously different from the mere "social" association between average selves in a mediocre society of mediocre needs and mediocre genius. But this point presents what must be seen as a relatively unremarkable refinement, given Nietzsche's philosophic disposition. If Nietzsche sought genius, he sought the possibilities of communal genius. Consonant with what we have seen of Nietzschean perspectivalism, the solitary or isolated individual possibility is meaningless and cannot exist.

This collective possibility beyond contemporary common humanity is, however, precisely what the sustained nihilism of the common culture works to anihilate. As we have seen in the previous chapter, the nihilistic point of the ascetic ideal preserves weakness; the excellence of the most recent manifestation of that ideal in science sustains nihilism. The possibility of communal genius, as a possibility for the *Übermensch* may not be a cultural or societal event. Yet once again: it is crucial to emphasise that Nietzsche is not a social, not a political thinker. And it remains relevant to recall that Nietzsche did not think that the possibility of the *Übermensch* (the possibility of communal genius) beyond current humanity had much chance. In a *Nachlaß* scheme outlining what Nietzsche calls his "new Rankordering," which he intended as a "*Vorrede zur Philosophie der ewigen Wiederkunft*," he expresses the inequality among human beings and among creative types as well.[44] After artists, philosophers, lawgivers, and value (religious) instigators, Nietzsche adds the ironic, ultimate distinction: "The missing type: the man, who commands most forcefully, posits new values, makes the most extensive judgements concerning all of humanity and understands thereby the means to its conformation—sacrificing the entirety for a higher configuration."[45] Seen from this perspective, together with what we have noted of Nietzsche's concinnity, the work *Thus Spoke Zarathustra*, must be considered eminently parodic or ironic: an ultimately postmodern piece of fancy.

Nevertheless, an explanation of Nietzsche's opposition to nihilism may be found in an expression of his final Dionysian affirmation. Dionysian affirmation (which recalls *amor fati*) is affectively the same as Dionysian pessimism and yet—and this is capital—it could be effectively different.

THE AESTHETICS OF ILLUSION

Nietzsche first expresses the role of truth as illusion in *The Birth of Tragedy*. This illusion covers over the reality of nature as the chaos of becoming, or, in the Schopenhauerian metaphor here employed, the Will. For Nietzsche:

> all that comes into being must be ready for a sorrowful end; we are forced to look into the terror so the individual existence . . . We are really for a brief moment primordial being itself, feeling its raging desire for existence and joy in existence: the struggle, the pain , the destruction of phenomena, now appear necessary to us, in view of the excess of countless forms of existence which force and push one another into life, in view of the exuberant fertility of the universal will.[46]

Here, Nietzsche traces the origin of the Apollonian ideal in a flight from the Real, taken in Lacan's usage as a flight from the horror of existence: "The Greek knew and felt the terror and horror of existence. That he might endure this terror at all, he had to interpose between himself and life the radiant dreambirth of the Olympians."[47] The truth of Apollo thus represents in the invention of the gods and the corresponding ideal of beauty and the individual (that is, the ideal of beauty in association with the individual emerges in the light of knowledge), the truth of the absolute. This is also the truth of the object: the nameable, knowable, and, most crucially of all, the countable ennumerable, mearureable, calculable thing. The possibility of reckoning and taking account of nature thus begins with individuation. The absolute, in its illusory guise as the theoretical, scientific measure can be taken to be identical with empirical reality: "And we, completely wrapped up in this illusion and composed of it, are compelled to consider this illusion as the truly nonexistent—i.e., as the perpetual becoming in time, space, and causality—in other words as empirical reality."[48] The attitude of this illusion is not identical to our own post-Galilean empirical reality. In the heroic worldview of the ancient Greek, the presentation of life as a satisfying spectacle from the Olympian viewpoint was meant as a justification of life as such (a life taken to be much less apparently reliable than it currently appears). The tragic view that makes life bearable is the viewpoint of the pleased company of the gods. Their delight, which for the Homeric and Sophoclean Greek is also a *human* delight, justifies the spectacle.

From the viewpoint of the ancient Greek, the living human being or hero presenting himself in such a way was an actor indeed: doing what the dramatic aesthetic of the scene required, and doing it in such a way as to pattern the scheme for once and always. The well-played scene is eternal. Evidently, it is this Shakespearean perspective, the *thing* of the spectacle that is recalled in the context of Nietzsche's thought of the Eternal Return. Moreover, in keeping with Heidegger's contention that the thought of the Eternal Return must always be thought in conjunction with the doctrine of the Will to Power (as characterizing the basis of existence), we may expect to find some correlate of this immortal scene-play in the notion of the Will to Power itself. And we do, insofar as the creative impulse of the Will to Power expresses its power from the side of power as an expression. In accordance with this active, creative emphasis, the "dramatic" or heroic view of life requires an extraordinary, originally powerful individual:[49] the type capable of surviving a life lived as a spectacle. The capacity to survive the agony of existence in its full tragic expression—that is also to say, with grace and style—is the cultured ideal of Dionysian ecstasy and the art of tragedy itself.

In its everyday impact, in a world already structured by the Apollinian vision, the Dionysian experience served to break the round of life by means of a conflict. This means that a time out of ordered time was established. This time out of time expanded the experience of the moment. The expansion of the moment, the ecstasy, disrupted the traditional Apollonian commemorative-anticipatory time of human events and rituals.[50] In this momentary and ecstatic way, the Dionysian ecstasy revealed life itself through an altered flow of time, a time apart from the past-prolonging (traditional) memory of human dissatisfaction. The human subject was offered a vision beyond the images of individuality—the masks of tragic life. We have here two temporal sides of art, the Apollonian in images or words, the Dionysian in music and intoxication: that is, again, the beautiful illusion and the maddening illusion.

The talk of ecstasy is often glossed as a state of intoxication: an ec-static state both in the Heideggerian sense of being beside oneself and in the normal usage, which indicates a state of drunkenness.[51] David Krell expresses the Heideggerian/Nietzschean image of ecstasy with the (for me still troublesome) translation of *Rausch* with the word *rapture*. This word with all its connotations evidently does philosophers wading in cold godless seas in a desolate postmodern landscape worlds of good, but it is not Nietzschean enough—not even when we read the word *rapture* through French or at least Lacanian lynx/Lynceus-eyes, thinking of St. Theresa. *Rausch* for Nietzsche is something else again; it is not rapture and it is not the oozy frivolity of a rush—although, it should be said, the English *rush* closes on the German *Rausch*. For the present consideration of the perspective upon existence and time in the original cult experience of *Rausch* it must be recalled that: (1) intoxication offered the organism a special vantage point on experience as such; (2) where this intoxication was a communal event, the altered character of experience had a special cultural significance; and (3) this collective frenzy had a specific focus (meaning) in the mystery of the god (we are speaking of a profound individual *and* cultural and, so, spiritual event). For these reasons, the kind of *Rausch* enjoyed by the Greek in Dionysian celebration is no longer possible. Lacking the culture, we cannot have the cult, no matter the stoutness of our assertion or the energy of our aspiration. This would seem to suggest that the category of the Dionysian must be meaningless today. This would be the case, if one were unhistorical enough or gullible enough to think Nietzsche's point simply nostalgic.

It has been argued that for the Homeric Greeks, the truth of tragedy required only that life make a good show of itself. In the arena of life, for the Pindaric as for the Sophoclean mind, the human player could be heroically acquitted through his or her own fair efforts, for the enjoyment of the Olympian spectators.[52] But where life does not justify itself in play (and this it can only do for a hero) life must be justified in some other way. The inadequate human being suffers from life as from an "illness" and demands that it be corrected or "justified" or "rectified." This justification is achieved through a significance reaching "behind the scenes" (i.e., a meta-physical significance). This need for a metaphysical account of life recalls one we have seen before: the impulse toward rationality manifest in a nonheroic, or weak culture. Unable to overcome its horror of life, such a culture is impelled to rationalize it. And inasmuch as it means to survive, it has no other option.

The difference between our culture and the culture that could permit a Dionysian world view is the difference between a plebeian culture (together with a democratic ideal) and a heroic culture. The difference between the impotent view and the potency of the Greek vision lies, then, in the differing response to what Nietzsche names the raw "excess" of nature. On the side of the

Dionysian and in opposition to the Apollinian, "*Excess* revealed itself as truth. Contradiction, the bliss born of pain, spoke out from the very heart of nature."[53] Distinguishing between measure or denial and excess or affirmation, Nietzsche's asks how the suffering essential to life is to be overcome.

We have seen that the noble individual is characterized by his or her capacity for suffering. So Nietzsche notes, "The *pride* of unhappiness must be *learned*."[54] This does not mean a brutish insensitivity, nor is it an ordinary asceticism: the heroic individual overcomes suffering through his or her own disposition toward life. The heroic individual is able to draw suffering into him or herself, consecrating his or her right to be. The focus here is not on pain but on life. This is Nietzsche's point when he writes that "there is a will to suffering at the foundation of all organic life (against 'happiness" as 'end')."[55] The emphasis on life reflects the essential life-expressive character of the Will to Power, which opposes self-preservation. Suffering is to be understood as stimulating life, but without necessarily charging the energy of defensive or conservative life. The personal disposition regarding suffering as a life-stimulus must have enough constitutionally appropriate power to treat pain as a stimulant: the decadent, on the other hand, can only retreat from life. Only one who has power to begin with can attain a benevolent, affirmative disposition toward the painful and tragic in life. Hence, in a *Nachlaß* gloss contrasting Dionysus with Apollo, Nietzsche explains, "The past can be interpreted as the enjoyment of a generative and destroying power, as continuous creation."[56] This is only possible from a certain perspective; we have already noted the sense of Nietzsche's rank ordering: "I teach: that there are higher and lower human beings and that by their very existence an individual can justify entire millienia."[57] Such a singular individual is the creative artist. Nietzsche's vision, we see once more, cannot be for everyone because it is in its nature conceived from a height as an esoteric scheme.

Before we can discuss this esoteric height however, we will need to return to the success of calculative representation, or the illusion of scientific truth. To summarize the devolution of the Real as calculable, manipulable reality, it should be said that in addition to the truth about truth as illusion there is also the dream of truth (originally Apollonian) or the illusion of truth (in its Socratic or Alexandrian, banal expression). This latter illusion is nothing but that which sustains life as the present culture of impotence lives the Socratized-Apollonian ideal of the subheroic Greek choice to represent the Real as manipulable reality, assuaging their still-creative, still-aesthetic sensibilities in the only way open to a decadent culture.[58]

This is a surface life-style. It is not easy to penetrate beneath this surface, because we still have a horror of the profound, and rightly so. What is deep is deeper than the day's knowledge: it is the Real or the abyss.

Thus, despite a contemporary preoccupation with the opposition between historical interests and absolute findings, that is, despite our sophisticated interest in undermining the rational aspect of the Real in order to impose our own constructive vision upon it, Nietzsche's philosophical undertaking must be admitted to remain unappealing. A serious consideration of the possibilities of a perspectivalist philosophy challenges science and current technological culture. The value of such a challenge is difficult to assess. As we have seen, religion and science have an associative equipollence in the telic orientation of the ascetic ideal. The sacrifice of this vision of things (that is, of progressive truth) is not possible for an understanding (that is, contemporary) that does not oppose the sovereignity of the technological manifestation of the ascetic ideal. But Nietzsche's vision of truth demands such a sacrifice. For Nietzsche writes that truth is unattainable. This is Nietzsche's truth, Dionysus's last truth, namely: "the truth, to be eternally condemned to untruth. But only the belief in an attainable truth, a trustworthy, self-confirming illusion is proper to humanity. Doesn't man actually live by virtue of being progressively deceived?"[59] For those who are human and so susceptible to a cultivation of impotence, the only way to live with the truth about the illusion of "truth" is through art: "The truth is ugly: *we have art* so that we are not undone by truth [*damit wir nicht an der Wahrheit zu Grunde gehn*]."[60] The redemptive art intended here is a deliberate art, aware of itself as art, "in which precisely the *lie* is sanctified and the *will to deception* has a good conscience."[61] Truth must be found in naming the illusion as illusion: "for all life is based on semblance, art, deception, points of view, and the necessity of perspectives and error.[62] Beyond that small shining employment of our truth-ideal, that is, beyond its critical, reflective application, it has no value.

CREATION AND AFFIRMATION

Nietzsche proposes the affirmation of art in place of the optimism characterizing science (that is really a denial of life), because only a perspectival (or Dionysian) art has the resources to both suggest and survive the sensitive vision into the heart of existence. What is possible in this vision is an affirmative (but not Stoic), tragic wisdom (thus the rigor of classic wisdom is conserved) beyond the science of calculable technology. The simple preservation of life as such holds no allure for a creative or an artist's science. Instead, what is to be affirmed is the classic mystery of the power of existence, beyond the individual and beyond the moment. This affirmation denies neither the individual nor the moment but, conceiving each as aesthetically necessary, "consecrates" both artistically. This affirmation is a will to life, connected with life's own artistry, or creative force. Yet this thrill of life-expression involves nothing less than an affirmation of the intimate anguish of every creation: "the future promised and consecrated in the past; the triumphant Yes to life beyond death and change . . . For the eternal joy in creating to exist, for the will to life eternally to affirm itself, the 'torment of the childbearer' *must* also exist eternally."[63] This association is not a cheerful making-do, an accomodation to reality, or a banal Nietzschean "reality-principle": for Nietzsche, there cannot be creation, there cannot be art, unless there is agony. This is not simply because the sensitivity required for the one leaves one open for the latter, and it is not because the ways of creation so far available are often painful and terrible. With the kind of pain and the kind of terror that forgets itself in creation, what is brought into the open comes into being as such. Thus we are not speaking of the conjunction of pity and terror, but we are speaking of the tragic. As cathartic origins, pain and artistic being share a common root, but not all pain, not all artistry ends in catharsis, that is, purification or expression.

What is created, what comes forth into being in nature and in art is brought into presence from nonbeing. Recall here the archaic insight given expression in Anaximander that in order for an individual to come to birth, all that prefigures what comes to be must be transformed and obscured as what had been. All other possibilities give way of necessity to what presences among beings: both herald and burgeoning crowd of contenders perish in the manifestation of what comes to be. What is, by its limitation in being, by its very form, violates the unlimited. In this way, every becoming is a slash across the face of nature's chaos that seizes form, passes it, plays it, and finally by this same necessity loses hold. The point of perspectivalism in the blind necessity of the expression of the Will to Power remains itself the balance of the Anaximandrian scales of existential judgment. The aesthetic conception of the Will to Power as the consecration of the Eternal Return of the Same reconciles presencing and perishing. The thought of the Eternal Return must be conceived as the challenge for conception and style of an aestheticisation of the rudeness of a point of view that blocks other viewpoints, to be closed off in its turn.

With this in mind, the artist's or creator's aesthetic is to be understood in terms of *amor fati*. If we consider the Heraclitean image Nietzsche uses at the end of his first book, comparing the "world-building force to a playing child that places stones here and there and builds sand hills only to overthrow them again,"[64] it is clear that this eternal play refers beyond the perceiver or individual human being as creative-constructive to the universal character of existence. This connection must not be overlooked. Nietzsche neither advocates a simple phenomenalism, nor a subjective perspectivism. Instead the origin of the emphasis may be found in Heraclitus: "αἰὼν παῖς ἐστι παίζων, πεσσεύων. παιδὸς ἡ βασιληίη." C. Kahn translates: "Lifetime is a child at play, moving pieces in a game." *Αἰών*, called "lifetime" here, is not the lifetime of human life but the lifetime of Being.[65] Not only is the "world-building power" of the child's play effective through our perceptions or perspectives, but this constructive character describes the interpretive play of life or existence as well. The interpretive world-building force of the Will to Power maps a description of the eternity of Nietzsche's Eternal Return.[66] The aorgic[67] aesthetic connection underlines both that the goal of the game (as a play of force) is the play and that a particular game is only actualized through the force or intensity of the play. The significance of gaming as such is less important in this context than Nietzsche's concept of nonadiaphoric or perspectival necessity. The world of soccer, then, is less paradigmatic for Nietzsche's notion of play than is the image of the self-absorbed world child playing on the beach or, even better, the casual roll of the die as the image of necessity where—rather than a human wager—a world is at stake. As *amor fati*, the roll of the die echoes Nietzsche's concept of affirmative necessity.

The seductive style of Nietzsche's presentation of his doctrine of the Eternal Return in *The Gay Science* and *Thus Spoke*

Zarathustra demonstrates that the generation of *amor fati* (or, to use Heidegger's strikingly suitable expression here, we may also speak of authentic resolve) is an aesthetic understanding of being, where aesthetic is meant as a sensitive conception, a disposition in being in which the artist transforms existence. What is wanted, as we shall see in the next chapter, is the resolute attitude from which it is possible to become, to affirm, or to express what one (already) is—to find the promise of the future through the past precisely by way of its determination in the past—transforming 'thus it was' to 'thus I willed it'! Thus: the project here is *not* to find the promise of the past (its redemption) in the future. Rather, it is the past willed backwards that is to redeem the future. If the transformative power of *amor fati* is a blessing, it is not for that anything like a salvation.

Let it be added here that an affirmative scientific aesthetic would be possible only through critique or daring thought. This kind of critique recognizes the nihilistic dimension of science in its illusion of truth. In this way, the illusion of truth becomes the willed illusion: "One must oneself will the illusion—in that lies the tragic."[68] An artist's science is essential to an aesthetics of truth or a tragic wisdom. But philosophy plays an essential, intermediary role,[69] just because "the problem of science cannot be recognized in the context of science."[70]

In sum, then, Nietzsche's aesthetic idea, that is, his critical canon for artistic evaluation, emphasizes the genealogy of creation. The original question, as we have noted before, asks whether need or superfluity has become creative. The distinction is important, not because art is more or less necessary in the one case than in the other—in both cases it may be said that the artist has no choice—but because the compulsion reaches beyond itself where one becomes creative out of overflowing power. This distinction is later expressed as an issue of self-overcoming, *amor fati*, and in still other terms, Nietzsche will speak of the *Übermensch*: of the post, transfigured, higher human being. Earlier, it was noted that the artist's aesthetic is opposed to the concept of traditional aesthetics as a spectator's or evaluator's aesthetic. The spectator must become a player and the artwork the field of play; the position of distant or pure judgment is not at stake, and appreciative delight is not at issue. For Nietzsche, the highest (creative) aesthetic value is *Rausch*, or the celebration of abandon—calculated, aesthetic abandonment: this is the artist's own pleasure.

Because Nietzsche seeks the aesthetic-creative view, the tragic view of Dionysian pessimism emerges in direct opposition to the denial of life in religion (or as the consequence of any celebration of the absolute for the nihilistic and banal benefit of everyday interests in science). At the end of *Twilight of the Idols*, Nietzsche invokes this opposition when he explains what is Dionysian: "Affirmation of life even in its strangest and sternest problems, the will to life rejoicing in its own inexhaustibility through the *sacrifice* of its highest types—*that* is what I called Dionysian, that is what I recognized as the bridge to the pscyhology of the *tragic* poet."[71] Nietzsche's aesthetic view differs from that of Schopenhauer, as it differs from the larger tradition of aesthetics, because it is affirmative-creative (seen from the artist's side) and not resigned-passive.[72] From this position, the claim that his critique of science is not nihilistic but aesthetic highlights the affirmative aspect of this critique. In Deleuze's expression: "But does not critique, understood as critique of knowledge itself, express new forces capable of giving thought another sense? A thought that would go to the limit of what life can do, a thought that would lead life to the limit of what it can do? A thought that would *affirm* life instead of a knowledge that is opposed to life."[73] For Deleuze, as for the present author, the realization of this possibility of life-affirmation necessarily "presupposes a completely different will" in the rule of Western culture. So Deleuze charges: "We have not understood that the tragic is pure and multiple positivity, dynamic gaiety. The tragic is affirmation: because it affirms chance and the necessity of chance; because it affirms multiplicity and the unity of multiplicity."[74] It will do to repeat again although no emphasis can ever be enough that this viewpoint is far from optimistic. Only a *pessimistic*, only a hard or tragic understanding is able to understand what Deleuze names the "dynamic gaiety" in the "necessity of chance" as the tragic. Such an understanding is nothing short of Nietzschean: it is Dionysian, if it is also literally titanic: "The will to life rejoicing in its own inexhaustibility through the expenditure (*in Opfer*) of its highest types."

NOTES

1. As Nietzsche's Zarathustra declares "Es giebt kein Aussen!" (*Zarathustra* III "Der Genesende 2" [KSA IV, p. 272]). For an extended

and consistently dogmatic (or one-sided) treatment of this theme, see Holger Schmid's *Nietzsches Gedanke der tragischen Erkenntnis*. Apart from the interpretive complexities of Zarathustra's rhetoric, it is clear that perspectivalism must be as nonsubjective as it is nonobjective. Or as Nietzsche's double-sided reflection on the dynamics of perspectival considerations concludes "Folglich ist die Aussenwelt *nicht* das Werk unsere Organe—?" (JGB 15, KSA V, p. 29) ("Consequently the external world is *not* the work of our organs?" [BGE, p. 27]).

2. "Meine Wahrheit ist *furchtbar*: denn man hiess bisher die Lüge Wahrheit" (KSA VI, p. 365), EH, p. 126.

3. "Das nicht-sehen-wollen um jeden Preis, wie im Grunde die Realität beschaffen ist" (KSA VI, p. 368), EH p. 128. Hence Nietzsche can write, "Die Überschätzung der Güte und Wohlwollens, ins Grosse gerechnet, mir bereits als Folge der décadence gilt, als Schwäche-Symptom" (ibid.).

4. "Der organische Prozeß setzt fortwährendes Interpretiren voraus" (KSA XII, p. 140).

5. van Fraassen, *Scientific Image*, p.5.

6. "Schlechte Interpretations-Künste . . . schlechten 'Philologie'" (JGB 22, p. 34, KSA V, p. 37), BGE, p. 34.

7. "Und wer da fragt, "aber zur Fiktion gehört ein Urheber?"—dürfte dem nicht rund geantwortet werden: *Warum*? Gehört dieses 'Gehört' nicht vielleicht mit zur Fiktion? Ist es denn nicht erlaubt, gegen Subject, wie gegen Prädikat und Objekt, nachgerade ein Wenig ironisch zu sein? Dürfte sich der Philosoph nicht über die Glaübigkeit an die Grammatik erheben?" (KSA V, p. 54; JGB 34), BGE, pp. 47–48.

8. "Man muss die Hypothese wagen, ob nicht überall, wo 'Wirkungen' anerkannt werden, Wille auf Wille wirkt—und ob nicht alles mechanische Geschehen, insofern eine Kraft darin thätig wird, eben Willens-kraft, Willens-Wirkung ist" (JGB 36, KSA V, p. 55), BGE, p. 79.

9. S. Rosen asks, why a multiplicity? One reply runs as follows: Either one or more than one or none. But at least one, but not only one (opposition). Therefore multiplicity.

10. "Das Leben ist nicht Anpassung innerer Bedingungen an äußere, sondern Wille zur Macht, der von innen her immer mehr 'Äußeres' sich unterwirft und einverleibt" (KSA XII, p. 295). This substance of this point is employed against Darwinian struggle and success. Cf WP 647.

11. E.g., KSA XII, p. 209 or FW 118. Cf. KSA XII, pp. 106, 296; KSA XI, p. 222.

12. WP 654. See also W. Müller-Lauter's discussion of Nietzsche and Roux.

13. Where the pseudopodia of an amoeba in its molecular structure are in constant construction and reconstitution, it is striking to note that

on this level the concepts of construction and constitution are not metaphorical.

14. "Der fundamental Hang, *gleichzusetzen, gleichzuschen* wird modifizirt . . . durch den *Erfolg*. Dieser Prozeß ist ganz entsprechend jenem äußeren mechanischen (der sein Symbol ist), daß das *Plasma* fortwährend, was es sich aneignet, sich gleich macht und in seine Formen und Reihen einordnet" (KSA XII, pp. 295–96), WP 510. Cf. KSA XI, pp. 221–22, where Nietzsche discusses the balance of power in the interpretative constellation of Will to Power: "das Auseinandertreten des Protoplasma im Falle, daß eine Form sich gestaltet, wo das Schwergewicht an 2 Stellen gleich vertheilt ist. Von jeder Stelle aus geschieht eine zusammenziehende, *zusammenschnürenden* Kraft: da *zerreißt* die Zwischen-Masse." For the metaphor here, cf. KSA XII, p. 106, where Nietzsche speaks of perception, projection, incorporation, and approximation in terms centering about the image of "Idioplasma."

15. There are several sizes, species or kinds of amoebae; the plasmodium is a biological curiosity because it is an association of cells where distinction among cells is unclear. A syncytium is often regarded as a "single" cell. There is a controversy here in connection with higher organisms, because certain cellular tissues, muscle and particularly cardiac tissue, display comparable syncytial characteristics.

16. P. Churchland, "The Ontological Status of Observables: In Praise of the Superempirical Virtues" in Churchland and Hooker, *Images of Science*, p. 45.

17. So Nietzsche assures us, "All honor to the ascetic ideal *insofar as it is honest!*" ("Alle meine Ehrfurcht des asketischen Ideals, sofern er ehrlich ist") (GM III:26, KSA V, p. 407), G, p. 158.

18. Breazeale, "Hegel-Nietzsche Problem," p. 163.

19. ibid.

20. Cf. JGB, "Vorreda"; GM III, 24–28.

21. "Ja, was zwingt uns überhaupt zur Annahme, dass es einen wesenhaften Gegensatz von 'wahr' und 'falsh' giebt? Genügt es nicht, Stufen der Scheinbarkeit anzunehmen und gleichsam hellere und dunklere Schatten und Gesammttöne des Scheins—verschiedene *valeurs*, um die Sprache der Maler zu reden?" (JGB 34, KSA V, pp. 53–54), BGE, p. 47.

22. Derrida, "on ne peut s'empêcher de les chercher" (*Éperons*, p. 57).

23. For Nietzsche, "Der Mann hat das Weib geschaffen" (G-D, "Sprüche und Pfeile," 13, KSA VI, p. 61).

24. "Jung: beblümtes Höhlenhaus. Alt: ein Drache fährt heraus" (JGB 237, KSA V, p. 174), BGE, p. 146.

25. Thus early in *Jenseits von Gut und Böse* as a simile for the emergence of a philosopher's "conviction" on the "stage," Nietzsche refers to the fortunes of Apuleius and quotes the Latin, "*adventavit assinus, pulcher et fortissimus*" (JGB 8).

26. "Man findet bei Zeiten gewisse Lösungen von Problemen, die gerade uns starken Glauben machen; vielleicht nennt man sie fürderhin seine 'Überzeugungen'. Später—sieht man in ihnen nur Fusstapfen zur Selbsterkenntnis, Wegweiser zum Probleme, das wir *sind*,—richtiger, zur grossen Dummheit, die wir sind, zu unserem geistigen Fatum, zum *Unbelehrbaren* ganz 'da unten'" (JGB 231, KSA V, p. 170), BGE, p. 144.

27. "Fait virer la vérité de la femme, la vérité de la vérité . . . La femme (la vérité) ne se laisse pas prendre. A la vérité la femme, la vérité ne se laisse pas prendre. Ce qui á la vérité ne se laisse pas prendre est—féminin" (Derrida, *Éperons*, p. 43).

28. This misogyny is ambivalent enough but it was real and as deep as Nietzsche was himself a child of his day. I do not think it necessary to trace it as some do to the predictable psychological result of a boyhood lost in "a household of women." Nietzsche could well have grown up among men, he could well have had a wife and three daughters, and he would likely have thought the same way about women. The male commentators who criticize Nietzsche's perspective on women play, without intending to do so, the same misogynistic game Nietzsche played. In different ways, feminist readings are vulnerable to the same trap laid by Nietzsche's style.

29. Here note that a contemporary, rather modishly feminist account speaking of woman's connection with truth in terms of her *essence* as "abyssal or "*jouissant*" does not differ from this more traditional reading. See Jean Graybeal, *Language and the Feminine in Nietzsche and Heidegger* (Bloomington: Indiana University Press, 1990).

30. "Les questions de l'art, du style, de la vérité ne se laissent donc pas dissocier de la question de la femme" (Derrida, *Éperons*, p. 57).

31. "Seine grosse Kunst ist die Lüge, seine höchste Angelegenheit ist der Schein und die Schönheit. Gestehen wir es, wir Männer: wir ehren und lieben gerade *diese* Kunst und *diesen* Instinkt am Weibe" (JGB 232, KSA V, p. 171), BGE, p. 145.

32. "Das, was Dante und Goethe vom Weibe geglaubt haben—jenes, indem er sang 'ella guardava suso, ed io in lei' dieser indem er es übersetzte, 'das Ewig-Weibliche zieht uns *hinan*'—ich zweifle nicht dass jedes edlere Weib sich gegen diesen Glauben wehren wird, denn es glaubt eben *das* vom Ewig-Männlichen" (JGB 236, KSA V, p. 173), BGE, p. 146.

33. "Devenir-femme de l'idée" (Derrida, *Éperons*, p. 71).

34. "Vorausgesetzt, dass die Wahrheit ein Weib ist" (KSA V, p. 11).

35. "Ich denke doch, das Sich-Putzen gehört zum Ewigweiblichen?" (JGB 232, KSA V, p. 171), BGE, p. 145.

36. "Le titre retenu pour cette séance aura été *la question du style*. Mais—la femme sera mon sujet" (Derrida, *Éperons*, p. 27).

37. "Voire à détruire un fétiche" (ibid., p. 88).

38. Nietzsche's term for this project is "higher egoism."

39. Cf. Löwith, Kaufmann, Dannhauser, etc. Also see McGinn, "Culture as Prophylactic: Nietzsche's Birth of Tragedy as Culture Criticism" for a balanced and sympathetic account.

40. Note the emphasis in JGB 37!

41. For example, in the doctrine of the *Übermensch* proclaimed by Zarathustra.

42. Thus Nietzsche's preoccupation with the idea of the friend in Zarathustra and the *Nachlaß*, and—although the conviction of Nietzsche's homosexuality is more in vogue than ever (*pace* Köhler!)—on the matter of marriage and companionship.

43. "NB. Es muß *viele* Übermenschen geben: alle Güte entwickelt sich nur unter seines Gleichen" (KSA XI, p. 541).

44. Cf. WP 287.

45. "Ein fehlender Typus: der Mensch, welcher am stärksten befielt, führt, neue Werthe setzt, am umfänglichsten über die ganze Menschheit urtheilt und Mittel zu ihrer Gestaltung weiß—unter Umständen sie *opfernd* für ein *höheres* Gebilde" (KSA XI, p. 213). Nietzsche adds the Platonic comment, "Erst wenn es eine Regierung der Erde giebt, werden solche Wesen Entstehen, wahrscheinlich lange *im höchsten Maaße mißrathend*" (KSA XI, p. 213). The dominion of the earth suggested here refers to the "naturalization of humanity" and—correlatively—the "dehumanisation of nature"—and not (contra Heidegger's reading in "Nietzsche's Wort 'Gott ist tot'") to a technological realm of ultimate world-disposal. Cf. FW 109.

46. "Wie alles, was entsteht, zum leidvollen Untergange bereit sein muss, wir werden gezwungen in die Schrecken der Individualexisten hineinzublicken . . . der Kampf, die Qual, die Vernichtung der Erscheinungen dünkt uns jetzt wie nothwendig, bei dem Uebermass von unzähligen, sich in's Leben drängenden und stossenden Daseinsformen, bei der überschwänglichen Fruchtbarkeit des Weltwillens" (GT 17, KSA I, p. 109).

47. Der Grieche kannte und empfand die Schrecken und Entsetzlichkeiten des Daseins: um überhaupt leben zu können, musste er vor sie hin die glänzende Traumgeburt der Olympischen stellen" (GT 3, KSA I, p. 35), BT, p. 42.

48. "Welchen Schein wir, völlig in ihm befangen und aus ihm bestehend, als das Wahrhaft-Nichtseiende, d.h. als ein fortwährendes Werden in Zeit, Raum und Causalität, mit anderen Worten, als empirische Realität zu empfinden genöthigt sind" (GT 4, KSA I, p. 38–39), BT, p. 45.

49. This is the dialectic between sickness and health (which might as well be expressed in terms of the nihilism of power (strength) and the nihilism of impotence (weakness): "for one who is typically healthy being sick can even be an energetic *stimulant* to life, to more life" (EK, p. 40) ("für ein typisch Gesunden kann . . . Kranksein sogar ein energisches Stimulans zum Leben, zum Mehr-leben sein" [EK, KSA VI, p. 266]).

50. Even so, Apollonian world-time is accommodated for the Dionysian ecstasy is itself a ritualized event.

51. Allison's "Nietzsche Knows No Noumenon," or "Nietzsche's Archilochus," to the contrary, shows the special significance of the *Rausch*. See also the discussion of this value in the next chapter.

52. In short and in other words: Anaximander did not invent the vision of justice typified in his philosophy.

53. "Das Übermaass enthüllte sich als Wahrheit, der Widerspruch, die aus Schmerzen geborene Wonne sprach von sich aus dem Herzen der Natur heraus" (GT 4, KSA I, p. 41), BT, pp. 46–47.

54. "Man muß den *Stolz* des Unglücks *lernen*" (KSA XI p. 103). This "unhappiness" must be contrasted with Nietzsche's characterization of ordinary (*not* noble) happiness as "viehisch-dummes" (ibid., p. 101). The "pride" to be learned here is the classic—that means: Greek—antipode of the Christian sin of pride. Cf. ibid., p. 105.

55. "Es giebt einen Willen zum Leiden im Grunde alles organischen Lebens (gegen 'Glück' als 'Ziel')" (KSA XI, p. 222).

56. "Die Vergänglichkeit könnte ausgelegt werden als Genuß der zeugenden und zerstörenden Kraft, als beständige Schöpfung" (KSA XII, p. 113).

57. "Ich lehre: daß es höhere und niedere Menschen giebt, und daß ein Einzelner ganzen Jahrtausenden unter Umständen ihre Existenz rechtfertigen kann" (KSA. XI, p. 278).

58. In an early *Nachlaß* note, Nietzsche writes: "Wir kommen über die Ästhetik nicht hinaus—ehemals glaubte ich, ein Gott mache sich das Vergnügen, die Welt anzusehen: aber wir haben das Wesen einer Welt, welche die Menschen allmählich geschaffen haben: ihre Ästhetik" (KSA IX, p. 581).

59. "Die Wahrheit, ewig zur Unwahrheit verdammt zu sein. Dem Menschen geziemt aber allein der Glaube an die erreichbare Wahrheit, an die zutrauensvoll sich nahende Illusion. Lebt er nicht eigentlich durch ein fortwährendes Getäuschtwerden?" (KSA I, p. 760).

60. "Die Wahrheit is häßlich: *wir haben die Kunst* damit wir nicht an der Wahrheit zu Grunde gehn" (KSA XIII, p. 500). Thus Nietzsche explains, "Über das Verhältniß der Kunst zur Wahrheit bin ich am frühesten ernst geworden: und noch jetzt stehe ich mit einem heiligen Entsetzen vor diesem Zweigespalt. Mein erstes Buch [war] auf ihm geweiht; die Geburt der Tragödie glaubt an die Kunst auf dem Hintergrund eines anderen Glaubens: daß es *nicht möglich ist mit der Wahrheit zu leben*" (KSA XIII, p. 500). In his last work, however, Nietzsche writes, on the same subject, that he had then discovered "die Moral selbst als décadence symptom." This revelation is not a moral or psychological insight but is rather "eine Neuerung, eine Einzigkeit ersten Rangs in der Geschichte der Erkenntniss" (EH, KSA VI, p. 311).

61. "Gerade die Lüge sich heiligt, der Wille zur Täuschung das Gute Gewissen zur Seite hat" (GM III: 25, KSA V, p. 402), G, p. 153.

62. "Denn alles Leben ruht auf Schein, Kunst, Täuschung, Optik, Nothwendigkeit des Perspektivischen und des Irrthums" (GT V:5, KSA I, p. 18), BT, p. 23.

63. "Die Zukunft in der Vergangenkeit verheissen und geweiht; das triumphirende Ja zum Leben über Tod und Wandel hinaus . . . Damit es die Lust des Schaffens giebt, damit der Wille zum Leben sich ewig selbst bejaht, muss es auch ewig die "Qual der Gebärerin" geben" (G-D, KSA VI, p. 159), TI, p. 110.

64. CT 24 (KSA I, p. 153, BT, p. 142.

65. It is of casual note that αἰών corresponds to *eon*. Kahn points out (with Liddell and Scott) that αἰών, defined as "vitality, life; lifetime, duration, time," is "cognate with *aiei*, forever" (*The Art and Thought of Heraclitus*, p. 71).

66. Although nontelic, the play of forces is not limited to static (pointless or adiaphoric) play but has a focus expressed Nietzsche's statement that "jede Macht in jedem Augenblicke ihre letzte Consequenz zieht" (JGB 22, KSA V, p. 37).

67. *Aorgic* translates Friedrich Hölderlin's coinage ("Aorgische") expressing an Empedoclean opposition.

68. "Man muß selbst die Illusion wollen—darin liegt das Tragische" (KSA VII, p. 428).

69. This should not taken to suggest that philosophy lacks its own aesthetic pathos.

70. Once again: "Denn das Problem der Wissenschaften kann nicht auf dem Boden der Wissenschaft erkannt werden" (GT ii, KSA I, p. 13), BT, p. 18.

71. "Das Jasagen zum Leben selbst noch in seinen fremdesten und härtesten Problemen; der Wille zum Leben, im Opfer seiner höchsten Typen der eignen Unerschöpflichkeit frohwerdend, etc. (G-D, KSA VI, p. 160), TI, p. 110.

72. Thus Nietzsche's aesthetics opposes any kind of contemplation, whether resigned or transcendent, in favor of a celebration of the artist's powers. The aesthetic of the artist, from Nietzsche's point of view, is the only aesthetic viewpoint. But it is necessary to concede that insofar as asethetics is traditionally defined from the position of the nonartist as spectator, or recently: speculator. Nietzsche cannot be said to present a typically defined "aesthetics" or theory of beauty. See my essay, "From Nietzsche's Artist to Heidegger's World: The Post-Aesthetic Perspective."

73. Deleuze, *Nietzsche and Philosophy*, p. 101.

74. Deleuze, *Nietzsche and Philosophy*, p. 36.

CHAPTER 7

A Dionysian Philosophy: Art in the Light of Life

THE ETERNAL RETURN OF THE SAME: INTERPRETATION AND WILL

Nietzsche's perspectivalist representation of the world comprises both cultural manifestations of the Will to Power and the Real or chaos beyond such interpretations. On the level of culture, Nietzsche's doctrine of the Eternal Return of the Same appears opposed to an ultimate perspectivalism. For unlike relativism, as we have seen, perspectivalism does not claim the same value or right for every perspective. Hence, even within Nietzsche's own project, the sameness of the Eternal Return is problematic. On the level of the Real, a vision of a nonuniform and dynamic nature is essential to a universe of Will to Power, characterized as chaos in all eternity. If the doctrine of the Eternal Return is interpreted as implying nothing but a repetitive, cyclic necessity, it must also be manifestly discordant with the notion of the Will to Power.[1]

To temper this discord, I propose to review the sense of the Eternal Return as Nietzsche offers it to us, both in the penultimate section of book 4 of *The Gay Science* and in *Thus Spoke Zarathustra*, as a Kantian "what if." This thought, which Nietzsche named in *Ecce Homo* "the highest formula of affirmation," corresponds to a weighting challenge, a question, or, once again: for Kantian ears: a crisis of judgment.[2] This "weight" or crisis of judgment must be heard in an aesthetic cosmic (life- and world-interpretive) dimension. Where Nietzsche's perspectivalism stresses the importance of physiological origins (because it is genealogical), the Eternal Return is not an axiological doctrine of what should be for all and "at any price" but a still esoteric docrine: that is, a selective device addressed to those rare and well-made human beings of Nietzsche's philosophic promise.[3]

The notion of the Eternal Return as the heaviest weight (*das grösste Schwergewicht*), expresses a stress on existence itself and as a question or formula of affirmation is meant to stress—emphasizing as well as weighing on—existence as such.

Conceived as a seal stamped on existence, the existential aspect of the thought of the Eternal Return of the Same is the expression of an ontological dimension in a historical context. This interpretation could be taken to recall the sense of resolute historicality as it is manifest in the latter part of Heidegger's *Being and Time*. Yet a simple conversion of Nietzsche's meaning to Heidegger's thinking is manifestly unworkable. As traced in the preceeding chapters, the historical context of contemporary human existence is one of decline, or decadence. Hence, thinking Nietzsche through Heidegger, the thought of the Eternal Return must seal or confirm decadence. And such a nihilistic seal or confirmation of decadence is not the point of Nietzsche's thought.

Likewise, in another reading of the axiological ideal in a pluralistic, almost postmodern context, the doctrine of the Eternal Return is taken as (far from instructive but still edifying) literature.[4] This reading follows a currently received diverticulum in academic thought: namely the fermentation of a vision conflating philosophy with literature.[5] But in what follows, recalling the terms of interactive perspectivalism as we have seen it thus far,[6] I will attempt to present the doctrine of Eternal Return as an aesthetic *aorgic* (that is, "*aorgisch*" as Hölderlin named the Empedoclean Real of the natural universe in contrast with the "*organisch*" world of human reason and order) principle of generative creation, on the positive basis of a therapeutic vision for creative philosophic types (Nietzsche's "free spirits"). To do this, I must first consider the significance of the Eternal Return as a doctrine confirming existence. If such a confirmation "fixes" existence, must it not repudiate becoming and change? For such a doctrine would be ultimately metaphysical, just as Heidegger supposes.

Yet what is most problematic following our extended consideration of Nietzsche's reflections on science is neither Heidegger's representation of Nietzsche's teaching of the Eternal Return as it imprints the image of the same on the flashing passage of time nor the axiological import of the doctrine of the Eternal Return as a program for life. What is problematic is the question of the scientific formulation and value of the idea of the Eternal Return of the Same. If any one notion could be singled out as stumbling block for interpreters of Nietzsche's thought and philosophers of science alike, it would be the doctrine of the Eternal Return.

The greater number of Nietzsche interpreters, even those who concentrate on the topic, are often content to ignore—with some hedging qualification—the scientific claims Nietzsche made for the Eternal Return.[7] A few authors, such as Alistair Moles and Gunter Abel as noted earlier in the present study, take the doctrine and Nietzsche's "scientific" proofs for it literally. It is hardly my contention that the issue of science in Nietzsche's published or unpublished work is unimportant. But Nietzsche is never to be taken on the matter of science in the way we might ordinarily take another author, even one of Nietzsche's era and formation. Accordingly when Lawrence Lampert at the conclusion of his masterful study, *Zarathustra's Teaching*, writes of the possibility of a "Nietzschean science" the Straussian optimism of this expression inevitably overreaches Nietzsche's project. We do better to hear Tracy B. Strong's brilliant and careful analysis of the dynamic of Nietzsche's ironic engagement with science (scholarship) as it moves through science to art and hence to the political.[8] Such a critical ironic perspective informs the present interpretation.

It is significant that Nietzsche's scientific "proofs" for the Eternal Return are not genuine proofs—by scientific standards or as judged in accord with Nietzsche's own reflective procedures. In fact, for all his rigorous philological formation (background), Nietzsche offers no scientific proofs or rigorous theoretical justifications for *any* of his doctrines. This scholarly reservation holds for Nietzsche's first and most earnest theoretical account of the origins of tragedy as for his lifelong concern with art and life, language and metaphor, morality and society, and so on. Thus the fact that such "proofs" of the Eternal Return of the Same are extant in his notes should be suspect. For while this last teaching may be properly represented as a keystone of his thinking, a reflective keystone is not a central issue or monumental conceptualization so much a formula locking in an entire thought structure.

Nietzsche's own autobiographical claim in *Ecce Homo* is that he "deliberately mislead(s)" the reader with such scientific formulations. Thus it may be said further that he does this throughout wherever he presents mechanical or thermodynamic or other "proofs" for the hypothesis of eternal recurrence, representing the Eternal Return of the same as "true." We have already traced the concinnous troping of truth in Nietzsche's texts. The topic of the

Eternal Return changes little in this resonant dimension. We recall: the ultimate danger of science is nothing but the promise of fixity; what scientific technology effects is sameness, the actual installation of what is called "happiness." Nietzsche's entire concern with science in its industrial technological, its social cultural, character refers to the levelling effect by which such a standard(ized) happiness achieved. The question raised by the issue of art in the service of life ("we have art so that we do not perish from the truth") raises the further question of the interconnectedness of things, a question we will need to leave for the conclusion.

A seal set on shifting, ever-becoming appearance fastens and marks it with the character of being. This fascination with static being characterizes the dominant, reactive Will to Power of Western culture. Thus Nietzsche writes, in the so-called Recapitulation *Nachlaß* note: "To *seal* Becoming with the character of Being— that is the supreme *Will to Power.*"[9] Heidegger employs this declaration to draw a line between Nietzsche's doctrine of the Eternal Return as an expression of Will to Power and the possibility of an illumination of the Eternal Return through the essence of technology.[10] The same line, of course, notoriously distinguishes Nietzsche for Heidegger as the "last Metaphysician" of Western culture. Yet, both orientations are deliberate misprisions. Heidegger does not expressly advert to the difference between the Will to Power as it is the reactive expression of impotence and need, and as it is an active expression of overflowing abundance. For this reason, it is easy to extend Heidegger's emphasis on the fixing-framing function of the Will to Power, concluding that the dominance of the technological worldview represents the single essence of the character of the Will to Power as the affirmation of existence in the Eternal Return of the (mechanical) Same for Nietzsche himself.[11] But this extension is illicit. Nietzsche explains that because "Knowledge-in-itself is impossible in a world of becoming," knowledge is possible only as "error concerning oneself, as will to power, as will to deception."[12] Where "*knowledge* and *becoming* exclude one another," knowledge depends on error.[13] In the scientific culture of the West, the starting point and end result of a will to know is the limited illusion of fixed, certain knowledge. Art too may be viewed as the "will to overcome becoming, as 'eternalization.'"[14] As the cult of appearance, art has special importance for Nietzsche. This importance, however, is derived from the basic perspectival circumstance of human existence, that is, "in a world where there is no Being."[15] In this latter Dionysian world of change, the project of mastering what becomes (i.e., as effective interpretation) has more than one ultimate expression.

Withal, the very idea of "stamping becoming with the character of being" must be conceived as fundamentally ambiguous. The project of setting a seal on becoming, as a weight to be set on existence, remains open to resolution either as calculative knowledge (together with—in its current scientific guise—its antiaesthetic, ascetic ideal), or else as art (with its own aesthetic, ascetic ideal). But these "settings" are equivocal seals or imprints; they may not be conceived as identical weights. For the selective power of art provides an interpretive determination that can be distinguished from the narrowing calculation used to effect the progress and technical advances of modern science. Nietzsche, attributes an unmistakably moral, unmistakably social interest to the latter, while the former corresponds to an ultimately asocial but ineluctably nonindividualistic tragic perception of the multiple realities of what is Real. In this measure, art transcends the moral, social ideal of science. Thus,

> To ascertain *what* is, *how* it is, seems something unspeakably higher and more serious than any "'thus it ought to be,'" because the latter, as a piece of human critique and presumption, appears ludicrous from the start. It expresses a need that desires that the structure of the world should correspond with our human well-being . . . [16]

For Nietzsche, "perhaps this 'thus it ought to be' is our desire to overcome the world."[17] The reflection on this possibility is a result of what he calls a *"Wille zur Wahrheit,"* here expressed as a fundamentally nihilistic orientation toward an absolute. Nietzsche reads such a Will to Truth *"as the impotence of the will to create."*[18] The Will to Truth, operating from a position of weakness and expressed as a will to permanence, is directed toward this-worldly things, particularly in its latest technologico-scientific expression. This ideal of happiness denies becoming for just as knowledge and becoming exclude one another, fixity is the aim not only of knowledge but of a contentment ideal that is increasingly realized in the current technological era as the happiness of the greatest possible number, even if this happiness is also the least kind of happiness. Such happiness requires fixity or guarantees, and "change and happiness exclude one another."[19] Such minimal but guaranteed happiness accords with the aim of scientific progress in the West. Thus in its latest and noblest manifestation, the ascetic ideal preserves its vulgar origin in the technologico-scientific project of fixing becoming. This fixation characterizes the violence of technology and the manipulative illusion of interpretive impotence.

RESSENTIMENT AND AMOR FATI

Grounded in technological world-interpretation, the illusion of reactive culture effectively fixes reality. This congealing achievement does not cancel the value of Nietzsche's reading of reactive culture as illusion. Regarding the nature of the Real as perpetual becoming, the process of chaos is yet amenable to interpretation (interpretively malleable, that is to say, suitable for fixation or constitution) according to human standards and is effected with logical filters; concept nets and selective, manipulative experimental procedures; and technical artifices. The interests of an interpretive expression of scientific reality are those of a decadent culture. Thus Nietzsche could declare,

> For seventeen years I have not wearied of exposing the *despiritualizing* influence of our contemporary scientific pursuits. The harsh Helot condition to which the tremendous extent of science has condemned every single person today is one of the main reasons why education and *educators* are no longer forthcoming.[20]

Nietzsche means to expose the social advantage served by science as that of a general, mediocre, or banal culture. But he means to express more than a merely personal disaffection by speaking of banal culture—the European culture of his day—in this way. It is worth considering a note entitled "On the Order of Rank" to understand the significance of mediocrity for Nietzsche:

> What is mediocre in the typical man? That he does not understand the necessity for the *reverse side of things*: that he combats evils as if one could dispense with them; that he will not take the one with the other—that he wants to erase and extinguish the *typical character of a thing*, a condition, an age, a person, approving of only one part of their qualities and wishing to

abolish the others. The "desirability" of the mediocre is what we others combat: the *ideal* conceived as something in which nothing harmful, evil, dangerous, questionable, destructive would remain.[21]

This emphasis recalls Nietzsche's "immoral" observation in *Twilight of the Idols*: "Nothing offends a philosopher's taste more than man *when he expresses desires.*"[22] It is of course finally unimportant that Nietzsche's immoralist despises humanity as such,[23] because this can obscure the affirmative insight that "What justifies man is his reality—it will justify him eternally."[24] It is factual human activity, considered as an active expression of the Will to Power, that alone justifies humanity. We can see the value of a distinction between factual action and mere human longing (moral ideals, in the Nietzschean sense), if we recall the difference Nietzsche articulates between the ordinary human being involved in a projected world of wishes and discontents (*Ressentiment* or the Spirit of Revenge) and the individual who has "turned out well"—taking care to emphasize that such good form is narrowed to Nietzsche's own sensibilities. To determine the difference between these two possibilities for human being "one must have a **measure**: I differentiate the *grand style*: I differentiate *activity* and and reactivity; I differentiate *overflowing profligates* and the afflicted-passionates (—the Idealists)."[25] We have already seen the importance of this differentiation between active and reactive, the expressive and the apprehensive. But a further reference can only augment this comprehension and its relevance to the idea of a "grand style." We turn once more to Heidegger.

In a section of the Nietzsche lectures entitled "The Grand Style," Heidegger cites Nietzsche's expression of the grand style as "the supreme feeling of power: [*das höchste Gefühl der Macht*]."[26] It is clear that Heidegger regards the grand style in terms of intimate expression in art, as distinguished from all banal drives: "art in the grand style is the simple tranquility resulting from the protective mastery of the supreme plenitude of life. To it belongs the original liberation of life, but one which is restrained; to it belongs the most terrific opposition, but in the unity of the simple; to it belongs fullness of growth, but with the long endurance of rare things."[27] Heidegger sees the challenge

and the effective edge of Nietzsche's esoteric criterion for life affirmation—"the original liberation of life"—just because the ideal
of art offers the artist a glimpse of what is perfect. In this way,
Heidegger expresses the problem of selection as Nietzsche—and,
indeed, we might add, as Plato in his dialogues on love—understood it: "who is to determine what the perfect is? It could only be
those who are themselves perfect and who therefore know what it
means. Here yawns the abyss of that circularity in which the
whole of human Dasein moves. What health is only the healthy
can say. Yet healthfulness is measured according to the essential
starting point of health."[28] The recognition of perfection, of
beauty, is only available to one who understands perfection, who
knows beauty. For this, one must oneself be perfect—beautiful.
The (re)cognitive circle here is only to be resolved by entering into
the contradiction; cognitively, of course, the circle remains a circle. In the context of Heidegger's thinking a circle, a cognitive
contradiction, as a possibility for thought that collides with or
assaults itself is the most profitable origin for thought, for those
willing to follow the question. In this case, the self-compelling
question is elicited in terms of Nietzsche's distinguishing creative
canon. Nietzsche's aesthetic canon articulates the difference
between the creative inspiration of the artist who is wanting and
the artist who is overfull. Heidegger repeats this canon, as it
expresses the possibility of a creative achievement other than one
born of neediness, or an insufficiency: "The contrary possibility is
that the creative is not a lack but plentitude, not a search but full
possession, not a craving but a dispensing, not hunger but super-
abundance."[29] Conversely, the creative power of insufficiency is
thoroughly indigent: "it is not active but always reactive, utterly
distinct from what flows purely out of itself and its own fullness."[30] This means that if Heidegger makes much of Nietzsche's
so-called Recapitulation note whereby the highest Will to Power
marks becoming with the impression of being, as that which per-
dures, he is far from misconceiving it in every sense. Heidegger
rightly saw that this expression can be understood either as the
highest Will to Power from the side of inadequacy or that from
the side of overabundance. It is the latter that is to be expressed as
art in the grand style.

Ressentiment-inspired projects (including science and its culture) may be defined and are themselves effective as instances of

the Will to Power (manifest in the ascetic ideal). Where we understand Will to Power as interpretation, each force expresses its being (its power) only as a force and so requires opposition: this opposition must be equal to or exceed the original power. Because an impotent expression has no excess power to be expressed, it can only be preservative: a will *to* power. The conservative project of weakness ultimately goes to ground because, in Nietzsche's words, it is a decadent project. Yet, such a weak going to ground is no description of death or dissolution: it is not an authentic "going under." The essence of decadence is the preservation of life even in decay. Thus, a reactive "going to ground" amasses substantial reserves and utilizes all its power and every trick in order to sustain and increase power. The result is nothing less than self-preservation. And although many interpreters have read this and its concommitant ethos of increase or material power enhancement that goes with it as Nietzsche's ultimate ideal, it should be contrasted with the Will to Power that expresses an abundance of power. This latter and best expression of the Will to Power forgets or loses itself.

The meaning of the Eternal Return is to be conceived with reference to the perception of existence as aesthetically justifiable in its affirmation or consecration as the field of illusion and loss. The yes of the "yes-sayer" corresponding to Nietzsche's longest self-ambition is the yes beyond yes and no, where chance itself "plays with us."[31] Yet such an affirmation of existence—glorious as it is, superficial as it is—is fundamentally tragic. And Nietzsche concedes and never ceases to emphasize even where the tone is affirmation, this conception of the world *"ist absonderlich düster und unangenehm."*[32]

Here, we can see that the aesthetic representation of the Eternal Return as the *heaviest* weight, as a seal set on existence, fixes each event as a benediction or Homeric consecration: affirming existence and its having been. This sanctifying affirmation is the critical difference between a weight set on the dynamic of existence in the thought of the Eternal Return and the projects of *Ressentiment* that delimit nature in the common interest. In Nietzsche's doctrine, the end of fixity is not sought; the "weighting" or setting of existence remains purely cosmetic—if we can understand cosmetic in its original, cosmic, ordering sense. This "weighting" works in a cosmic, aesthetic sense to set off and to offset the individual.[33] An affirmation of becoming and change celebrates the individual being as participating in the project of existence, as reflecting in itself the same eternally creative, eternally destructive transitoriness that belongs to life. That things are just what they are: coming into and going out of being, at once sources of delight and occasions for suffering, articulates the same irreversible round that inspires the fear and the flight of ordinary idealisation. Yet, it is this very repetition heightened as the same that Nietzsche seeks to affirm. Consequently, eternity is alienated from the ideal of an insurmountable project of permanence. The latter is merely the quest for the persistence of the individual, opposing time and its dissolution in the becoming of being. So far from the ideal of a system or structure of permanence, the dynamic nature of existence reveals nothing but finitude. Such a dynamic does not express life in a perpetual round but articulates its essential timeliness, the time of mortal existence instead. It is this finitude that is affirmed, confirmed, and so sealed in the thought of the Eternal Return in the artistic life-attitude that Nietzsche calls the "Dionysian." The notion of *amor fati* as such an aesthetic view point embraces more than a universal will or nature opposing the individual: it is not a simple-minded love of the fate to which one must submit. Hence, despite religious overtones, the blessing of becoming is *not* Job's dumb-willing assent to the tragedy of his life—for the love of God and His divine purpose. Nietzsche's worldview is full of gods, but it is not the Judeo-Christian vision of the divine.

The notion of *amor fati* here conceived works to confirm or affirm one's individualized opposition to the universe highlighted in transience as intransigence, as Anaximander or Schopenhauer had conceived it. In this universal opposition, the illusion of the individual ego offends life, transgressing what is perpetually transforming being. From this perspective, as the facticity of a persistent individual or existence, life itself is a cosmic offence: a continual transgression of becoming into being, which, in the eventual and unavoidable dissolution of being, is ultimately redeemed in the perishing of each individual configuration. If we take this in Judeo-Christian terms, we hear the message of *vanitas*. Sounded from Schopenhauer's position, the justice proclaimed has a bitter and resentful tone. Nonwilling, the renunciation of Buddhism is better than not willing at all. As Nietzsche begins and closes the third essay of *On the Genealogy of Morals*, "man would rather will nothingness than not will."[34] Nietzsche differs from this last perspective—but in a deeply concordant fashion—because Nietzsche defines human nature as identical with that of what Lacan calls the "Real." Closed off from the remotest possibility for supernatural alignment and so for either justification or redemption, and excluded from the alienation of difference needed for a metaphysical offence, human beings enter the world of illusion, becoming themselves yet another mask for existence. In this way, Nietzsche teaches the Dionysian and its aesthetic affirmation of tragic life as a project "beyond pity and terror, *to realize in oneself* the eternal joy in becoming—the joy that also encompasses *joy in destruction.*"[35]

This late reflection recalls Nietzsche's earliest book and first expression of this aesthetic ideal. It is not gratuitous here to observe that it is *because* a thinker thinks his original thought that he can be underway in the thinking of this thought to the end.

THE PERSPECTIVAL DOMINANCE OF DECADENCE

More than once it has been necessary to repeat that for Nietzsche, science must be construed as a *moral* phenomenon, indeed, as the latest manifestation of the ascetic ideal. In a Heideggerian context, Nietzsche's genealogical perspective on the moral phenomenon of science (its psychology) implies that the (ontic) danger inherent for Heidegger in the essence of technology cannot be avoided. The threatening danger of technology is illuminated in its historical expression. If it is part of what it is to be human to manifest a violent and reactive Will to Power, that is, without comprehension and in a measure cut only to the standard of the (most reactive) human Will to Power itself, then the danger that threatens in the age of technology is not only the ambivalence or uncanniness of an invitation to respond to Being, but also contemplated from the side of weakness, the very vulnerability of human beings. Inauthenticity, for Heidegger, is an extended flight from the challenging call to pose and harbor the question of Being.

But the conflict between authenticity and inauthenticity is not at stake in a juxtaposition of Nietzsche and Heidegger. The violence of human being in the world belongs to the human condition, as its danger and its saving power. This is the point of contact that we seek. In *Introduction to Metaphysics*, Heidegger expresses the original nature of this violence after Heraclitus: "In the conflict [*Aus-einandersetzung*, setting apart] a world comes into being."[36] In a Nietzschean context, this may be read as an expression of the interactional chaos of multiple Wills to Power. This interpretation must be seen to be especially germane where we read further in Heidegger's account of the origins of the violence that characterizes human being in the world and expresses the dynamic of the Will to Power in a world that is essentially chaos: "It is this conflict [*Kampf*] that first projects and develops what had hitherto been unheard of, unsaid, and unthought. The battle is then sustained by the creators, poets, thinkers, statesmen. Against the overwhelming chaos they set the barrier of their work, and in the work they capture the world thus opened up."[37] The battle of Wills to Power is sustained, among other ways, through culture. In this way, Western culture may be said to be the cultivation of conflict that imposes the beings of cultural interest (as *Seienden*) in interpretational expressions wrested from the scope of the "overpowering power." For Heidegger, the original receptive preservation and interpretive expression of Being as beings is poietic, that is, genuinely vital and creative, in the original, tragic culture of ancient Greece. This poietic character is altered when the world that is made (represented, interpreted) is no longer preserved as such but represented as "merely finished and as such available to everyone, already there, no longer embodying any world." This last disposition expresses the essence and interest of popular culture: "now man does as he pleases with what is available. The essent becomes an object, whether to be beheld (view, image) or to be acted upon (product and calculation). The original world-making power, φύσις, degenerates into a prototype to be copied and imitated."[38] Heidegger's description of the reproductive prototype for efficient manipulation recalls Nietzsche's image of the fanciful logical and scientific units that permit practical computation and control: "Nature thus becomes a special field, differentiated from art."[39] Heidegger explains the same historical (and for us here: perspectival) consequences of such a separation, where the human being "turns round and round in his own circle. He can ward off whatever threatens this limited sphere. He can employ every skill in its place. The violence that originally creates the paths engenders its own mischief of versatility, which is intrinsically issueless, so much so that it bars itself from reflection about the appearance in which it moves."[40]

The dead-end for thought and life that threatens here is expressed with a word taken from (and offered in the context of a deliberate reference to) Nietzsche: "But where is nihilism really at work? Where men cling to familiar essents and suppose that it suffices to go on taking essents as essents, since after all that is what they are . . . To forget being and to cultivate only the essent—that is nihilism. Nihilism thus understood is the ground of the nihilism which Nietzsche exposed."[41] The analysis of the devolution of Being that Heidegger offers throughout his work shows that the reference to nihilism, as a reference to nothing and to the reductive danger threatening in the essence of technology, is not just a dramatic flourish but a deep meditation of Nietzsche's criticism of values and morality in intimate connection with mechanically manipulative scientific culture.

How is the sheerly manipulative culture of science (as an expression of βᾱναυσία rather than τέχνη) relevant to Heidegger's reflections on the essence of technology (for Heidegger an essence sharing in the source of ποίησις)? My suggestion here has been simply that the ποίησις characterizing modern τέχνη is effectively structured or calculated in βᾱναυσία. This dangerous banausic or utilitarian expression of technology works on a mechanical, that is, a regulated, repetitive level. This suggestion of technological efficacy initiates the question of artifactualism in the context of technology and science. The question of artifactualism arises from the recognition that an originally mechanical orientation elicits mechanical results. Likewise, chemical, physical, or historical orientations can call forth *correspondingly* artificial results. Every physical seeking grasps or apprehends what is sought. The seeking as such is fashioned as a specific template, used to cut out what is sought from what is. The point of this intentional analysis of seeking is not to impugn the validity of science (a contextual validity in any case) but merely to illuminate the artifactual origins, the formed, factitious, or fashioned source of its efficiency. This genealogical understanding is essential because efficiency is exactly what is wanted (and is precisely what is attained) by (banausic) science. Heidegger explains in "The Question Concerning Technology" that "the modern physical theory of nature prepares the way first not simply for technology but for the essence of modern technology. For already in physics the challenging gathering together into ordering revealing holds sway."[42]

He concludes that this kind of calculating-informational orientation may not be abandoned, not even in the light of the interpretive approach proposed by quantum mechanics:

> It is challenged forth by the rule of Enframing, which demands that nature be orderable as standing-reserve. Hence physics in all its retreating from the representation turned only toward objects that has alone been standard until recently, will never be able to renounce this one thing: that nature reports itself in some way or other that is identifiable through calculation and that it remains orderable as a system of information.[43]

The objectifying approach to mastering nature and setting its reality to work is fundamentally banal, that is, dependent upon the foreclosure of the question of Being. When Heidegger explains that "Enframing (*Das Ge-stell*) is the gathering together that belongs to that setting-upon which sets upon man and puts him in position to reveal the real, in the mode of ordering, as standing-reserve,"[44] he employs the illuminative insight of the interpretive character of the Will to Power as it simultaneously expresses the character of the interpreting force, along with the interpreted object. In this case the scientific, technological human being is the interpreted interpreter:

> As soon as what is unconcealed no longer concerns man even as object, but does so, rather, exclusively as standing-reserve, and man in the midst of objectlessness is nothing but the orderer of the standing reserve, then he comes to the very brink of a precipitous fall; that is he comes to the point where he himself will have to be taken as standing-reserve.[45]

We have seen that Nietzsche's different—and less "timely"— analysis of the essential reductive or ontic nihilism of the scientific, technological interpretation (the ascetic ideal as it expresses a reactive Will to Power), articulates the interpreter within the same Will to Power.

If the point of theory making and computation in science and technology is the production of calculable results, the presentation or utility of such results is characterized by an irrecusable banality. There can be no ambiguity and no esoteric rarity or genius in such a scientifically reproducible, technologically manageable account of nature. This banality is unavoidable in a culture that seeks calculable, predictable results in the first place. From Nietzsche's genealogical perspective, one must know *who* expresses a desire to know (or to be good, or to be moral), in order to evaluate the nature of what is achieved as the known (the good, the moral). The Heideggerian analysis of technology reveals one sense of the ultimate danger of technology as calculative knowing and the discovery/covering over of the concealed/unconcealed, in the representation of human being as standing reserve. But, in this expression Heidegger proposes the saving hope of a re-covery of the sense of Being as emergent (un)concealment, as ἀλήθεια. This position is—in a sense opposing Nietzsche's elitist interest—preeminently given over to think the "humanistic" importance of human correspondence (that is, as *ge-eignet*) to the event of *Ereignis* as such. This seems to be a humanism in the worst scholastic or academic sense. But as William J. Richardson emphasized in his classic essay on Heidegger and science, the very idea of humanism is inherently ambiguous in the modern context.[46] Heidegger observes that the incorporation of human beings into the scientific, technological domain threatens to limit human value to mere physical, chemical, biological, or economic value. This observation should not be reduced to the pronouncement of ordinary humanism: the degradation of human value forecloses the realization of the exclusively human possibility of thinking Being beyond beings as such. This is a humanism subordinate to and responsible for the provenance of Being. Heidegger's humanism—if we call it that against Heidegger's own expression and merely in order to distinguish between Heidegger and Nietzsche—is not of the reactive kind, so deplored in Nietzsche's criticism of decadent, ascetic culture. For Heidegger observes that the tendency to treat human beings as so much standing reserve works to install select human knowers in a supreme position vis-à-vis all beings: "Meanwhile man, precisely as the one so threatened, exalts himself to the posture of lord of the earth."[47] This (what would be merely banal) lordship is, for Heidegger, an illusion. This illusion is a part of the threatening danger from the human side that closes off the possibility of a responsive openness to Being. Heidegger's concern is with the illusion of value at stake and the transcendence denied in the scientific ideal of truth. Although my description of the notion of a banausic essence of technology is better suited to articulating Nietzsche's critique of technology, Heidegger too may be taken to oppose the banausic expression of technology, favoring as he does a recollection of the primordial essence of technology as τέχνη. In our present, limited context, we can do no more than note the parallel between Nietzsche's expression of the ascetic ideal and its self-diminution in the service of an ultimate power-project with the threat posed by the technologically inspired Heideggerian "Lord of the Earth" to the (authentic) possibility of an appropriate response to the call of Being. Relevant too, in a further but still limited context, would be the belief, shared by Heidegger and Nietzsche but differently expressed, in the saving, transformative power of art. But the greatest difference between Nietzsche and Heidegger turns upon Nietzsche's fearful sensitivity to the dominance of the banal, or the mediocre. This dominance is the deadly obstacle to the recollection of the original sense of $\tau\acute{\epsilon}\chi\nu\eta$ in productive, creative art—which Heidegger, fearlessly for his part, invokes and attends.

DIONYSIAN AESTHETIC PESSIMISM

If tragedy is born out of the spirit of music and if unlike the other arts born of the muses, tragedy dies at its own hand, according to Nietzsche, then in this oldest of romances, as it may be brought to life again in the Eternal Return, we may ask about the future of art. We may ask about the philosophers of the future and the future as such. As we have seen, Nietzsche seeks to instigate a new kind of affirmative response to life—for the (well-constituted, select) individuals he imagined as "philosophers of the future." We will have an opportunity to consider the point of such a limiting qualification after recalling the character of this affirmative response to life. Nietzsche calls this response, variously, *"amor fati,"* the "tragic," the "Dionysian"; and, reflecting on his first expression of Dionysian affirmation in *The Birth of Tragedy*, he explains it as:

> a formula of *supreme* affirmation born out of fullness, of superfluity, an affirmation without reservation, even of suffering, even of guilt, even of all that is strange and questionable in existence . . . This ultimate, joyfullest, boundlessly exuberant Yes to life is not only the highest insight, it is also the *profoundest*, the insight most strictly confirmed and maintained by truth and knowledge. Nothing that is can be subtracted, nothing is dispensable.[48]

Life-affirmation requires power, it needs overwhelming, overabundant strength. But let us not forget here, as we have seen and as we have underlined as the fundamental difference between Heidegger and Nietzsche, that for Nietzsche, "Only excess of strength is proof of strength."[49] Dionysian affirmation is nothing less than an expressive abundance.[50] Such a richly exuberant strength says Yes! to life, in "all its realities," especially in its transitory destructiveness and says Yes! not out of bland new-age confidence or traditional other-wordly religious hopes but Yes! because of its own power-in-excess. So as we noted at the end of the preceeding chapter, Nietzsche declares

> Affirmation of life even in its strangest and sternest problems; the will to life rejoicing in its own inexhaustibility through the *sacrifice* of its highest types—*that* is what I called Dionysian, *that* is what I recognized as the bridge to the psychology of the *tragic* poet. *Not* so as to get rid of pity and terror, not so as to purify oneself of a dangerous emotion through its vehement discharge . . . but, beyond pity and terror, *to realize in oneself* the eternal joy of becoming.[51]

We have already seen that this kind of affirmation is *not* available to the ordinary individual. This critical restriction must never be discounted in Nietzsche's doctrine of *amor fati*, or tragic, Dionysian life-affirmation. The radicality of this limitation calls for further discussion as we are here attempting to see why such a limit applies.

To express an artistic affirmation of life it is essential that one live what Nietzsche calls an "ascending life." An ascending life differs from a decadent life by virtue of its original power, it is *higher*. Full life-expression is a matter of physiological and spiritual endowment. Nietzsche writes, "After all, no one can spend more than he has—that is true of individuals, it is also true of nations."[52] If this is preeminently true of artists, it is essential to recall Strong's provocative gloss on Nietzsche's own related aesthetico-economic reserve. For Strong, the fundamentally diffident reservation must move in what Strong names a "circle of praxis" betraying Nietzsche's own concern not only for himself and his readers but also for the very possibility of engagement. For Strong, "it is essential to ask what anxiety is expressed in the passage"[53] from scholarship to art, and it is clear that Strong regards this anxiety in terms of the political dimension. But this same political anxiety must also correspond to what I have named the esoteric *difficulty* of a Dionysian philosophy of life. In this special Nietzschean sense, the Dionysian artist is intoxicated with life. The artist is filled to excess with power, and drunk with the feeling of that abundance: "The essence of intoxication [*Rausch*] is the feeling of plenitude and increased energy."[54] Such intoxication is of the essence of art; the *Rausch* is its creative impulse:

> In this condition one enriches everything out of one's own abundance: what one sees, what one desires, one sees swollen, pressing, strong, overladen with energy. The man in this condition transforms things until they mirror his power— until they are reflections of his perfection. The *compulsion* to transform into perfection is—art.[55]

This expression is ecstatically styled, hyperbolic to an extreme. But this description of the intoxicated visionary and master of aesthetic transformation accords with Nietzsche's understanding of "the artist." This is nothing like a fluffy romanticism—and Nietzsche tells us that himself. Instead, Nietzsche's tragic aesthetic is more dangerous and much crueler:

> In the face of tragedy the warlike in our soul celebrates its Saturnalias; whoever is accustomed to suffering, whoever seeks out suffering, the *heroic* man extols his existence by means of tragedy—for him alone does the tragic poet pour this draught of sweetest cruelty.[56]

The tragic possibility of this *süssesten Grausamkeit* requires a perspectival position of tremendous power. This is the position of ascendent life, and from this position: "much may be risked, much demanded, and much *squandered [vergeudet]*."[57] From the preceeding chapters, it is clear that this is not the position of our cultural life, as actual for us today as it was for Nietzsche.[58] Thus Nietzsche immediately offers an illustrative reference implying that ascendent life is a past, but perhaps also a future possibility: "What was formerly a spice of life would be *poison* to us."[59] It is a poisonous life-stimulus, an affirmation in the face of what is questionable and horrible in life that determines what is rare and great for Nietzsche. Can such an overtly offensive ideal have any relevance for us? Better, perhaps, to ask if we can find any (future or possible) relevance for it at all (past or future)?

Nietzsche's original image highlights the obliquity of his position. For Nietzsche, the primordially affirmative artist-player of life is the world-child Zeus. The play of life is represented in a preface to an unwritten book, taken according to Heraclitus's world insight as: "the eternal jest of world-destruction and world-creation."[60] In *Philosophy in the Tragic Age of the Greeks*, this insight is singularly "understandable to one who, like Heraclitus, is related to the contemplative god. Before his fire-gaze not a drop of injustice remains in the world poured all around him; . . . In this world only play, play as artists and children engage in it, exhibits coming-to-be and passing away, structuring and destroying, without any moral additive, in forever equal innocence."[61] In Heraclitus' innocently creative vision, "The world is the *game* Zeus plays."[62] But what kind of play, what kind of game is this? A game that is primordially for a god cannot be a game for everyone. In the natural world, as we have seen it to be a perspectival world of power, such playing is possible only for children and artists.[63] Does this mean that any of us, insofar as we are artistic or childlike, play the original world-game? Are all artists, together with all children, special, Dionysian beings? To be Dionysian—that is, to be an artist or to be a child? The answer to these questions must be no for, in fact, we are interested only in a special qualification of the being of the artist and the child, any artist, any child.

Both "artist" and "child" embody becoming. Insofar as this is fully and genuinely the case, that is, insofar as becoming holds innocent sway, the child's game, the play of art is creation:

> Such is the game the aeon plays with itself. Transforming itself into water and earth, it builds towers of sand like a child at the seashore, piles them up and tramples them down. From time to time it starts the game anew. An instant of satiety—and again it is seized by its need, as the artist is seized by his need to create. Not hybris (*Frevelmmuth*) but the ever self-renewing impulse to play calls new worlds into being.[64]

Nietzsche's conception of the world-play defies both straightforward and sophisticated exposition. Thus as an instance of the latter, Lampert's (politically scientistic and marvellously "correct") optimistic possibility of a "Nietzschean science": a kind of "play" reconciled with (an egregiously underexamined conception of) "love" must fall woefully short of the fundamentally aorgic play of the eon essential to Nietzsche's conception of the world-child.

Such childish, loving world playfulness cannot at all be said to

hold for every child, neither does it hold metaphorically (or ideally) for every child at every moment—and it is even more naive if
utterly fashionable to think of practicing artists in this way except
(and this exception is important) "on good days." On such "good
days" of easy and unrestrained creation, the artist's creative play
reveals a world of possibility which is offered to us as an artistic
articulation of the "*innocence* of becoming."[65] This expression
affirms life in the utter extension of a good day gone well. This is
the classical, arcadian meaning of the perfect: a disposition ready
to bless everything, a transfiguring moment, the moment of great
and tragic style. The affirmative disposition of the artist here is
the singular achievement of the *Übermensch* promoted by Nietzsche's Zarathustra. Nietzsche describes this ideal:

> the species of man he delineates delineates reality as *it is*: he is
> strong enough for it—he is not estranged from or entranced by
> it, he is *reality itself*, he still has all that is fearful and question-
> able in reality in him, *only thus can man possess greatness*[66]

This is the Nietzschean meaning of *Gelassenheit*:

> One is necessary, one is a piece of fate, one belongs to the
> whole, one *is* in the whole—there exists nothing which could
> judge, measure, compare, condemn our being, for that would be
> to judge, measure, compare, condemn the whole . . . *But noth-
> ing exists apart from the whole.*[67]

There is no Archimedian, absolute, objective standpoint in the
world seen as Will to Power (and nothing besides). But the world
seen as Will to Power and nothing else is the whole and nothing
besides where "nothing exists apart from the whole."

In other words, to see the world in this way is to see the world
as Will to Power *from within*. This perspective remains alien to us
as modern academic thinkers, as the inheritors of Socratic rationality in the ascetic tradition of science, where our ambition is to
stand apart from the world. As scientific decadents in this sense
we are unequal to a Dionysiac's life-affirmative vision. While,
conversely,

> Recognition, affirmation of reality is for the strong man as great
> a necessity as is for the weak man, under the inspiration of
> weakness, cowardice and *flight* in the face of reality —the
> "ideal" . . . They are not at liberty to know décadents *need* the
> lie—it is one of the conditions of their existence.[68]

Compared with the Renaissance, a classic time of prodigal power for Nietzsche,

> we moderns with our anxious care for ourselves and love of our neighbour, with our virtues of work, of unpretentiousness, of fair play, of scientificality—acquisitive, economical, machine-minded—appear as a *weak* age.[69]

Beyond contemporary decadence, Nietzsche seeks a superior possibility. In this he knowingly opposes the democratic ideal of equality. So he claims:

> "Equality," a certain actual rendering similar of which the theory of "equal rights" is only the expression, belongs essentially to decline . . . *Declining* life, the diminution of all organizing power, that is to say the power of separating, of opening up chasms, of ranking above and below, formulates itself in the sociology of today as the *ideal*.[70]

For Nietzsche, different from the modern decadent ideal: "the chasm between man and man, class and class, the multiplicity of types, the will to be oneself, to stand out—that which I call *pathos of distance*—characterizes every *strong* age."[71] The articulation of types in the pathos of distance does not articulate social standing, for social distinctions belong to the herd typology. Instead the *pathos of distance* delineates the expression of power. As an active, creative manifestation of the Will to Power, the expression of power is the expression of life in unstinting, unflinching expenditure and sacrifice, not out of or in terms of an ascetic ideal. Here we can see, as before, an important difference between the kind of extraordinary possibility Nietzsche evokes and the ordinary possibilities and goals of everyday life. Thus Nietzsche defines his conception of freedom: "The value of a thing sometimes lies not in what one attains with it, but in what pays for it—what it costs us."[72] When greatness is at stake, when genius, art, and life in the grand style are at stake, nothing can be withheld, because nothing is to be reserved beyond the endeavor itself. Rather is it its own justification:

> The great human being is a terminus; the great epoch, the Renaissance for example, is a terminus. The genius—in his works, in his deeds—is necessarily a prodigal: his great ness lies in the fact that *he expends himself*. . . . The instinct of self-preservation is as it were suspended; the overwhelming pressure

of the energies which emanate from him forbids him any such care and prudence.[73]

What Nietzsche means by progress must always be understood in the light of this conception of genius. And genius is always to be understood as distinct from the ordinary way of human being. So he explains, "I too speak of a 'return to nature,' although it is not really a going-back but a *going-up*."[74] Common humanity must be overcome just because its overall tendency is to decline:

> Mankind does *not* represent a development of the better or the stronger or the higher in the way that is believed today. "Progress" is merely a modern idea, that is to say a false idea. . . . onward development [*Fortentwicklung*] is *not* by any means, by any necessity the same thing as elevation, advance, strengthening.[75]

Nietzsche interprets talk of equal rights as a self-dissimulating species of injustice and as a covert attack against rarer possibilities:

> there exists no more poisonous poison: for it *seems* to be preached by justice itself, while it is the *end* of justice. . . . "Equality for equals, inequality for unequals"—*that* would be the true voice of justice: and, what follows from it, "Never make equal what is unequal."[76]

Nietzsche protests further in a later note in the same text: "Injustice *never* lies in unequal rights, it lies in the claim to '*equal*' rights."[77]

The tendency to ignore rank-order in the extension of possibilities for life-affirmation and life-negation leads to a misunderstanding of the essence of Dionysian affirmation and muddles the significance of Nietzsche's doctrine of the Eternal Return. Once again one may not apply Nietzsche's categories or possibilities to Everyman. For if one does, one easily concludes that Everyman could well "go under," or, more commonly and understandably, one concludes that anyone might be capable of becoming an *Übermensch*. This erring assessment is equivalent to conceiving Nietzsche's philosophic project as a kind of exotic preparation for today's various self-realization programs. Almost every commentator plays negative theologian here and says that such a popularizing program is not what Nietzsche means. I have found it essential to go further than this gnomic denial to say why Nietzsche may not be read in this way. The principle obstacle is the unpleasantness, the difficulty of Nietzsche's esotericism, not because the doctrine may not be communicated (a difficulty intrinsic to the doctrine), but precisely because the general reader cannot be expected to have ears for what is said (a difficulty instrinsic to the possibility of reception). But no more should we as general readers, *need* we as general readers, wish to have such ears. For where there are fine little ears, like those Lou Salomé fancied, there are also the ears of the braying yea-ah, ears that can be pulled like our own. The expressly asinine extension of Nietzsche's thought conception inevitably levels the quality of tragic genius and denies even the possibility of a grand style. Opposed to such democratic tastes, Nietzsche's depiction of the great individual (Goethe) emphasises the tragic or what is dark and cruel, and the scene of senseless and abysmal suffering:

> A spirit thus *emancipated* stands in the midst of the universe with a joyful and trusting fatalism, in the faith that only what is separate and individual may be rejected, that in the totality everything is redeemed and affirmed—*he no longer denies.* . . . But such a faith is the highest of all possible faiths: I have baptised it with the name *Dionysus*.[78]

However, in reference to Everyman, Nietzsche's Dionysian consecration echoes little more than a promise of death. The promise of death, the opposite of eternal life, does not appeal to everyday or ordinary spiritual prospects. Nietzsche suggests the average faith in the future (the desire to keep life for oneself, no matter what and no matter the cost), in rough terms when he writes in *The Antichrist* against the doctrine of immortality as a principle of equality:

> such a raising of every sort of egoism to infinity, to *impudence*, cannot be branded with sufficient contempt. And yet it it to *this* pitiable flattery of personal vanity that Christianity owes its *victory*—it is with this that it has persuaded over to its side everything ill-constituted [*Missrathene*], rebellious-minded, underpriveleged [*Schlechtweggekommene*], all the dross and refuse of mankind. 'Salvation of the soul'—in plain words: "The world revolves around *me*."[79]

Against this equality and its pretensions, Nietzsche is more than blunt, but nasty, effusively unkind: "'Immortality' granted to every Peter and Paul has been the greatest and most malicious outrage on *noble* mankind every committed."[80] Thereby, so Nietzsche claims, the possibility of the aristocratic is undermined. To be sure, the aristocratic is not undermined because the doctrine of the immortality of the soul opposes the oppression of lower orders. That is not the point. We may not forget the uncanny self-preservative power of the lower order. In fact, in real life, this lower order is not at all oppressed or prevented from glorifying itself: to the contrary. Hence Nietzsche's interest in this context is to reverse such a claim.[81] The standard projection of the ascetic ideal seeks its own preservation. It seeks power in the same dynamic we have analyzed before in this context: "the will to the end, the *nihilistic* will wants power."[82] For Nietzsche, the danger in the nihilistic expression of the Will to Power (from the side of the ascetic ideal) is the loss of the possibilities for greatness. Thus the matter of rights must be refined: "A right is a privilege [*Vorrecht*]. The privilege of each is determined by the nature of his being."[83]

THE TROPING OF THE ETERNAL RETURN:
AN APOSEMATIC APOSIOPESIS

I have suggested that the projection of the thought of the Eternal Return of the Same addresses the character of a philosophic elite. Thus it is expressed in its inexpress charac ter. As Eugen Fink has it, with a wink to Habermas who for his own part stumbled fatally both on this doctrine and on the related principle of the Will to Power: the "teaching of the Eternal Return is not 'teachable'; it cannot be presented in a communicative discourse. It is rather Nietzsche's 'esoteric wisdom.'"[84] This esoteric challenge calls forth the particular (dis-)position of an eternal affirmation of individual being in perpetual becoming, that is, in the becoming of what is as it is. An eternal affirmation is indiscriminate, comprising every event, great and small, beyond yes and no. But this breadth impels Nietzsche to circumscribe an eternal affirmation as the most dangerous perspective. It is in this gigantic perspective that, "What I do or do not do now is as important for *everything that is yet to come* as is the greatest event of the past."[85] Such an active perspective anticipates the thought of the Eternal Return.

Nietzsche's conception of the Eternal Return of the Same is commonly taken to literally involve the factual recurrence of the events of one's life, exactly ordered, the same as they have already and will come to pass. Life, like the demon's hourglass, in the image Nietzsche invents as a spectre for one's loneliest loneliness, is to be turned upon itself; the sand-flow of events begins, passes, and turns ever again. But should we accept this reading? How credulous would we need to be! Better to ask, how naive *should* we get? As Nietzsche teases in a section entitled *"Vita Femina,"* the antepenultimate section of book 4 of *The Gay Science* (recalling the elusiveness of the right perspective on the beautiful), even the admirable ancient Greek prayer: "Everything beautiful twice and even three times!" (*Zwei und drei Mal alles Schön*) importunes too much in longing for repetition. Nietzsche could hardly be said to be unsympathetic (and whoever would be?) to the noble desire to experience the beautiful not only once, but once more and still again. Nietzsche's reflection of rarity here is that the needed coincidence of the same lucky accidents that brought forth the original experience of beauty suggests the very consummate impossibility of its repetition:

> Not only do we have to stand precisely in the right spot in order to see this, but the unveiling must have been accomplished by our own soul because it needed some external expression and parable, as it it were a matter of having something to hold on to and retain control of itself. But it is so rare for all of this to coincide that I am inclined to believe that the highest peaks of everything good, whether it be a work, a deed, humanity, or nature, have so far remained concealed and veiled from the great majority and even from the best human beings. But what does unveil itself for us, *unveils itself for once only.*[86]

The consistency characterizing Nietzsche's thought recurs in this reflection and in the thought of the Eternal Return. From this recollection, then, the thought of the Eternal Return will not be expressed by the *claim*—material, scientific, or (meta)physical—that one *is* indeed to live again, in the same way, with the same stars and spiders, moonlight and evening conversations. This (impossible) possibility is not the point. And stylistically, it is inevitably beside the point. What is at issue in the thought of the Eternal Return is suggested by the dangerous perspective noted above. For Nietzsche, everything that is has a fundamental and valuable necessity in its being and its becoming. Thus, in the (usually autobiographically construed) declaration of *amor fati* that begins the fourth book of *The Gay Science*, Nietzsche can assert, "I want to learn more and more to see as beautiful what is necessary in things."[87] The meaning of *amor fati* admits necessity as an expression of chaos *and* the musical play of chance ($\tau \acute{\upsilon} \chi \eta$).

To understand necessity in this way, let us proceed to the thought of the Eternal Return by contrasting it with Nietzsche's observation of voyagers—in life:

> how much enjoyment, impatience, and desire, how much thirsty life and drunkenness of life comes to light every moment. And yet silence will soon descend on all these noisy, living, life-thirsty people. How his shadow stands even now behind everyone, as his dark fellow traveller![88]

Here Nietzsche describes the threat of death, the constant companion, striding like T. S. Eliot's wasteland shadow to meet the traveller embarked upon any and every venture. In this context, the contest of the wayfarer, the passenger or traveler, he observes,

> It is always like the last moment before the departure of an emigrants' ship: people have more to say to each other than ever, the hour is late, and the ocean and its desolate silence are waiting impatiently behind all of this noise—so covetous [*begierig*] and certain of its prey.[89]

He finds the emigrant's ship analogy compelling because it describes the flight of life-anticipation that is nothing like an appreciation of the fullness of any moment because, so far from the moment, this flight is intent only upon what is to come:

> And all and everyone of them suppose that the heretofore was little or nothing while the near future is everything; and that is the reason for all this haste, this clamor, this outshouting and overreaching each other. Everyone wants to be first in this future—yet death and deathly silence alone are certain and common to all in this future.[90]

The only sure anticipation is death. But to note this point is to imply its denial:

> How strange it is that this sole certainty and common element makes almost no impression on people and nothing is *further* from their minds that the feeling that they form a brotherhood of death.[91]

Nietzsche then avows his own life-direction, opposing the general flight of life-anticipation and life-recollection: "It makes me happy that men do not want at all to think the thought of death! I should like very much to do something that would make the thought of life yet a hundred times more *thought provoking.*"[92] Kaufmann's translation of this passage includes a cautionary gloss noting an important contrast with existentialism. This distinction is certainly useful if the meaning of existentialism is taken to correspond to that of a Kierkegaard or a Dostoievsky or even a Sartre. But Kaufmann is wrong if he does not hear in Nietzsche's words an understanding anticipation of Heidegger's prototypically existential understanding of Dasein's uttermost possibility as "Being-unto-death." For Heidegger, it is inauthenticity as the very flight from this same utmost possibility that characterizes human Dasein, "proximally and for the most part." Nietzsche describes this flight in its trajectory and ordinary expression and its antithesis for Nietzsche, just as for the spirit of Heideggerian authenticity and *Gelassenheit*, is life. Behind the thought of the Eternal Return stands a positive emphasis: each living moment is of incomparable weight and inestimable value. By the thought of the Eternal Return, conceived as *amor fati*, Nietzsche would like to make the "thought of life even a hundred times more appealing." If the thought of the Eternal Return retains its melancholy, this must be referred to the improbability of its general success as an invitation to think the thought of life as such:

> The question in each and every thing: "Do you desire this once more and innumerable times more?" would lie upon your actions as the greatest weight. Or how well disposed would you have to become to yourself and to life to *crave nothing more fervently* than this ultimate eternal confirmation and seal.[93]

Thus the thinking of the Eternal Return of the same is directed against the seekers of the future, the men of the future, the last men. The thought of the Eternal Return must oppose—indeed it threatens to crush—what is small in us, discontented, longing, and, in a Heideggerian word, inauthentic.[94] What is inauthentic is everything that is in us that repudiates what we are and thereby repudiates the cyclic limitation of life, that is, the Same. But what is inauthentic is, in Nietzsche's words, merely all-too-human. What is inauthentic in us is everything in us. Thus we do not affirm finitude but deny it. And the whole of our desire is in the service of this denial. In this what we want is always something more, something else, we want—in the very impotence and bile of our desire—something other than life's eternal round. Against those caught in the past in their hope for future compensation, Nietzsche writes:

> Let us leave such chatter and such bad taste to those who have nothing else to do but drag the past a few steps further through time and who never live in the present—which is to say the many, the great majority. *We, however, want to become those we are [Wir aber wollen Die werden, die wir sind].*[95]

As we know, the motto from Pindar is key to Nietzsche: he subtitles his autobiography, "How one becomes what one is" (*Wie man wird, was man ist*). The exhortation, to become what one is, is more than a favorite saying but suggests the extended sense of the concept of *amor fati* and the projective image of the thought of the Eternal Return. In this, one can never forget the aspect of cyclical (recurrent) life as cruel and triumphantly benign, as horrifying and beautiful, as tragic and joyous-affirmative. Reviewing this listing, the paired aspects appear top-heavy to us: the cruel, the horrifying, the tragic—these are the possibilities that weigh on us. The perspectival illusion responsible for this "weighting" is natural, and despair is the quietistic answer to tragic fate. And apart from tragic despair and revulsion at the blatantly horrible or cruel, the more common responses of boredom and diffidence answer the rest of life's indecisive recurrence (its eternal round). So Nietzsche explains the effects of thinking to the deepest registers of life as feeling:

> in an enormously generalized way all the grief of an invalid who thinks of health, of an old man who thinks of the dreams of his youth, of a lover deprived of his beloved, of the martyr whose ideal is perishing, of the hero on the evening after a battle that has decided nothing but brought him wounds and the loss of his friend.[96]

All this belongs to the many faces and the many numbing hours of the human tragedy in its ordinary depths much more than in its heroic heights. To think the projected life of these moments leads to pessimism in every profound sense. Yet Nietzsche goes on to speak of the project of bearing the pain of human existence

> But if one endured, if one could endure this immense sum of grief of all kinds while yet being the hero who, as the second day of battle breaks, welcomes the dawn and his fortune, being

a person whose horizon encompasses thousands of years past and future, being the heir of all nobility of all past spirit—an heir with a sense of obligation, the most aristocratic of old nobles and at the same time the first of a new nobility—the like of which no age has yet seen or dreamed of; if one could burden one's soul with all of this —the oldest, the newest, losses, hopes, conquests, and the victories of humanity.[97]

In this way, Nietzsche expresses the horizon of eternity. It is this horizon that promises in the moonlight of the thought of the Eternal Return, and it is this horizon that glitters in the Great Midday of its realization. To see oneself as the heir and crown of eternity, Nietzsche writes, is the goal:

if one could finally contain all this in one soul and crowd it into a single feeling—this would surely have to result in a happiness that humanity has not known so far: the happiness of a god full of power and love, full of tears and laughter, a happiness that like the sun in the evening, continually bestows its inexhaustible riches, pouring them into the sea, feeling richest, as the sun does, only when even the poorest fisherman is still rowing with golden oars! This godlike feeling would then be called—humanity![98]

The god in question, the godlike feeling, is what Nietzsche names Dionysian.

I introduced this consideration of the Eternal Return with Nietzsche's reflections on death. When I mentioned pessimism in the interim I did mean to underline the possibility of a Dionysian response to life when one understands its depths because the sight into the depths—as a sight into the abyss—leads to despair. The moment of tragic despair is more than the time for discontent—resignation or resentment—but can be the prelude to an affirmation. Beyond despair, in the dawn that can follow the darkest night, another ideal may break. It is not impossible to think that Schopenhauer himself, as Nietzsche's symbol for the tragic hero of the pessimism of strength, could have been thus inclined, perhaps despite himself, perhaps only in Nietzsche's fancy, to an ideal opposing Schopenhauer's weak pessimism, with its world-denying conclusions:

to the ideal of the most exuberant, most living and most world-affirming man, who has not only learned to get on and treat with all that was and is but who wants to have it again *as it was and is* to all eternity, insatiably calling out *da capo* not only to himself but to the whole piece and play but fundamentally to him who needs precisely this play—and who makes it necessary: because he needs himself again and again—and makes himself necessary—What? And would this not be: *circulus vitiosus deus?*[99]

The god again, the vicious circle, the labyrinth entered here, is, again, Dionysian.

But this possibility can be understood only in terms of what Nietzsche means by eternity. How is affirmation to be suggested by repeating to one's own life, in its great and petty moments, again and again, to eternity—musically, as it were: *da capo*? I have earlier had occasion to declare and more than once to indirectly echo my concord with Philonenko's interpretation of the Eternal Return, citing his commentary as poised against both Scho enhauer and the Pythagoras represented in mystical guise, where Philonenko claims that "the eternal return cannot be the eternal return of the events."[100] Well and good. Yet, what sense can remain as eventual *content* for the cry, "*da capo*"? Is it only the philosopher's arbitrary embellishment? Is the thought of the Eternal Return to be nothing more than an "existential/axiological" imperative? But what of the cosmological/ontological dimension so integral to perspectivism?

These are of course rhetorical questions. But given the terms of such questions, it should prove useful to consider the whole promise of the Eternal Return as it is introduced in *The Gay Science* as what is in literary terms and for usually abstruse philological purposes technically called an *aposematic aposiopesis*. That is, a kind of incomplete suggestion, deliberately left incomplete (that is, aposiopetic) *in order* to suggest a certain danger (and thus aposematic). To be sure, the expression hardly remains uncompleted in Nietzsche's notes for he writes: "My teaching says: the task is to live so that you must wish to relive your life—which you will in any event!"[101] This putative completion of the experimental thought of the Eternal Return (following Nietzsche's pregnant dash), appears to obviate the "what if" character of the published statements. The essence of the afterthought, which is in turn followed by an ellipsis and echoed by a paraphrase, claims the actual, physical return of the events of one's life as in fact inevitable: "*du wirst es jedenfalls . . . Es gilt die Ewigkeit!*" To resolve this difficulty, I propose to consider its patent (published) concinnity with the sense of the thought of Eternal Recurrence as a leitmotif (concurrent with the ideal of the *Übermensch*) in Nietzsche's *Zarathustra*.

Zarathustra recounts a story from the depths of a two-day silence to the bored sailors in "Of the Vision and the Riddle," telling the riddle that is "the vision of the most solitary man."[102] Zarathustra's story is the dispositional courage needed for the path upward, in spite of "the Spirit of Gravity" riding upon the climber's shoulder. This is not the same as the *Das grösste Schwergewicht* but is nonetheless related to it as pessimism is related to the life-affirmative ideal: *amor fati*.

The courageous disposition that opposes the Spirit of Gravity, surpassing discouragement, overcomes both pain and pity and proves the human unique among the animals. Furthermore, Nietzsche writes, "Courage also destroys giddiness at abysses: and where does man not stand at an abyss? Is seeing itself not—seeing abysses? . . . Pity . . . is the deepest abyss: as deeply as man looks into life, so deeply does he look also into suffering."[103] The connection that stands between the opposition to the Spirit of Gravity and the opposition to the pessimistic world-nihilating ideal remains: "Courage, however, is the best destroyer, courage that attacks: it destroys even death, for it says was *that* life? Well then! Once more!"[104] This is not a universalizable maxim, and it is of capital importance to emphasize that Nietzsche does not mean it to be and says so expressly: "He who has ears to hear, let him hear."

At this point, the second part of the vision begins, the second half of the riddle. At this juncture, the decision is made between the spirit of courage and that of gravity. More important, however, is the appended differentiation that is not an excuse for Zarathustra's refusal to carry the dwarf further: "'But I am the stronger of us two—you do not know my abysmal thought! That thought—you could not endure!'"[105]

This distinguishing mark, Zarathustra's abysmal thought, is the thought of the Eternal Return. The dwarf, the spirit of scientific gravity, is not equal to the thought of the Eternal Return but would be crushed by it, so says Zarathustra, for the dwarf cannot see the significance of the doorway, *Augenblick* (Moment); that is to say, he cannot see the conflict or the collision between past and future. The *Augenblick* tells this riddle:

Two paths come together here: no one has ever reached their end.

"This long lane behind us: it goes on for an eternity. And that long lane ahead of us—that is another eternity.

"They are in opposition to one another, these paths; they abut on one another: and it is here at this gateway that they come together."[106]

Zarathustra's challenge to the dwarf asks him if he believes that the opposition between an eternal future and an eternal past is likewise an eternal opposition: "'Everything straight lies . . . All truth is crooked, time itself is a circle.'"[107] Thus in its dwarfed expression, its crippled pithiness, we can see that the gravely scientific doctrine of the circularity of time, the claim concerning the crookedness of truth, cannot be the same as Zarathustra's teaching—and it is not Nietzsche's. Nor does it provide an answer to the riddle. Even as Zarathustra descibes what appears to be a circle, he speaks to something beyond the tangential moment between the linear and the circular. The opposition in question is the collision of past and future. The solution required is better seen in the light of the next vision.

The illumination of the riddle is transmitted through the puzzling image of nihilistic transfiguration. The inevitably repellent opening tableau shows a young shepherd fallen asleep, somehow bitten fast on his tongue by a snake and now choking (the wierdness of the shepherd's predicament routinely passes without remark). This is the riddle of the "most solitary man"; that is: a man who sleeps, if he does not also conduct his life, with his mouth open.

In his vision, Zarathustra first attempts but fails to pull the snake free and at last cries: "'Bite! Bite! Its head off! Bite!'"[108] The riddle Zarathustra asks the listening sailors is this: "*Who* is the shepherd into whose mouth the snake thus crawled? *Who* is the man into whose throat all that is heaviest, blackest will thus crawl?"[109]

We recall that the young shepherd, for his part, upon hearing Zarathustra "bit . . . spat . . . and sprang up."[110] Zarathustra himself as Everyman, standard interpretations tell us, is the shepherd speaking to himself, calling to himself to bite the serpent of black memory. Beyond standard readings, more basically, the open mouth is the vulnerability of the dreamer; as an open mouth is the ordinary sign of distracted fascination. The heaviest and blackest threat is the snake of nihilism, black despair is the threat of what is small in men; like the dwarfmole, it is crippled and crippling at the same time.[111] Elsewhere, Nietzsche calls it the Spirit of Revenge, and it takes many forms: pity, resentment, longing. And of all the things that it is, the Spirit ruled by Revenge, conscious or otherwise, contrary to the fascination of superficial distraction, cannot be wholly absorbed in the moment. For the Everyman of Being, such whole absorption—like the decision, the bite of the young shepherd severing the serpent of time's remorse—would have to kill it. Instead the Spirit of Revenge lives in the future, suspended as a ransom for the past. Even where the Spirit of Revenge surfaces in the present, as the contemporary modern world will have it, it lives for the moment only. That is it lives but once—just this once—with no weight, no nobility, and no eternal value.

This last superficiality is the opposite of profoundity and given another edge it comprehends the thought of the Eternal Return. In this way the proximity and generality of the menacing Spirit of Revenge has another saving side: when one has overcome, when one has come to face the dwarfing fear of the abyss, one can dance around it, one can laugh at it. Then one looks deep into life, then it becomes possible to embrace the end of, the edge of life —superficial now *out of* profoundity—absorbing the abyss that looks back into one and, so transfigured by the lightness of light, by the gold of existence, one is "a transformed being, surrounded with light, *laughing*! Never yet on earth had any man laughed as he laughed!"[112] This is "not human" but Dionysian laughter.

The question here in the thought of the Eternal Return is the weight of the future and the past and the value of this weight will be its worth for the present. The Eternal Return is the recurrence of the Same. This is the Same in the eternal round of existence that applies to every life and every moment. The problem Nietzsche calls the "Spirit of Gravity," the "Spirit of Revenge," is the problem of memory and small fearful (human) desires. It is a phenomenological commonplace that one is never done with the past, and it is a further observation that if one longs for the future that is as a relief from or a compensation for, a redemption, from the past. But this desire is eternally frustrated where the future brings only the Same, in events and in memories.[113] Because the weight of the past is primary, Nietzsche's Zarathustra could ask the earnest dwarf: "And are not all things bound fast together in such a way that this moment draws after it all future things? *Therefore*—draws itself too?"[114] The present, the ec-static temporal moment, is the single moment of decision between past and future. What is capital in this expression is the insight that the (interpretive) anticipation of what comes into being draws it forth into (or interpreting it in) being and in this draws itself forth, exposing the generative power of the past in the same evocation.

The completed aposematic sense of the earlier cited doctrine of the Eternal Return: "to live so that you must *wish* to relive your life—which you will *in any event!*"[115] is now revealed in its significance as a warning to us. For Nietzsche never claimed that one repeats one's future, merely that one's future is condemned to repeat one's past.[116] One can never have done with the past: what was, was. The implications of this insight, which Zarathustra called the *stone fact*, inspired a generation of existentialist thought—although Kaufmann is right to maintain that that emphasis was not intended by Nietzsche himself. And yet, if Nietzsche's doctrine of Eternal Recurrence is not an existentialist doctrine, it does not become for that an imperative. The thought of the Eternal Return of the Same expresses a phenomenological hermeneutic of life. Thus its existential (not existentialist) value can approximate the Heideggerian expression of Dasein's ec-static temporality: "Dasein authentically exists as futural in resolutely disclosing a possibility which it has chosen."[117] Heidegger suggests the significance of the Eternal Return as (expressed) affirmation in (untimely—unhistorically) Nietzschean terms: "Coming back resolutely to itself, it is, by repetition, open for the 'monumental' possibilities of human existence."[118]

When the Demon steals upon one in one's ownmost, lowest loneliness, intoning the consequences of one's historical, protentional because retentional nature, or in other words, the consequences of history or human memory, the damning moment inspires the Spirit of Revenge in one who has turned out badly. This is the one who like most of us curses the demon, as one curses the past and resents what even the future may not bring.[119] For what the demon says is:

This life as you now live it and have lived it, you will have to live once more and innumerable times more; and there will be nothing new in it, but every pain and every joy and every thought and sigh and everything unutterably small or great in your life will have to return to you, all in the same succes sion and sequence.[120]

The Spirit of Revenge desires that what was not have been as it was. Unable to overcome time in its passing, and effectively impotent against what has been, the Spirit of Revenge can only hope for something more, something better. We can recall what Nietzsche heard in the promise of traditional salvation in the West: "Principle of 'Christian love': it wants to be well *paid*."[121] Against this kind of Revenge, Nietzsche describes the benevolent life- and self-disposition that is needed in order to respond to the brutal inflexibility of what was, in the Eternal Return (wherein all that was remains eternally the Same—and all that is and is to come repeats the Same), "*to crave nothing more fervently* than this ultimate eternal confirmation and seal?[122] That is to affirm what has been, so much to say, "That was life? Well then! Once more!" In this affirmation one does not seek another chance, one does not seek eternal compensation but affirms exactly what has been, exactly as it was, transfigured only with one's blessing: but thus I willed it, I would have it thus! So Nietzsche reflects, repeating Zarathustra, "To redeem the past and to transform every 'It was' into an 'I wanted it thus!' that alone would I call redemption."[123]

What is required for this salvific transfiguration of the past (and hence of what the future can be in the present repetition) is (to be able) to respond positively and joyously to the demonic question of the Eternal Return in the proposition that places the heaviest weight on one's life and activity: "Do you desire this once more and innumerable times more?" An affirmative response of this kind is offered at the conclusion of the late work *Twilight of the Idols:* "The psychology of the orgy as as an overflowing feeling of life and energy within which even pain acts as a stimulus."[124] In this, Nietzsche was able to understand the power of ancient Greek tragedy, and, in that, the Dionysian secret of life affirmation that does not hold back but, as an expression of power, gives itself out.

The possibility of such a transfiguring, affirmative vision is derived from an interpretive vision of the world as multiple, mutually interactive, Wills to Power. Yet this is only to say that for the reflective (recollective, anticipating) human being, as interpretive and historical there is not, never has been, and never will be a possibility of life, lived once and only once. What remains, apart from an eternal life beyond this life, is rather to live *for once and always.*

To propose that life be lived in the light of the eternity that returns means that this light is not the light of a world beyond the world or a time beyond time. This view of eternity offers a redemption through deliberation or courage. Such is the deliberation of the creative action characterizing the resolute hero the day after an undecided battle, in the light of suffering, of physical and spiritual pain. The difference is the one glory makes in the long light of (worldly or pagan) eternity. All pains and all joys are shortened in the Great Midday. The one thing that casts no shadow is the moment, and it is the moment that is to be affirmed in life, in what is as it is because it is, in the thought of the Eternal Return.

In this way, Nietzsche could explain that Zarathustra's ideal (the *Übermensch*) was an individual who saw flux and death and birth and life—all as reality, as many realities, as the Real. The only human redemption possible must be attained through this aesthetic sight of the Real. This vision teaches how one is to go under, and how one overcomes oneself, because it teaches the ineluctability (it teaches the fierce creator's or artist's joy) of death in life. Yet, as we have seen, it would be wrong to claim that the redemptive world- and life-affirmative power of this thought is meant for everyone.

Most people, and that still means most of us, that means most of the time, remain intent upon nothing else but nihilism in the ascetic, decadent sense Nietzsche describes. That is, most people are intent upon what is not, as it is not, because it is not, because what they desire to have they do not have. Thus Zarathustra names people in this banal dimension of inauthentic life characteriztic of all us (proximally Heidegger says and for the most part) tarantulas. Where Heidegger's expression of inauthenticity is meant to carry no axiological censure, Nietzsche's expression is not so limited. Tarantulas are spiders, they *spin.* In fact, of course, tarantulas do nothing of the kind. And, just as self-evidently, Nietzsche was more interested in the philology of *spinnen*, of weaving inventions, fancies, lies than he was in the physiological differences between spiders hanging in Nietzsche's own moonlit northern doorways and those found in southern deserts and tropics. What matters in such fanciful spinning is that, overcome by the Spirit of Revenge, the spirit that is all too human undercuts all self-overcoming. Thus the thought that "man is something that should be overcome" cannot occur, cannot be thought, cannot be borne by the common person.

The disposition toward life and living required to bear and to incorporate the thought of the Eternal Return of the Same is not possible for the average individual. One lives and works precisely so that things will get better. According to individual fortune, one remembers the halcyon, forgets the dismal past—but one never blesses it. Even when one lives historically, one seeks the future and its redemption: a future focus remains a human reality. Thus expressed, the progress ideal, the ascetic ideal of rationality and attainable happiness appears as a statement of living discontent, and it is an index of displacement and alienation.[125]

The average man or woman lacks, as most of us do, the resources—economically, socially, politically, physiologically—necessary for a benign disposition toward the moment. The reason for this may well be a matter for reproach where the individual may not be said to have been in the moment to begin with. But it is a trick of the oldest and basest version of the ascetic ideal to claim that it is only one's inner spirit that matters, to assert that the material world is irrelevant for those whose (real) lives are elsewhere. In answer to this, it is enough to affirm that the achievement of the moment remains an ideal for ascetic as for enthusiast. For the sake of this achievement Nietzsche could admire Pindar's melancholy joyful admonition to the hero sent upon a glistening track: the champion needing and summoning every last effort to achieve the balance of the moment, and as reward for his temporary accomplishment, envied by the gods. Such admonitions are, such admiration is, justly reserved for heroes. Most of us, most of the time, like the dwarf perched upon Zarathustra's shoulders, or squatting beside him in the dust, see only backward and forward, but do not see the moment. For this reason we remain proximally and for the most part, inauthentic, to employ Heideggerian language in an approximative context once again.

But, separate from most of us, from what we are for the most part, there are those rare moments that do turn out well—and there are those exceptional individuals (heroes of whatever kind) who have "turned out well." The harmony of such an individual delights as a work of art, as such an individual interprets his or her own delight. Nietzsche explains:

> And in what does one really recognize that someone has *turned out well*! In that a human being who has turned out well does our senses good: that he is carved out of wood at once hard, delicate and sweet-smelling. He has a taste only for what is beneficial to him; his pleasure, his joy ceases where the measure of what is beneficial is overstepped. He divines cures for injuries, he employs ill chances to his own advantage; what does not kill him makes him stronger. Out of everything he sees, hears, experiences, he instinctively collects together *his* sum: he is a principle of selection . . . he does honor when he chooses, when he *admits*, when he *trusts* . . . He knows how to *forget*—he is strong enough for everything to *have* to turn out for the best for him.[126]

This is meant to be Nietzsche's own self-description, and it is the story of an artist. Such an individual could well find a moment of overpowering, self-overpowering benevolence. In this moment one looks upon everything with a divine beaming glance that sees through to the depths. This glance consecrates without envy, without pity, and without hope—yet it is not resignation, it is not wanting. Thus distinguished, we have the Dionysian blessing of an affirmative disposition beyond affirmation and negation, the yes beyond yes and no.

And if this affirmation does not conduce to a Nietzschean science (it does not) just as it cannot engender a Nietzschean modernity, it is not useless as—for it is the substance of—a Nietzschean philosophy, indeed a Nietzschean philosophy of science or life (or art). For as philosophers we do need to know what in us wants the truth, what wants scientific regularity, technological mastery. Knowing that about ourselves, as philosophers of this most dangerous "perhaps," we are not automatically healed or redeemed. But where more lies in the question mark we place after ourselves than in any other insight, we raise the saving virtue, the grace of suspicion. It is the modesty that is the precondition for heroic valor, the very ironic, tragic meaning of the Delphic "know thyself." This is science on the ground of art—in the service of life.

NOTES

1. The notion of discord here forms the core of Heidegger's first volume on Nietzsche. Note however that for a classical reading this same discord remains, even where one followed a Pythagorean or Stoic initiative, taking a time-line of Heraclitean "Great Years" and counting the cycles of recurrence in aeons.

2. See Howard Caygill's sensitive treatment of this theme in his essay in Keith Ansell-Pearson, ed. *Nietzche and Modern German Thought*, pp. 216–39.

3. Thus I disagree with the major thesis in B. Magnus, *Heidegger's Metahistory of Philosophy: Amor Fati*, if it is drawn out into a generally applicable (exoteric) context as Magnus seems to intend in his more recent work, yet I would also hold that Magnus's original thesis can be read esoterically. See Fink.

4. Cf. Nehamas, *Nietzsche: Life as Literature* and Magnus, Stewart, and Mileur, *Nietzsche's Case: Philosophy as/and Literature*.

5. This intestinal worldview is taken after the image provided by Rorty, who, for his part, follows Derrida and more recently Foucault, but always in the (to me, questionable) spirit of Dewey.

6. And this means leaving (for the moment) the question of (eso/exoteric) applicability in reserve.

7. Such as, for one example, Lawrence Hatab's game admission that such inattention "may be precarious because Nietzsche himself dabbled in 'scientific justifications' for eternal recurrence)." Hatab, *Nietzsche and Eternal Recurrence*, p. 93.

8. Strong, p. 168.

9. Nietzsche, "Dem Werden den Charakter des Seins *aufzuprägen*—das ist der höchste *Wille zur Macht*" (KSA XII, p. 312), WP 330.

10. Heidegger, "Who is Nietzsche's Zarathustra" in *Nietzsche: Volume 2*, Trans. Krell; "Wer ist Nietzsche's Zarathustra" (VA, p. 116). See *Nietzsche I*, pp. 27, 466ff., 656.

11. Heidegger, it is true, invites this very extension.

12. "Erkenntniß an sich im Werden unmöglich [ist] . . . Als Irrthum über sich selbst, als Wille zur Macht, als Wille zur Täuschung" (KSA XII, p. 313). As we have seen knowledge is only possible on the basis of error because "*Erkenntniß* und *Werden* schließt sich aus" (KSA XII, p. 382).

13. "*Erkenntniß* und *Werden* schließt sich aus" (KSA XII, p. 382).

14. "Wille zur Überwindung des Werdens, als 'Verewigen'" (KSA XII, p. 313), WP 330.

15. "In einer Welt, wo es kein Sein giebt" (KSA XIII, p. 271).

16. Nietzsche, "Feststellen, *was* ist, *wie* es ist, scheint etwas

unsäglich Höheres, Ernsteres als jedes 'so sollte es sein': . . . Es drückt sich darin ein Bedurfniß aus, welches verlangt, daß unsrem menschlichen Wohlbefinden die Einrichtung der Welt entspricht" (KSA XII, p. 300), WP 182.

17. "Vielleicht ist jenes 'so sollte es sein', unser Welt-Über-wälti-gungs-Wunsch" (ibid).

18. *"Als Ohnmacht des Willens zum Schaffen"* (KSA XII, p. 365), WP 317.

19. In its full context: "Das *Glück* kann nur im Seienden verbürgt sein: Wechsel und Glück schließen sich aus" (KSA XII, p. 365), WP 317. Cf. note 12, "Erkenntnis und Werden schließt sich aus." In effect then, and in opposition to Heidegger's parsing of Nietzsche's "recapitulation," showing Nietzsche positive recognition of the value of becoming and its endangerment by tendencies of nihilism: "Die logische Weltverneinung und Nihilisirung folgt daraus, daß wir Sein dem Nichtsein entgegensetzen müssen, und daß der Begriff 'Werden' geleugnet wird" (KSA XII, p. 369).

20. "Ich bin seit siebzehn Jahren nicht müde geworden, den *entgeistigenden* Einfluss unsres jetzigen Wissenschafts-Betriebs an's Licht zu stellen. Das harte Helotenthum, zu dem der ungeheure Umfang der Wissenschaften heute jeden Einzelnen verurtheilt, ist ein Hauptgrund dafür, dass voller, reicher, *tiefer* angelegte Naturen keine ihnen gemässe Erziehung und *Erzieher* mehr vorfinden" (G-D, KSA VI, p. 105), TI, p. 62.

21. "Daß er nicht die *Kehrseite der Dinge* als nothwendig versteht: da er die Übelstände bekämpft, wie als ob man ihrer entrathen könnte; daß er das Eine nicht mit dem Anderen hinnehmen will,—daß er den *typischen Charakter eines Dinges*, eines Zustandes, einer Zeit, einer Person verwischen und auslöschen möchte, indem er nur einen Theil ihrer Eigenschaften gutheißt und die andern *abschaffen* möchte. Die 'Wunschbarkeit' der Mittelmäßigen ist das, was von uns Anderen bekampt wird: das *Ideal* gefaßt als etwas, an dem nichts Schädliches, Böses, Gefährliches, Fragwürdiges, Vernichtendes übrig bleiben soll" (KSA XII, p. 519), WP 470.

22. "Einem Philosophen geht Nichts mehr wider den Geschmack als der Mensch, *sofern er wünscht*" (G-D, KSA VI, pp. 130–31), TI, p. 85.

23. The preface to *The Antichrist* affirms a kind of transcendent contempt: "Man muss der Menschheit überlegen sein durch Kraft, durch *Höhe* der Seele,—durch Verachtung"(A, KSA VI, p. 168). One needs to understand Nietzsche's contempt in this elevated dimension as little more than his esotericism.

24. "Was den Menschen rechtfertigt, ist seine Realitat,—sie wird ihn ewig rechtfertigen" (G-D, KSA VI, p. 131), TI, p. 85.

25. "Man muß einen **Maaßstab** haben: ich unterscheide den *großen*

Stil; ich unterscheide *Aktivität* und Reaktivität; ich unterscheide die *Überschüssigen Verschwenderischen* und die Leidend-Leiden-schaftlichen (—die 'Idealistin')" (KSA XII, p. 520).

26. Heidegger, *Nietzsche I*, p. 147.

27. Heidegger, *Nietzsche: Volume 1*, p. 126. ("Die Kunst des großen Stils ist die einfache Ruhe der bewahrenden Bewältigung der höchsten Fülle des Lebens. Zu ihr gehört die ursprungliche Entfesselung des Lebens, aber gebändigt; die reichste Gegensätzlichkeit, aber in der Einheit des Einfachen; die Fülle des Wachstums, aber in der Dauer des Langen und Wenigen" [*Nietzsche I*, p. 148–49].)

28. Heidegger, *Nietzsche: Volume 1*, p. 127. ("Wer soll festsetzen, was Vollkommene ist? Das können nur jene, die es selbst sind und es deshalb wissen. Hier öffnet sich der Abgrund jenes Kreisens, in dem das ganze menschliche Dasein sich bewegt. Was Gesundheit sei, kann nur der Gesunde sagen. Doch das Gesunde bemißt sich nach dem Wesensansatz von Gesundheit" [*Nietzsche I*, p. 150].)

29. Heidegger, *Nietzsche: Volume 1*, p. 132. ["Die Gegenmöglichkeit liegt darin, daß das Schaffende nicht der Mangel ist sondern die Fülle, nicht das Suchen, sondern der volle Besitz, nicht das Begehre, sondern das Schenken, nicht der Hunger sondern der überfluß" (*Nietzsche I*, p. 156.)]

30. Heidegger, *Nietzsche: Volume 1*, p. 132 ("Es ist nicht aktiv, sondern immer reaktiv im Unterschied zum rein aus sich selbst und seiner Fülle Quellenden" [*Nietzsche I*, p. 156].)

31. FW 277.

32. " . . . is particularly gloomy and unpleasant" (KSA XIII, p. 193).

33. This is not to be confused with the Heideggerian image of technicity as a "setting upon," as we shall see below.

34. "(eher will er) lieber will noch der Mensch das *Nichts* wollen, als *nicht* wollen" (GM III: 1 and 28, respectively, KSA V[p. 339], p. 412).

35. "Uber Schrecken und.Mitleid hinaus, die ewige Lust des Werdens selbst zu sein,—jene Lust, die auch noch die Lust am Vernichtung in sich schliesst" (G-D, KSA VI, p. 160), TI, p. 110.

36. Heidegger, *Introduction to Metaphysics,* trans. Ralph Manheim (New Haven: Yale University Press, 1959), p. 62. *Einführung in die Metaphysik* (Tübingen: Niemeyer, 1976), p. 47.

37. Heidegger, *Introduction to Metaphysics,* p. 61. "Der Kampf entwirft und entwickelt erst das Un-erhörte, bislang Un-gesagte und Un-gedachte. Dieser Kampf wird dann von den Schaffenden, den Dichtern, Denkern, Staatsmannern getragen. Sie werfen dem überwältigenden Walten den Block des Werkes entgegen und bannen in dieses die damit eröffnete Welt." *Einführung in die Metaphysik,* p. 47.

38. Heidegger, *Introduction to Metaphysics,* p. 63.

39. Ibid.

40. Heidegger, *Introduction to Metaphysics,* p. 158.

41. Heidegger, *Introduction to Metaphysics,* p. 203.

42. Heidegger, *The Question Concerning Technology,* p. 22. ("Die neuzeitliche physikalische Theorie der Natur ist die Wegbereiterin nicht erst der Technik, sondern des Wesens der modernen Technik. Denn das herausfordernde Versammeln in das bestellende Entbergen waltet bereits in der Physik" [VA, p. 25].)

43. Heidegger, *Question Concerning Technology,* p. 22. ("Er ist vom Walten des Ge-stells herausgefordert, das die Bestellbarkeit der Natur als Bestand verlangt. Darum kann die Physik bei allem Rückzug aus dem bis vor kurzem allein maßgebend, nur den Gegenständen zugewandten Vorstellen auf eines niemals verzichten: daß sich die Natur in irgendeiner rechnerisch feststellbaren Weise meldet und als ein System von Information bestellbar bleibt" [VA, p. 26].)

44. Heidegger, *Question Concerning Technology,* p. 24. ("Das Gestell ist das Versammelnde jenes Stellens, das den Menschen stellt, das Wirkliche in der Weise des Bestellens als Bestand zu entbergen" [VA, p. 27].)

45. Heidegger, *Question Concerning Technology,* pp. 26–27. ("Sobald das Unverborgene nicht einmal mehr als Gegenstand, sondern ausschließlich als Bestand den Menschen angeht und der Mensch innerhalb des Gegenstandlosen nur noch der Besteller des Bestandes ist,— geht der Mensch am äußersten Rand des Absturzes, dorthin nämlich, wo er selber nur noch als Bestand genommen werden soll" [VA, p. 30].)

46. See Richardson, "Heidegger's Critique of Science."

47. Heidegger, *Question Concerning Technology,* p. 22. ("Indessen preizt sich gerade der so bedrohte Mensch in die Gestalt des Herrn der Erde auf" [VA, p. 30].)

48. "Eine aus der Fülle, der Überfülle geborene Formel der *höchsten* Bejahung, ein Jasagen ohne Vorbehalt, zum Leiden selbst, zur Schuld selbst, zu allem Fragwürdigen und Fremden des Daseins selbst . . . Diese letzte, freudigste, überschwänglich-übermüthigste Ja zum Leben ist nicht nur die höchste Einsicht, es ist auch die *tiefste*, die von Wahrheit und Wissenschaft am strengsten bestätigte und aufrecht erhaltene. Es ist Nichts, was ist, anzurechnen, es ist Nichts entbehrlich" (EH, KSA VI, p. 311), E, p. 80.

49. "Das Zuviel von Kraft erst ist der Beweis der Kraft" (G-D, KSA VI, p. 57) TI, p. 21.

50. "Jenes wundervolle Phänomene . . . das den Namen des Dionysos trägt: es ist einzig erklärbar aus linem *Zuviel* von Kraft" (G-D, KSA VI, p. 158), TI, p. 21.

51. "Das Jasagen zum Leben selbst noch in seinen fremdesten und

härtesten Problemen; der Wille zum Leben, im *Opfer* seiner höchsten Typen der eignen Unerschöpflichkeit frohwerdend—*das* nannte ich dionysisch, *das* errieth ich als die Brücke zur Psychologie des *tragischen* Dichters. *Nicht* um von Schrecken und Mitleiden loszukommen . . . sondern um, über Schrecken und Mitleiden hinaus, die ewige Lust des Werdens *selbst zu sein*" (G-D, KSA VI, p. 160), TI, p. 110.

52. Ibid. "Niemand kann zuletzt mehr ausgeben als er hat—das gilt von Einzelnen, das gilt von Völkern" (G-D, KSA VI, p. 106), TI, p. 62.

53. Tracy B. Strong, "Nietzsche's Political Aesthetics," in *Nietzsche's New Seas: Explorations in Philosophy, Aesthetics, and Politics*, ed. Michael Allen Gillespie and Tracy B. Strong (Chicago: University of Chicago Press, 1988), p. 168.

54. "Das Wesentliche am Rausch ist das Gefühle der Kraftsteigerung und Fülle" (G-D, KSA VI, p. 116), TI, p. 72.

55. "Man bereichert in diesem Zustande Alles aus seiner eignen Fülle: was man sieht, was man will, man sieht es geschwellt, gedrängt, stark, uberladen mit Kraft. Der Mensch diese Zustandes verwandelt die Dinge, bis sie seine Macht wiederspiegeln—bis sie Reflexe seiner Vollkommenheit sind. Dies Verwandeln*müssen* in's Vollkommne ist—Kunst" (G-D, KSA VI, pp. 116–17), TI, p. 72.

56. "Vor der Tragödie feiert das Kriegerische in unserer Seele seine Saturnalien; wer Leid gewohnt ist, wer Leid aufsucht, der *heroische* Mensch preist mit der Tragödie sein Dasein,—ihm allein kredenzt der Tragiker den Trunk dieser süssesten Grausamkeit" (G-D, KSA VI, p. 128), TI, p. 82.

57. "Darf auch Viel gewagt, Viel herausgefordert, Viel auch *vergeudet* werden" (G-D, KSA VI, p. 137), TI, p. 90.

58. Nor—if it may [*with a surety!*] be anticipated that this tired point requires yet another sharpening—is such a life characterized by the life of German culture during the twelve years following 1933.

59. "Was Würze ehedem des Lebens war, für uns ware es *Gift*" (KSA VI, p. 137).

60. "Das Spiel des großen Weltenkindes Zeus und den ewigen Scherz einer Weltzertrümmerung und einer Weltenstehung" (KSA I, p. 758).

61. "Ein Werden und Vergehen, ein Bauen und Zerstören, ohne jede moralische Zurechnung, in ewig gleicher Unschuld" (PG 7, KSA I, p. 830), PT, p. 62.

62. "Die Welt ist das *Spiel* des Zeus" (PG 6, KSA I, p. 828), PT, p. 58.

63. KSA I, p. 830.

64. "Diese Spiel spielt der Aeon mit sich. Sich verwandelnd in Wasser und Erde thürmt er, wie ein Kind Sandhaufen am Meere, thürmt auf und zertrümmert; von Zeit zu Zeit fängt er das Spiel von Neuem an.

Ein Augenblick der Sättigung: dann ergreift ihn von Neuem das Bedürfniß , wie den Künstler zum Schaffen das Bedürfniß zwingt. Nicht Frevelmuth, sondern der immer neu erwachende Spieltrieb ruft andre Welten ins Leben" (PG 7, KSA I, pp. 830–31), PT, p. 62.

    65. "*Unschuld* des Werdens" (G-D, KSA VI, p. 97), TI, p. 53. See also p. 54.

    66. "Diese Art Mensch, die er concipirt, concipirt die Realität, wie *sie ist*: sie ist stark genug dazu—, sie ist ihr nicht entfremdet, entrückt, sie ist *sie selbst*, sie hat all deren Furchtbares und Fragwürdiges auch noch in sich, *damit erst kann der Mensch Grösse haben*" (EH, KSA VI, p. 370), E, p. 130.

    67. "Man ist nothwendig, man ist ein Stuck Verhängniss, man gehört zum Ganzen, man ist im Ganzen—es giebt Nichts, was unser Sein richten, messen, vergleichen, verurtheilen könnte, denn das hiesse das Ganze richten, messen, vergleichen, verurtheilen . . . *Aber es giebt Nichts ausser dem Ganzen!*" (G-D, KSA VI, p. 96), TI, p. 54.

    68. "Die Erkenntniss, das Jasagen zur Realität ist für den Starken eine ebensolche Nothwendigkeit als für den Schwachen, unter der Inspiration der Schwäche, die Feigheit und *Flucht* vor der Realität—das 'Ideal' . . . Es steht ihnen nicht frei, zu erkennen: die décadents haben die Lüge *nöthig*, sie ist eine ihrer Erhaltungs-Bedingungen" (EH, KSA VI, pp. 311–12), E, p. 80.

    69. "Wir Modernen, mit unsrer ängstlichen Selbst-Fürsorge und Nächstenliebe, mit unsren Tugenden der Arbeit, der Anspruchslosigkeit, der Rechtlichkeit, der Wissenschaftlichkeit—sammelnd, ökonomisch, machinal—als eine *schwache* Zeit" (G-D, KSA VI, p. 138), TI, p. 91.

    70. "Die 'Gleichheit,' eine gewisse thatsächliche Anähnlichung, die sich in der Theorie von 'gleichen Rechten' nur zum Ausdruck bringt, gehört wesentlich zum Niedergang: . . . Das *niedergehende* Leben, die Abnahme aller organisirenden, das heisst trennenden, Klüfte aufreissenden, unter- und überordnenden Kraft formulirt sich in der Soziologie von heute zum *Ideal*" (G-D, KSA VI, pp. 138–39), TI, p. 91.

    71. "Die Kluft zwischen Mensch und Mensch, Stand und Stand, die Vielheit der Typen, der Wille, selbst zu sein, sich abzuheben, Das, was ich *Pathos der Distanz* nenne, ist jeder *starken* Zeit zu eigen" (ibid.).

    72. "Der Werth eine Sache liegt mitunter nicht in dem, was man mit ihr erreicht, sondern in dem, was man fur sie bezahlt,—was sie uns kostet" (G-D, KSA VI, p. 139), TI, p. 92. Cf. AC 45.

    73. "Der grosse Mensch ist ein Ende; die grosse Zeit, die Renaissance zum Beispiel, ist ein Ende; Das Genie—in Werk, in That—ist nothwendig ein Verschwender: *dass es sich ausgiebt*, ist seine Grösse . . . Der Instinkt der Selbsterhaltung ist gleichsam ausgehängt; der übergewaltige Druck der ausströmenden Kräfte verbietet ihm solche Obhut und Vorsicht . . . Er strömt aus, er strömt über, er verbraucht sich, er schont sich

nicht—mit Fatalität, verhängnissvoll, unfreiwillig, wie das Ausbrechen eines Flusses über seine Ufer unfreiwillig ist" (G-D, KSA VI, p. 146), TI, p. 98.

74. "Auch ich rede von 'Ruckkehr zur Natur', obwohl es eigentlich nicht ein Zurückgehn, sondern ein *Hinaufkommen* ist" (G-D, KSA VI, p. 150), TI, p. 101.

75. "Die Menschheit stellt *nicht* eine Entwicklung zum Besseren oder Stärkeren oder Höheren dar, in der Weise, wie dies heute geglaubt wird. Der 'Fortschritt' ist blosse eine moderne Idee, das heisst eine falsche Idee . . . Fortentwicklung ist schlechterdings *nicht* mit irgend welcher Nothwendigkeit Erhöhung, Steigerung, Verstärkung" (AC 4, KSA VI, p. 171), A, p. 116.

76. "Es giebt gar kein giftigeres Gift: denn sie *scheint* von der Gerechtigkeit selbst gepredigt, während sie das *Ende* der Gerechtigkeit ist . . . 'Den Gleichen Gleiches, den Ungleichen Ungleiches—*das* wäre die wahre Rede der Gerechtigkeit: und was daraus folgt, Ungleiches niemals Gleich machen'" (G-D, KSA VI, p. 150), TI, p. 102.

77. "Das Unrecht liegt *niemals* in ungleichen Rechten, es liegt im Anspruch auf '*gleiche*' Rechte" (AC 57, KSA VI, p. 244), A, p. 179.

78. "Ein solcher *freigewordner* Geist steht mit einem freudigen und vertrauenden Fatalismus mitten im All, im *Glauben*, dass nur das Einzelne verwerflich ist, das im Ganzen sich Alles erlöst und bejaht—*er verneint nicht mehr* . . . Aber eine solcher Glaube ist der höchste aller möglichen Glauben: ich habe ihn auf den Namen des *Dionysos* getauft" (G-D, KSA VI, p. 152), TI, p. 103.

79. "Eine solche Steigerung jeder Art Selbstsucht ins Unendliche, ins *Unverschämte* kann man nicht mit genug Verachtung brandmarken. Und doch verdankt das Christenthum *dieser* erbarmungswürdigen Schmeichelei vor der Personal-Eitelkeit seinen *Sieg*—gerade alles Missrathene, Aufständische Gesinnte, Schlechtweggekommene, den ganzen Auswurf und Abhub der Menschheit hat es damit zu sich überredet. Das 'Heil der Seele'—auf deutsch: 'die Welt dreht sich um *mich*'" (AC 43, KSA VI, p. 217), A, p. 156.

80. "Die 'Unsterblichkeit' jedem Petrus und Paulus zugestanden war bisher das grösste, das bösartigste Attentat auf die vornehme Menschlichkeit" (AC 43, KSA VI, p. 218), A, p. 156.

81. "Die Realität ist dass hier der bewussteste *Auserwählten-Dünkel* die Bescheidenheit spielt" (A 44, KSA IV, p. 220). In this case Nietzsche is referring to religious ideal of the Judeo-Christian elect.

82. "Der Wille zum Ende, der *nihilistische* Wille will zur Macht" (AC 9, KSA IV, p. 176), A, p. 120.

83. "Ein Recht ist ein Vorrecht. In seiner Art Sein hat Jeder auch sein Vorrecht" (AC 57, KSA VI, p. 243), A, p. 178.

84. Fink, *Nietzsche*, p. 217.

85. "Was ich jetzt thue oder lasse, ist fur *alles Kommende* so wichtig, als das grösste Ereigniss der Vergangenheit" (FW 233, KSA III, p. 512).

86. "Nicht nur müssen wir gerade an der rechten Stelle stehen, diess zu sehen: es muss gerade unsere Seele selber den Schleier von ihren Höhen weggezogen haben und eines dusseren Ausdruckes und Gleichnisses bedürftig sein, wie um einen Halt zu haben und ihrer selber mächtig zu bleiben. Diess Alles aber kommt so selten gleichzeitig zusammen, dass ich glauben möchte, die höchsten Höhen alles Guten, sei es Werk, That, Mensch, Natur, seien bisher für die Meisten und selbst für die Besten etwas Verborgenes und Verhülltes gewesen:—was sich aber uns enthüllt, *das enthüllt sich uns Ein Mal!*" (FW 339, KSA III, pp. 568–69), GS, p. 271.

87. "Ich will immer mehr lernen, das Nothwendige an den Dingen als das Schöne sehen" (FW 276, KSA III, p. 521), GS, p. 223.

88. "Wieviel Geniessen, Ungeduld, Begehren, wieviel durstiges Leben und Trunkenheit des Lebens kommt da jeden Augenblick an den Tag! Und doch wird es für alle diese Lärmenden, Lebenden, Lebensdurstigen bald so stille sein! Wie steht hinter Jedem sein Schatten, sein dunkler Weggefährte!" (FW 278, KSA III, p. 523), GS, p. 225.

89. "Es ist immer wie im letzten Augenblicke vor der Abfahrt eines Auswandererschiffes: man hat einander mehr zu sagen als je, die Stunde drängt, der Ozean und sein ödes Schweigen wartet ungeduldig hinter alle dem Lärme—so begierig, so sicher seiner Beute" (ibid.).

90. "Und Alle, Alle meinen, das Bisher sei Nichts oder Wenig, die nahe Zukunft sei Alles: und daher diese Hast, diess Geschrei, dieses Sich-Uebertäuben und Sich-Uebervortheilen! Jeder will der Erste in dieser Zukunft sein,—und doch ist Tod und Todtenstille das einzig Sichere und das Allen Gemeinsame dieser Zukunft!" (ibid.).

91. "Wie seltsam, dass diese einzige Sicherheit und Gemeinsamkeit fast gar Nichts über die Menschen vermag und dass sie am *Weitesten* davon entfernt sind, sich als die Brüderschaft des Todes zu fühlen!" (ibid.).

92. "Es macht mich glücklich, zu sehen, dass die Menschen den Gedanken an den Tod durchaus nicht denken wollen! Ich möchte gern Etwas dazu thun, ihnen den Gedanken an das Leben noch hundertmal *denkenswerther* zu machen" (ibid.).

93. "Die Frage bei Allem und Jedem 'willst du diess noch einmal und noch unzählige Male?' würde als das grösste Schwergewicht auf deinem Handeln liegen! Oder wie müsstest du dir selber und dem Leben gut werden, um nach Nichts *mehr zu verlangen*, als nach dieser letzten ewigen Bestätigung und Besiegelung?" (FW 341, KSA III, p. 570), GS, p. 274.

94. Cf. *Sein und Zeit*, pp. 383–84.

95. "Welche nicht mehr zu thun haben, als die Vergangenheit um ein kleines Stuck weiter durch die Zeit zu schleppen und welche selber niemals Gegenwart sind—den Vielen also, den Allermeisten! Wir aber *wollen Die werden, die wir sind*" (FW 335, KSA III, p. 563), GS, p. 266.

96. "Jenen Gram des Kranken der an die Gesundheit, des Greises, der an den Jugendtraum denkt, des Liebenden, der der Geliebten beraubt wird, des Märtyrers, dem sein Ideal zu Grunde geht, des Helden am Abend der Schlacht, welche Nichts entschieden hat und doch ihm Wunden und den Verlust des Freundes brachte" (FW 337, KSA III, p. 565), GS, p. 268.

97. "Aber diese ungeheure Summe von Gram aller Art tragen, tragen können und nun doch noch der Held sein, der beim Anbruch eines zweiten Schlachttages die Morgenröthe und sein Gluck begrüsst, als der Mensch eines Horizontes von Jahrtausenden vor sich und hinter sich, als der Erbe aller Vornehmheit alles vergangenen Geistes und der verpflichtete Erbe, als der Adeligste aller alten Edlen und zugleich der Erstling eines neuen Adels, dessen Gleichen noch keine Zeit sah und träumte: diess Alles auf seine Seele nehmen, Aeltestes, Neuestes, Verluste, Hoffnungen, Eroberungen, Siege der Menschheit" (ibid.).

98. "Diess Alles endlich in Einer Seele haben und in Ein Gefühl zusammendrängen:—diess müsste doch ein Glück ergeben, das bisher der Mensch noch nicht kannte,—eines Gottes Glück voller Macht und Liebe, voller Thränen und voll Lachens, ein Glück, welches, wie die Sonne am Abend, fortwährend aus seinem unerschöpflichen Reichthume wegschenkt und in's Meer schuttet und, wie sie, sich erst dann am reichsten fühlt, wenn auch der ärmste Fischer noch mit goldenem Ruder rudert! Dieses göttliche Gefühl hiesse dann—Menschlichkeit!" (ibid.).

99. "Für das Ideal des übermüthigsten lebendigsten und weltbejahendsten Menschen, der sich nicht nur mit dem, was war und ist, abgefunden und vertragen gelernt hat, sondern es, so wie es war und ist, wieder haben will, in alle Ewigkeit hinaus, unersättlich da capo rufend, nicht nur zu sich, sondern zum ganzen Stücke und Schauspiele, und nicht nur zu einem Schauspiel, sondern im Grunde zu Dem, der gerade dies Schauspiel nothig hat—und nöthig macht—Wie? Und dies wäre nicht—circulus vitiosus deus?" (JGB 56, KSA V, p. 75), BGE, p. 64.

100. "L'éternel retour ne peut pas être l'éternel retour du contenus" (Philonenko, "Mélancholie et consolation chez Nietzsche," p. 95). As Philonenko observes, "César ne sera pas tué deux fois. Cette idée n'est bonne que pour les astronomes devenues astrologues."

101. "Meine Lehre sagt: so leben, da du wünschen mußt, wieder zu leben ist die Aufgabe—du wirst es jedenfalls!" (KSA IX, p. 505).

102. This is firstly the story of Zarathustra's courage: "Nicht nur Eine Sonne war mir untergegangen." Z, KSA IV, p. 198.

103. "Der Muth schlägt auch den Schwindel todt an Abgründen:

und wo stünde der Mensch nicht an Abgründen! Ist Sehen nicht selber—Abgründe sehen? . . . Mitleiden aber ist der tiefste Abgrund: so tief der Mensch in das Leben sieht, so tief sieht er auch in das Leiden" (Z, KSA IV, p. 199), TSZ, p. 177.

104. "Muth aber ist der beste Todtschläger, Muth, der angreift: der schlägt noch den Tod todt, denn er spricht: 'War das das Leben? Wohlan! Noch Ein Mal! . . . Wer Ohren hat, der höre'" (ibid.), TSZ, p. 178.

105. "Ich aber bin der Stärkere von uns Beiden—: du kennst meinen abgründlichen Gedanken nicht! Den—könntest du nicht tragen!" (Z, KSA IV, p. 199), Z, p. 178.

106. "Zwei Wege kommen hier zusammen: die gieng noch Niemand zu Ende. Diese lange Gasse zurück: die währt eine Ewigkeit. Und jene lange Gasse hinaus—das ist eine andre Ewigkeit. Sie widersprechen sich, diese Wege; sie stossen sich gerade vor den Kopf:—und hier, an diesem Thorwege, ist es, wo sie zusammen kommen" (KSA IV, p. 199–200), Z, p. 178.

107. Ibid. ("Alles Gerade lugt, murmelte [er] verächtlich . . . Alle Wahrheit ist Krumm, die Zeit selber ist ein Kreis" [p. 200].)

108. "Den Kopf ab! Beiss zu!" (Z, KSA IV, p. 201), TSZ, p. 180.

109. "Wer ist der Hirt, dem also die Schlange in den Schlund kroch? Wer ist der Mensch, dem also alles Schwerste, Schwärzeste in den Schlund kriechen wird?" (Z, KSA IV, p. 201), TSZ, p. 180.

110. "Biss mit gutem Bisse! Weit weg spie er den Kopf der Schlange—und sprang empor" (ibid., p. 202).

111. So Nietzsche confesses "*Der Ekel* am Menschen ist meine Gefahr"(EH, KSA VI, p. 371).

112. "Nicht mehr Hirt, nicht mehr Mensch,—ein Verwandelter, ein Umleuchteter, welche *lachte*! Niemals noch auf Erden lachte je ein Mensch wie er lachte!" (Z, KSA IV, p. 202), TSZ, p. 180.

113. Cf. UB 2:1.

114. "Sind nicht solchermaassen fest alle Dinge verknotet, dass dieser Augenblick alle kommenden Dinge nach sich zieht! Also—sich selber noch?" (Z, KSA IV, p. 200), TSZ, p. 179.

115. "So Leben da du *wünschen* mußt, wieder zu leben ist die Aufgabe—du wirst es *jedenfalls*!" (KSA IX, p. 505).

116. So the concinnous echo in the *Nachlaß* text explains the claim "du wirst es jedenfalls!": "Wem das Streben das höchste Gefühl giebt, der strebe: wem Ruhe das höchste Gefühl giebt, der ruhe; wem Einordnung Folgen Gehorsam das höchste Gefühl giebt, der gehorche. Nur möge *er bewußt daruber werden* was ihn das höchsten Gefühl giebt und kein Mittel scheuen. Es gilt *die Ewigkeit*!" (KSA IX, p. 505).

117. "Das Dasein existiert als Zukunftiges eigentlich im entschlossenen Erschlieben einer gewählten Möglichkeit" (*Sein und Zeit*, p. 396; translated at *Being and Time*, p. 448).

118. "Entschlossen auf sich zurückkommend ist wiederholend offen für die 'monumentalen' Möglichkeit menschlicher Existenz" (*Sein und Zeit*, p. 396).

119. Resenting so much that, as Nietzsche says, the revengeful need an eternity of compensation. Nietzsche explains this need in *On the Genealogy of Morals*: "ja man hat das ewige Leben nöthig, damit man sich auch ewig im 'Reiche Gottes' schadlos halten kann für jenes Erden-Leben 'im Glauben, in der Liebe, in der Hoffnung'" (GM I:15, KSA V, p. 283).

120. "Dieses Leben, wie du es jetzt lebst und gelebt hast, wirst du noch einmal und noch unzählige Male leben müssen; und es wird nichts Neues daran sein, sondern jeder Schmerz und jede Lust und jeder Gedanke und Seufzer und alles unsäglich Kleine und Grosse deines Lebens muss dir wiederkommen, und Alles in der selben Reihe und Folge" (FW 341, KSA III, p. 570), GS, p. 273.

121. "Princip der 'christlichen Liebe': sie will zuletzt gut *bezahlt* sein" (AC 45, KSA VI, p. 222), A, p. 160.

122. "Nach Nichts *mehr zu verlangen*, als nach dieser letzten ewigen Bestätigung und Besiegelung?—" (FW 341, KSA III, p. 570), GS, p. 273.

123. "Die Vergangnen zu erlösen und alles 'Es was' umzuschaffen in ein 'So wollte ich es!'—das hiesse mir erst Erlösung" (EH, KSA VI, p. 348), E, p. 110.

124. "Die Psychologie des Orgiasmus als eines überstromenden Lebens- und Kraftgefühls, innerhalb dessen der Schmerz noch als Stimulanz wirkt" (G-D, KSA VI, p. 160), TI, p. 110.

125. So Nietzsche notes, "Alle fragen: 'Warum ist das Leben nicht so, wie wir es wunschen und *wann* wird es so sein?'" (KSA XIII, p. 100).

126. "Und woran erkennt man im Grunde die *Wohlgerathenheit*! Dass ein wohgerathner Mensch unsern Sinnen wohlthut: dass er aus einem Holze geschnitzt ist, das hart, zart und wohlriechend zugleich ist. Ihm schmeckt nur, was ihm zuträglich ist; sein Gefallen, sein Lust hört auf, wo das Maass des Zuträglichen überschritten wird. Er erräth Heilmittel gegen Schädigungen, er nützt schlimme Zufälle zu seinem Vortheil aus; was ihn nicht umbringt, macht ihn stärker. Er sammelt instinktiv aus Allem, was er sieht, hört, erlebt, *seine* Summe: er ist ein auswählendes Princip . . . er ehrt, in dem er *wählt*, indem er *zulässt*, indem er *vertraut* . . . er ist stark genug, dass ihm Alles zum Besten gereichen *muss*" (EH, KSA VI, p. 267), E, pp. 40–41.

BIBLIOGRAPHY

PRIMARY SOURCES: NIETZSCHE

Citations from Nietzsche's works are given in English in the body of the present study. Footnotes to these citations first list the original German text, followed by the relevant volume of the *Kritische Studienausgabe* (KSA). When possible, the German reference will be supplemented by a further reference to an English language translation. Nietzsche's published works (in both original and English versions) are cited according to title abbreviations, as listed in the bibliography of Nietzsche's works below.

Without denying the enormous services to English Nietzsche scholarship rendered by Walter Kaufmann, wherever possible I have chosen to cite R. J. Hollingdale's translations. Hollingdale's achievement not only renders Nietzsche's German but even, *ceteris paribus* as it must be, Nietzsche's style into good English prose. I do however retain responsibility for all translations cited as in cases too numerous to distinguish I have found it necessary to make changes both substantive and minor.

In the case of many of Nietzsche's unpublished notes (referred to as the *Nachlaß* in the present study), translations given are for the most part my own. To facilitate critical use, wherever possible I also give references to the corresponding section of the *Will to Power*, using the English title abbreviation WP, followed by the section number, in footnotes after citing the original German text. Additionally, references to Nietzsche's unpublished essay "Über Wahrheit und Lüge im aussermoralischen Sinne" (in KSA I) are augmented with page references to Daniel Breazeale's translation "On Truth and Lie" in his *Philosophy and Truth*.

Nietzsche Werke: Jugendschriften. Dichtung/Aufsätze/Vorträge/Aufze-ichnungen und Philologische Arbeiten 1858–1868. Munich: Musarion Verlag, 1923. (JS)

Nietzsche Werke: Kritische Studienausgabe. Ed. Giorgio Colli and Mazzino Montinari. Berlin: Walter de Gruyter, 1967. (KSA)

KSA Works Cited in the Text

Also Sprach Zarathustra. (Z; KSA IV)
Der Antichrist. (A; KSA VI)
Ecce Homo. (EH; KSA VI)
Die fröhliche Wissenschaft. (FW; KSA III)
Die Geburt der Tragödie. (GT; KSA I)
Götzen-Dämmerung. (G-D; KSA VI)
Jenseits von Gut und Böse. (JGB; KSA V)
Menschliches, Allzumenschliches. (MA; KSA II)
Morgenröthe. (M; KSA III)
"Die Philosophie im tragischen Zeitalter der Griechen." (PG; KSA I)
"Über Wahrheit und Lüge im aussermoralischen Sinne." (UWL; KSA I)
Unzeitgemässe Betrachtungen. (UB; KSA I)
Zur Genealogie der Moral. (GM; KSA V)

ENGLISH TRANSLATIONS CONSULTED

Beyond Good and Evil. Trans. R. J. Hollingdale. Harmondsworth: Penguin, 1967. (BGE)

The Birth of Tragedy with *The Case of Wagner.* Trans. W. Kaufmann. New York: Vintage, 1966. (BT)

Daybreak. Trans. R. J. Hollingdale. Cambridge: Cambridge University Press, 1982. (D)

Ecce Homo. Trans. R. J. Hollingdale. Harmondsworth: Penguin, 1979. (E)

Friedrich Nietzsche on Rhetoric and Language. Trans. and ed. Sander L. Gilman, Carole Blair, and David J. Parent. Oxford: Oxford University Press, 1989. (TL)

The Gay Science. Trans. W. Kaufmann. New York: Vintage, 1974. (GS)

Human, All Too Human. Trans. R. J. Hollingdale. Cambridge: Cambridge University Press, 1986. (H)

Nietzsche: A Self-Portrait from His Letters. Trans. and ed. Peter Fuss and Henry Shapiro. Cambridge, Mass.: Harvard University Press, 1971.

On the Genealogy of Morals with *Ecce Homo.* Trans. W. Kaufmann and R. J. Hollingdale. New York: Vintage, 1968. (G)

Philosophy and Truth: Selections from Nietzsche's Notebooks of the Early 1870's. Trans. and ed. Daniel Breazeale. Atlantic Highlands, N.J.: Humanities Press, 1979. (TL)

Philosophy in the Tragic Age of the Greeks. Trans. Marianne Cowan. South Bend: Gateway, 1962. (PT)

Selected Letters of Friedrich Nietzsche. Trans. and ed. Christopher Middleton. Chicago: University of Chicago Press, 1969.

Thus Spoke Zarathustra. Trans. R. J. Hollingdale. Harmondsworth: Penguin, 1969. (TSZ)

Twilight of the Idols with *Antichrist*. Trans. R. J. Hollingdale. Harmondsworth: Penguin, 1968. (TI)

Untimely Meditations. Trans. R. J. Hollingdale. Cambridge: Cambridge University Press, 1983. (UM)

The Will to Power. Trans. W. Kaufmann and R. J. Hollingdale. New York: Vintage, 1968. (WP)

SECONDARY SOURCES: NIETZSCHE

Abel, Günter. "Interpretationsgedanke und Wiederkunftslehre." *Zur Aktualität Nietzsches*, Vol. 2, ed, Mihailo Djuric and Josef Simon. Wurzburg: Königshausen & Neumann, 1984.

————. *Nietzsche: Die Dynamik der Willen zur Macht und die ewige Wiederkehr*. Berlin, New York: Walter de Gruyter, 1984.

Alderman. Harold G. *Nietzsche's Gift*. Columbus: Ohio University Press, 1977.

Allison, David B. "A Diet of Worms: Aposiopetic Rhetoric in *Beyond Good and Evil*." *Nietzsche-Studien* 19: 43–58 (1990).

————, ed. *The New Nietzsche*. New York: Delta, 1977; Cambridge, Mass.: MIT Press, 1985.

————. "Nietzsche, Archilochus and the Question of the Noumenon." In *The Great Year of Zarathustra*, ed. David Goicoechea, 297–315. Lanham: University Press of America, 1983.

————. "Nietzsche Knows No Noumenon." *Boundary 2* 9/3 and 10/1: 295–310 (1981).

Alwast, Jendris. *Logik der dionysische Revolte. Nietzsches Entwurf einer aporetische dementierten "Kritischen Theorie."* Miesenheim am Glan: Anton Hain, 1975.

Amstutz, Jakob. "Nietzsche and Andre Gide." *The Great Year of Zarathustra*, 232–256. See Goicoechea, 1983.

Andler, Charles. *Nietzsche: sa vie et sa pensée*. Paris. Gallimard. 1920. (1958).

Ansell-Pearson, Keith, ed. *Nietzsche and Modern German Thought*. London: Routledge, 1991.

————, and Howard Caygill, eds. *The Fate of the New Nietzsche*. London: Avebury Press, 1993.

————. "The Significance of Michel Foucault's Reading of Nietzsche: Power, the Subject, and Political Theory." *Nietzsche-Studien* 20: 267–83 (1991).

Arras, J. D. "Art, Truth and Aesthetics in Nietzsche's Philosophy of Power." *Nietzsche-Studien* 9: 239–59 (1982).

Babich, Babette E. "From Nietzsche's Artist to Heidegger's World: The Post-Aesthetic Perspective." *Man and World* 22: 3–23 (1989).

———. "Heidegger on Nietzsche and Technology: Cadence, Concinnity, and Playing Brass." *Man and World.* 26: 239–60 (1993).

———. "Nietzsche and the Condition of Postmodern Thought: Post-Nietzschean Post-Modernism." In *Nietzsche as Postmodernist*, 249–66. See Koelb 1990.

———. "Nietzsche and the Philosophy of Scientific Power: Will to Power as Constructive Interpretation." *International Studies in Philosophy* 22/2: 79–92 (1990).

———. "On Nietzsche's Concinnity: An Analysis of Style." *Nietzsche-Studien* 19: 59–80 (1990).

———. "Self-Deconstruction: Nietzsche's Philosophy as Style." *Soundings* 73/1: 105–16 (1990).

Bachman, Matthias Lutz, ed. *Über Friedrich Nietzsche: Eine Einführung in seine Philosophie.* Frankfurt: Josef Knecht, 1985.

Baeumer, Max L. "Das moderne Phänomen des Dionysischen und seine 'Entdeckung' durch Nietzsche." *Nietzsche-Studien* 6: 123–53 (1977).

———. "Nietzsche and the Tradition of the Dionysian." In *Studies in Nietzsche and the Classical Tradition*, ed. James C. O'Flaherty, et al., 165–89. Chapel Hill: University of North Carolina Press, 1976.

Baier, Horst. "Die Gesellschaft—ein langer Schatten des toten Gottes: Friedrich Nietzsche und die Entstehung der Soziologies aus dem Geist der décadence." *Nietzsche-Studien* 10/11: 6–22 (1981–82).

Balmer, Hans Peter. *Freiheit statt Teleologie. Ein Grundgedanke von Nietzsche.* Freiburg: Alber, 1977.

———. "Nietzsche als Wissenschaftskritiker." *Zeitschrift für philosophische Forschung* 21: 130–43 (1966).

Barbera, Sandro, and Giuliano Campioni. "Wissenschaft und Philosophie der Macht bei Nietzsche und Renan." *Nietzsche-Studien* 13: 279–315. 1984.

Barrack, Charles M. "Nietzsche's Dionysus and Apollo: Gods in Transition." *Nietzsche-Studien* 3: 115–29 (1974).

Bataille, Georges. *On Nietzsche.* Trans. Bruce Boone. New York: Paragon House, 1992.

———. *Sur Nietzsche.* Paris: Gallimard, 1945.

Bauer, Martin. "Zur Genealogie von Nietzsches Kraftbegriff: Nietzsches Auseindandersetzung mit J. G. Vogt." *Nietzsche-Studien* 13: 211–27 (1984).

Behler, Ernst. *Derrida-Nietzsche: Nietzsche-Derrida.* Munich: Ferdinand Schöningh, 1988.

Bergmann, Peter. *Nietzsche, "The Last Antipolitical German.* Bloomington: Indiana University Press, 1987.

Bergoffen, Debra. Nietzsche's Madman: Perspectivism without Nihilism." In *Nietzsche as Postmodernist*, 57–71. See Koelb 1990.

———. "Posthumous Popularity: Reading, Privileging, Politicizing Nietzsche." *Soundings* 73/1: 31–60 (1990).

Bertman, M. A. "Hermeneutic in Nietzsche." *The Journal of Value Inquiry* 7: 254–60 (1973).

Bertram, Ernst. *Nietzsche: Versuch einer Mythologie*. Berlin: G. Bond, 1918.

Bindschedler, Maria. *Nietzsche und die Poetische Lüge*. Berlin: Walter de Gruyter, 1966.

Biser, Eugen. "Das Desiderat einer Nietzsche-Hermeneutik: Der Gang der Wirkungsgeschichte." *Nietzsche-Studien* 9: 1–38 (1980).

Bittner, Rüdiger. "Nietzsche's Begriff de Wahrheit." *Nietzsche-Studien* 16: 70–90 (1987).

Blanchot, Maurice. "Nietzsche et l'Écriture fragmentaire." *La Nouvelle Revue Française* 14: 967–83 (1966); 15: 19–34 (1967).

Bolz, Norbert W. "Nietzsches Genealogie der Wissenschaften" *Literaturmagazin* 12 (Nietzsche): 262–86 (1980).

———. "Von Nietzsche zu Freud: Sympathy for the Devil." In *Spiegel und Gleichnis. Festschrift für Jacob Taubes*, ed. Norbert W. Bolz and Wolfgang Hübner, 388–403. Würzburg: Königshausen & Neumann, 1983.

Böning, Thomas. "Das Buch eines Musikers ist eben nicht das Buch eines Augenmenschen. Metaphysik und Sprache beim frühen Nietzsche." *Nietzsche-Studien* 15: 72-106 (1986).

Boudot, Pierre. *L'ontologie de Nietzsche*. Paris: Presses Universitaires de France, 1971.

Brandes, Georg. "Aristokratischer Radikalismus." In *90 Jahre philosophischer Nietzsche-Rezeption*, ed. Alfred Guzzoni, 1–15. Meisenheim: Hain, 1979. Reprinted from: *Deutscher Rundschau* 63: 67-81 (1890).

Brassard, Werner. *Untersuchung zum Problem des Übermenschen bei Friedrich Nietzsche*. Diss. Freiburg im. Briesgau, 1962.

Bräutigam, Bernd. "Verwegene Kunststücke Nietzsches ironischer Perspektivismus als schriftstellerisches Verfahren." *Nietzsche-Studien* 6: 45 (1977).

Breazeale, Daniel. "Aground on the Ground of Values: Friedrich Nietzsche." *Analecta Husserliana*, ed. A-T. Tymieniecka and C. O. Schrag. 15: 335-349 (1983).

———. "The Meaning of the Earth." In *The Great Year of Zarathustra (1881–1981)*, 113–41. See Goicoechea 1983.

———. "The Hegel-Nietzsche Problem." *Nietzsche-Studien* 4: 146–64 (1975).

————. *Toward A Nihilist Epistemology: Hume and Nietzsche*. New Haven: Yale University Press. Diss. 1971.

Brock, Bazon. "Lebensmusik gegen Trommelwirbel und Trompetensignal." *Literaturmagazin* 12 (Nietzsche): 38–56 (1980).

Brock, Werner. *Nietzsches Idee der Kultur*. Bonn: F. Cohen, 1930.

Bröcker, Walter. *Das was kommt. Gesehen von Nietzsche und Hölderlin*. Stuttgart: Neske, 1963.

Brown, Richard S. G. "Does the 'True World' Not Remain a Fable?" In *The Great Year of Zarathustra,*, 97–112. See Goicoechea 1983.

Bruder, Kenneth. "Necessity and Becoming in Nietzsche." *Journal of the British Society for Phenomenology* 14/3: 225–39 (1983).

Bueb, Bernhard. *Nietzsches Kritik der praktischen Vernunft*. Stuttgart: Ernst Klett Verlag, 1970.

Clark, Maudemarie. *Nietzsche on Truth and Philosophy*. Cambridge: Cambridge University Press, 1990.

————. "Nietzsche's Perspectivist Rhetoric." *International Studies in Philosophy* 18/2: 35–43 (1986).

Colli, Giorgio. *Distanz und Pathos. Einleitung zu Nietzsches Werken*. Trans. Ragni Mani Gschwend and Reimer Klein. Frankfurt: Europaische Verlagsanstalt, 1982.

————. *Dopo Nietzsche*. Milan: Adelphi, 1974.

Combie, Jerry H. "Nietzsche as Cosmologist." *Interpretation* 1974–75: 38–47.

Congdon, Lee. "Nietzsche, Heidegger and History." *Journal of European Studies* 3/3: 211–17 (1973).

Dannhauser, Werner J. *Nietzsche's View of Socrates*. Ithaca: Cornell University Press, 1974.

Danto, Arthur. *Nietzsche as Philosopher*. New York. Columbia University Press, 1980.

Davey, Nicholas. "Hermeneutics and Nietzsche's Early Thought." In *Nietzsche and Modern German Thought*, 88–118. See Ansell-Pearson 1991.

————. "Nietzsche's Doctrine of Perspectivism." *Journal of the British Society for Phenomenology* 14/3: 240–57 (1983).

Deleuze, Gilles. *Nietzsche and Philosophy*. Trans. Hugh Tomlinson. New York: Columbia University Press, 1983.

————. *Nietzsche et la philosophie. Paris: Presses Universitaires de France, 1962*.

Derrida, Jacques. *Éperons: Les Styles de Nietzsche*. Venezia: Corboe Fieri, 1976. Paris: Flammarion, 1978.

————. Güter Wille zur Macht." In *Text und Interpretation,* ed. Phillipe Forget, 62–77. Munich: Find, 1984.

————. *L'Oreille de l'autre, otobiographies, transferts, traductions, textes et débats.* Ed. Claude Lévesque, Christie V. McDonald. Montreal: ULB Editions, 1982.

———. *Otobiographies: L'Enseignement de Nietzsche et la politique du nom propre.* Paris: Galilée, 1984.

———. *Positions.* Trans. Alan Bass. Chicago: University of Chicago Press, 1981.

———. "The Question of Style." In *The New Nietzsche*, 176–79. See Allison 1984.

———. *Spurs: Nietzsche's Styles.* Trans. Barbara Harlow. Chicago: University of Chicago Press, 1979.

———. "White Mythology." in Derrida, *Margins of Philosophy*. Trans. Alan Bass. Chicago: University of Chicago Press, 1982.

Dickopp, Karl-Heinz. "Nietzsches Kritik des Ich-denke." Diss., Bonn, 1965.

Dittberner, Hugo. "Tänzelnd und Böse. Über Nietzsche." *Literaturmagazin* 12 (Nietzsche): 24–37. Hamburg: Rowohlt/Reinbek, 1980

Den Ouden, Bernard. *Essays on Will, Creativity, and Time: Studies in the Philosophy of Friedrich Nietzsche.* Washington: University Press of America, 1982.

Eden, Robert. *Political Leadership and Nihilism: A Study of Weber and Nietzsche.* Tampa: University Press of Florida, 1983.

Eisler, Rudolph. *Nietzsches Erkenntnistheorie und Metaphysik: Darstellung und Kritik.* Leipzig: Hermann Haacke, 1902.

Emmerich, Erika. *Wahrheit und Wahrhaftigkeit in der Philosophie Nietzsches.* Halle, 1933.

Etterich, Walter. "Die Ethik Friedrich Nietzsches im Grundriß, im Verhältnis zu Kantischen Ethik betrachtet." Diss., Dortmund, 1914.

Fell, Albert P. "The Excess of Nietzsche's 'Amor Fati.'" In *The Great Year of Zarathustra*, 81–96. See Goicoechea 1983.

Figl, Johann. *Interpretation als philosophisches Prinzip: Friedrich Nietzsches universale Theorie der Auslegung im späten Nachlass.* Berlin: Walter de Gruyter, 1982.

Fink, Eugen. *Nietzsches Philosophie.* Stuttgart: Kohlhammer, 1960.

Fischer, Anton. *Die philosophischen Grundlagen der wissenschaftliche Erkenntnis.* Vienna: Springer, 1967.

Fischer, Kurt Rudolph. "The Existentialism of Nietzsche's 'Zarathustra.'" *Daedalus.* Summer 1974: 998–1016.

———. "Is Nietzsche a Philosopher?" *Bucknell Review* Winter 1970: 117–30.

Fisher-Dieskau, Dietrich. *Wagner znd Nietzsche.* New York: Seabury Press, 1976.

Flemming, Siegbert. *Nietzsches Metaphysik und ihr Verhältnis zur Erkenntnistheorie und Ethik.* Berlin: Simion, 1914.

Friedrich, Rainer. "*Euripidarisophanizein* and *Nietzschesokratizein:* Aristophanes, Nietzsche, and the Death of Tragedy." *Dionysius* (Halifax) 4: 5–36 (1980).

Funke, Monike. *Ideologiekritik und ihres Ideologie bei Nietzsche.* Stuttgart: Frommann-Holzboog, 1974.

Furness, Raymond. "Nietzsche and Empedocles." *Journal of the British Society for Phenomenology* 2: 91–94 (1971).

Gadamer, Hans-Georg. "Dichten und Deuten." *Kleine Schriften.* Tübingen, 1967.

———. "The Drama of Zarathustra." In *The Great Year of Zarathustra,* 339–69. See Goicoechea 1983.

———. "Das Drama Zarathustras." *Nietzsche-Studien* 15:1–15 (1986).

———. "Text und Interpretation." In *Text und Interpretation,* ed. Philippe Forget, 24–55. Munich: Fink, 1984.

———. "Und dennoch: Macht des Guten Willens." In *Text und Interpretation,* ed. Philippe Forget, 59–61. Munich: Fink, 1984.

Gebhard, Walter. "Erkennen und Entsetzen: Zur Tradition der Chaos-Annahmen im Denken Nietzsches." In *Friedrich Nietzsche. Strukturen der Negativität,* ed. Walter Gebhard, 13–47. Frankfurt am Main: Peter Lang, 1984.

———, ed. *Friedrich Nietzsche: Perspektivität und Tiefe.* Frankfurt am Main: Peter Lang, 1982.

———. *Friedrich Nietzsche. Strukturen der Negativität.* Frankfurt am Main: Peter Lang, 1984.

———. *Nietzsche's Totalismus: Philosphie der Natur Zwischen Verkärums und Veshïngust.* Berlin: Walter de Guyter, 1983.

———. "Zur Gleichnissprache Nietzsches: Problem der Bildlichkeit und Wissenschaftlichkeit." *Nietzsche-Studien* 9:61–90 (1980).

———. "Der Zusammenhang der Dinge": Weltgleichnis und Naturverklärung im Totalitätsbewustsein des 19. Jahrhunderts. Tubingen: 1983.

Gelven, Michael. "From Nietzsche to Heidegger: A Critical Review of Heidegger's Work on Nietzsche." *Philosophy Today.* Spring, 68–80 (1981).

———. "The Meaning of Evil." *Philosophy Today.* 200–21. 1983.

———. "Nietzsche and the Question of Being." *Nietzsche-Studien* 9: 209–23 (1980).

Gerhardt, Volker. "Macht und Metaphysik: Nietzsches Machtbegriff im Wandel der Interpretation." *Nietzsche-Studien* 10/11: 193–221 (1982/83).

———. *Pathos und Distanz: Studien zur Philosophie Friedrich Nietzsches.* Stuttgart: Reclam, 1988.

———. "Zum Begriff der Macht bei Friedrich Nietzsche." *Perspectiven der Philosophie* 7: 73–88 (1981).

Gillespie, Michael Allen, and Tracy B. Strong. *Nietzsche's New Seas: Explorations in Philosophy, Aesthetics, and Politics.* Chicago: University of Chicago Press, 1988.

Gilman, Sander. "Nietzsche's Reading on the Dionysian: From Nietzsche's Library." *Nietzsche-Studien* 6: 292–94 (1977).

Giesz, L. *Nietzsche, Existenzialismus und Wille zur Macht.* Stuttgart: Deutsche-Verlags Anstalt, 1950.

Goicoechea, David, ed. *The Great Year of Zarathustra (1881–1981).* Lanham: University Press of America, 1983.

Goth, Joachim. *Nietzsche und die Rhetorik.* Tübingen: Max Niemayer, 1970.

Granier, Jean. "Nietzsche's Conception of Chaos." In *The New Nietzsche*, 135–41. See Allison 1985.

————. "La Pensée nietzschienne du chaos." *Revue de la Metaphysique et de la Morale.* 1977: 129–66.

————. "Perspectivism and Interpretation." In *The New Nietzsche*, 190–200. See Allison 1985.

————. *Le Probléme de la vérité dans la philosophie de Nietzsche.* Paris: Seuil, 1966. 1969.

————. "Le Statut de la philosophie selon Nietzsche et Freud." *Nietzsche-Studien* 8: 210–24 (1979).

Grau, Gerd Günther. *Ideologie und Wille zur Macht. Zeitgemäße Betrachtungen über Nietzsche.* Berlin: Walter de Gruyter, 1984.

Graumann, C. F. *Grundlagen einer Phänomenologie und Psychologie der Perspektivität.* Berlin: Walter de Gruyter, 1960.

Grimm, Ruediger Hermann. "Circularity and Self-Reference in Nietzsche." *Metaphilosophy* 10: 289–305. (1979).

————. *Nietzsche's Theory of Knowledge.* Berlin: Walter de Gruyter, 1977.

Gustafsson, Lars. *Sprach und Lüge—Drei sprachphilosophische Extremisten: Friedrich Nietzsche, A. B. Johnson, Fritz Mauthner.* Trans. Susanne Seul. Hamburg: Fischer, 1982.

Haar, Michel. "La Critique nietzschiene de la Subjectivité." *Nietzsche-Studien* 12: 80–110 (1983).

————. "Nietzsche and Metaphysical Language." Trans. Cyril and Liliane Welch, 5–36. In *The New Nietzsche.* See Allison, 1985.

Habermas, Jürgen. "Nachwort." Foreword to *Friedrich Nietzsche: Erkenntnistheoretische Schriften* Frankfurt am Main: Suhrkamp, 1968.

Haller, Friedrich. *Mandragora: Bemühungen um Nietzsche.* Frankfurt am Main: Peter Lang, 1982.

Hamacher, Werner. "Das Versprechen der Auslegung: Überlegungen zum hermeneutischen Imperativ bei Kant und Nietzsche." In *Spiegel und Gleichnis: Festschrift für Jacob Taubes,* ed. Norbert W. Bolz and

Wolfgang Hübner, 252–387. Wurtzburg: Königshausen & Neumann, 1983.

Hayman, Ronald. *Nietzsche: A Critical Life.* Oxford: Oxford University Press, 1980.

Heftrich, Eckhard. *Nietzsches Philosophie: Identität von Welt und Nichts.* Frankfurt am Main: Vittorio Klostermann, 1962.

Heidegger, Martin. *Nietzsche.* Vols. 1, 2. Pfullingen: Neske, 1961.

———. *Nietzsche Volume 1: The Will to Power as Art.* Trans. David Farrell Krell. New York: Harper and Row, 1979.

———. *Nietzsche Volume 2: The Eternal Recurrence of the Same.* Trans. David Farrell Krell. New York: Harper and Row, 19784.

———. *Nietzsche Volume 3: The Will to Power Knowledge and as Metaphysics.* Trans. Joan Stambaugh, David Farrell Krell, and Frank A. Capuzzi. New York: Harper and Row, 1987.

———. *Nietzsche Volume 4: Nihilism.* Trans. Frank A. Capuzzi. New York: Harper and Row, 1982.

———. "Nietzsches Wort 'Gott ist tot.'" In *Holzwege,* 15 ed., 193–247. Frankfurt am Main: Klosterman, 1963. Also in *Heidegger, The Question Concerning Technology and Other Essays.* Trans. Wilkin Lovitt, 53–112. New York: Harper and Row, 1977.

Heidemann, Ingeborg. "Nietzsches Kritik der Metaphysik." *Kant-Studien* 53: 507–43 (1961–62).

Heller, Erich. *The Importance of Nietzsche.* Chicago: University of Chicago Press, 1988.

Heller, Peter. *Studies on Nietzsche.* Bonn: Bouvier, 1980.

Henke, Dieter. *Gott und Grammatik. Nietzsches Kritik der Religion.* Tubingen: Pfullingen 1981.

———. "Nietzsches Darwinismuskritik aus der Sicht gegenwärtiger Evolutionsforschung." *Nietzsche-Studien* 13: 189–210 (1984).

———. "Wer ist Nachfolger des Homo Sapiens—der Heerdenmensch, der Übermensch? Nietzsches Bedeutung für Anthropologie und Naturwissenschaften." In *Ist Gott tot? Über Friedrich Nietzsche,* ed. W. Böhme, 42–67. Freiburg: Alber, 1989.

Hennemann, Gerhard. "Friedrich Nietzsche als Naturphilosoph." *Philosophia Naturalis* 11: 490–501 (1969).

Hilpert, Konrad. "Die Überwindung der objektiven Gültigkeit." *Nietzsche-Studien* 9: 97–121 (1980).

Hollingdale, R. J. *Nietzsche: The Man and His Philosophy.* Baton Rouge: Louisiana State University Press, 1985.

Howey, John. *Heidegger and Jaspers on Nietzsche.* The Hague: Martinus Nijhoff, 1973.

Howey, Richard L. "Some Difficulties about Reading Nietzsche." *Nietzsche-Studien* 8: 378–88 (1979).

Hoy, David. "Philosophy as Rigorous Philology? Nietzsche and Poststructuralism." *New York Literary Forum* 8–9: 171–85 (1981).

Jacobs, Carol. *The Dissimulating Harmony.* Baltimore: Johns Hopkins University Press, 1978.

Jähnig, Dieter. "Die Befreiung der Kunsterkenntnis von der Metaphysik in Nietzsches Geburt der Tragödie." In *Welt-Geschichte: Kunst-Geschichte*, 122–60. Cologne: M. DuMont Schauberg, 1975.

———. "Nietzsches Kritik der historischen Wissenschaften." *Praxis* 6: 223–36. Zagreb, 1970.

Janz, Curt P. "Friedrich Nietzsche's Verhältniss zur Musik seiner Zeit." *Nietzsche-Studien* 7: 308–26 (1978).

Jaspers, Karl. *Nietzsche: An Introduction to the Study of His Philosophical Activity.* Trans. Charles F. Wallraff and Frederick J. Schmitz. Tucson: University of Arizona Press, 1965.

Joos, Walter. *Die desperate Erkenntnis. Ein Zugange zur Nihilismusproblematic bei Friedrich Nietzsche.* Frankfurt am Main: Peter Lang, 1983.

Juranville, Alain. *Physique de Nietzsche.* Paris: Denoël/Gonthier, 1973.

Kabermann, Friedrich. "Fragezeichen Für Solche, Die Antwort Haben: Zu Nietzsches 'historischer Philosophie' und der 'historischen Philosophie über Nietzsche.'" *Nietzsche-Studien* 6: 75–115 (1977).

Kaulbach, Friedrich. "Ästhetische und philosophische Erkenntnis beim frühen Nietzsche." In *Zur Aktualität Nietzsches*, ed. M. Djuric and J. Simon, 63–80. Wurzburg: Königshausen & Neumann, 1984.

———. *Nietzsches Idee seiner Experimentalphilosophie.* Cologne: 13 öhlau, 1980.

———. "Nietzsches Kritik an der Wissensmoral und die Quelle der philosophischen Erkenntnis: die Autarkie der perspektivischen Vernunft in der Philosophie." In *Nietzsche kontrovers*, vol. 4, ed. Rudolph Berlinger and Wiebke Schrader, 71–90. Wurzburg: Königshausen & Neumann, 1984.

Kaufmann, Walter. *Nietzsche: Philosopher, Psychologist, Antichrist.* Princeton: Princeton University Press, 1974.

Kaulhausen, Marie Hed. *Nietzsches Sprachstil: Gedeutet aus seinem Lebensgefühl und Weltverhältnis.* Munich: R. Oldenbourg, 1977.

Kein, Otto. *Das Apollinische und Dionysische bei Nietzsche und Schelling.* Berlin: Junker und Dünnhaupt, 1935.

Kirchhoff, Jochen. "Zum Problem der Erkenntnis bei Nietzsche." *Nietzsche-Studien* 6: 16–44 (1977).

Kittler, Friedrich A. "Vergessen." *Texthermeneutik* 195–221.

Kittman, Siegried. *Kant und Nietzsche. Darstellung und Vergleich ihrer Ethik und Moral.* Frankfurt am Main: Peter Lang, 1984.

Klages, Ludwig. *Die psychologischen Errungenschaften Nietzsches.* Leipzig, 1926; Bonn: Bouvier, 1979.

Koelb, Clayton. *Nietzsche as Postmodernist: Essays Pro and Contra.* Albany, N.Y.: SUNY Press, 1990.

Köhler, Joachim. "'Die fröhliche Wissenschaft': Versuch über die sprachliche Selbstkonstitution Nietzsches." Diss., Wurzburg, 1977.

————. *Zarathustras Geheimnis: Friedrich Nietzsche und seine ver-schlüsselte Botschaft.* Nördingen: Greno, 1989.

Kofman, Sarah. *Nietzsche et la métaphore.* Paris: Payot, 1972.

Köster, Peter. "Nietzsches Beschwörung des Chaos." *Tübingen Theological Quartalshrift* 153: 132–63 (1973).

Krell, David Farrell. "Heideger Nietzsche Hegel: An Essay in Descensional Reflection." *Nietzsche-Studien* 5: 255–62 (1976).

————. "Heidegger's Reading of Nietzsche: Confrontation and Encounter." *JBSP* 14: 271–82 (1983).

————. *Nietzsche and the Task of Thinking.* Ann Arbor, Mich.: University Microfilms, 1971.

————. "Nietzsche in Heidegger's Kehre." *Southern Journal of Philosophy.* 13/2: 197–204 (1975).

Künzli, Rudolf E. "Nietzsche's Zerography: Thus Spoke Zarathustra." *Boundary* 2 9/3 and 10/1: 99–117 (1981).

————. "Nietzsche und die Semiologie: Neue Ansätze in Der Französischen Nietzsche-Interpretation." *Nietzsche-Studien* 5: 263–88 (1976).

Kutzner, Heinrich. "(S)Tä(h)lernes Schreiben." *Literatur-Magazin* 12: 141–52 (1980).

Lacoue-Labarthe, Philippe. "Le Détour (Nietzsche et la rhétorique)." *Poetique* 5: 53–62 (1971).

Lea, F. A. *The Tragic Philosopher: Friedrich Nietzsche.* London: Methuen, 1977.

Lingis, Alphonso. "The Last Form of the Will to Power." *Philosophy Today* 22: 192–205 (1978).

————. "Theory and Idealization in Nietzsche." In *The Great Year of Zarathustra*, 257–78. See Goicoechea 1983.

Löb, Walther. "Naturwissenschaftliche Elemente in Nietzsches Gedanken." *Deutsche Rundschau* 137: 261-69 (1908).

Love, Frederick R. *Nietzsche's Saint Peter: Genesis and Cultivation of an Illusion.* Berlin: Walter de Gruyter, 1981.

Löwith, Karl. *From Hegel to Nietzsche.* Trans. D. E. Green. New York: Doubleday, 1967.

————. *Nietzsches Philosophie der ewigen Wiederkehr des Gleichen.* Berlin, 1935; Stuttgart: Kohlnahmer, 1956.

————. *Von Hegel zu Nietzsche: Der revolutionäre bruch im Denken des neunzehnten Jahrhunderts.* Stuttgart, 1964.

Löw, Reinhard. "Die Aktualität von Nietzsches Wissenschaftskritik." *Merkur* 38: 399–409 (1984).

————. *Nietzsche, Sophist und Erzieher: Philosophische Untersuchungen zur systematischen ort von Friedrich Nietzsches Denken.* Weinheim: Acta Humaniora, 1984.

Lukács, Georg. *Die Zerstörung der Vernunft.* 2d ed. Berlin: Luchterhand, 1962.

Lyotard, Jean-François. *The Post-Modern Condition: A Report on Knowledge.* Trans. Geoff Bennington and Brian Massumi. Minneapolis: University of Minnesota Press, 1984.

Mauer, Reinhart. "Nietzsche und das Experimentelle." In *Zur Aktualität Nietzsches,* vol. 1, ed. M. Djuric and J. Simon, 7–28. Wurzburg: Königshausen & Neumann, 1984.

———. "Nietzsche und die kritische Theorie." *Nietzsche-Studien* 10/11: 34–56 (1981–82).

———. "The Origins of Modern Technology in Millenarianism." In *Philosophy and Technology,* ed. Paul T. Durbin & Friedrich Rapp, 253–65. Dordrecht: D. Reidel, 1983.

———. "Wird Nietzsche wieder aktuell?" *Redliches Denken Festschrift für G. G. Grau,* ed. F. W. Kerff, 78–91. Stuttgart: Frommann Holzboog, 1981.

MacIntyre, Alasdair. *After Virtue.* Notre Dame: Notre Dame University Press, 1981.

———. "Philosophy and Sanity: Nietzsche's Titanism." *Encounter* 32: 79–82 (1969).

Magnus, Bernd. *Heidegger's Metahistory and Philosophy: Amor Fati, Being and Truth.* The Hague: Martinus Nijhoff, 1970.

———. "Nietzsche's Eternalistic Counter-Myth." *Review of Metaphysics* 26: 604–16 (June 1973).

———. *Nietzsche's Existential Imperative.* Bloomington: Indiana University Press, 1978.

———. "Overman: An Attitude or an Ideal?" In *The Great Year of Zarathustra,* 142–61. See Goicoechea 1983.

———, Stanley Stewart, and Jean-Pierre Milew. *Nietzsche's Case: Philosophy as/and Literature.* New York and London: Routledge, 1993.

de Man, Paul. *Allegories of Reading: Figural Language in Rousseau, Nietzsche, Rilke, and Proust.* New Haven: Yale University Press, 1979.

May, Keith M. *Nietzsche and the Spirit of Tragedy.* New York: St. Martin's Press, 1990.

McGinn, Robert E. "Culture as Prophylactic: Nietzsche's Birth of Tragedy as Culture Criticism." *Nietzsche-Studien* 4: 75–138 (1975). See especially 113–35.

Megill, Allan. *Prophets of Extremity: Nietzsche, Heidegger, Foucault, Derrida.* Berkeley and Los Angeles: University of California Press, 1985.

Minson, Jeffrrey. *Genealogies of Morals: Nietzsche, Foucault, Donzelot and the Eccentricity of Ethics.* New York: St. Martin's Press, 1985.

Mittasch, Alwin. *Friedrich Nietzsche als Naturphilosoph.* Stuttgart: Alfred Kröner, 1952.

———. *Friedrich Nietzsches Stellung zur Chemie.* Berlin, 1944.

———. *Unvergänglichkeit? Naturförschergedanken über Unsterblichkeit.* Heidelberg, 1947.

Moles, Alistair. *Nietzsche's Philosophy of Nature and Cosmology.* New York: Peter Lang, 1990.

Mongré, Paul. *Sant' Ilario. Gedanken aus der Landschaft Zarathustras.* Leipzig: C. G. Naumann, 1897.

Montinari, Mazzino. *Nietzsche lesen: Einführung in die Lektüre von Nietzsches Werken und Nachlass.* Berlin: Walter de Gruyter, 1982.

———. "Nietzsches Philosophie als Leidenschaft der Erkenntnis." *Studi Germani* 7: 337–52 (1969).

Morgan, George. *What Nietzsche Means.* New York: Harper & Row, 1965.

Müller, Severin. "Perspektivität der Erkenntnis und Perspektivität des Willens: Zur Pluralität des Wirklichen bei Leibniz und Nietzsche." In *Friedrich Nietzsche: Perspektivität und Tiefe,* ed. W. Gebhard, 15–70. Frankfurt am Main: Peter Lang, 1982.

Müller-Lauter, Wolfgang. "Der Geist der Rache un die ewige Wiederkehr: Zu Heideggers später Nietzsche Interpretation." In *Redliches Denken,* ed. F. W. Korff, 92–113. Stuttgart: Frommann-Holzborg, 1981.

———. "Der Organismus als innere Kampf: Der Einfluss von Wilhelm Roux auf Friedrich Nietzsche." *Nietzsche-Studien* 7: 189–223 (1978).

———. *Nietzsche: Seine Philosophie der Gegensätze und die Gegensätze seiner Philosophie.* Berlin: Walter de Gruyter, 1971.

———. "Nietzsches Lehre vom Willen zur Macht." *Nietzsche-Studien* 3: 1–60 (1974).

Nassen, Ulrich, ed. *Texthermeneutik: Aktualität, Geschichte, Kritik.* Paderborn: Schöningh, 1979.

Nehamas, Alexander. *Nietzsche: Life as Literature.* Cambridge: Harvard University Press, 1985.

Neumann, Harry. "Socrates and History: A Nietzschean Interpretation of History." *Nietzsche-Studien* 6: 64–74 (1977).

Oehler, Richard. *Freidrich Nietzsche und die Vorsokratiker.* Leipzig, 1904.

Oehler, Max, ed. *Nietzsches Bibliothek.* Vierzehnte Jahresgabe der Gesellschaft der Freunde des Nietzsche-Archivs. Weimar: R. Wagner Sohn, 1942. (Contains a list of the books in Nietzsche's personal library gathered from collections of such libraries edited by Arthur Berthold in 1900 and 1913. Includes an appendix of the books Nietzsche borrowed from from Pforta and Basel [though, in the absence of records, not those borrowed during his student years at Bonn or Leipzig]. The appendix itself was first discovered where one would least expect to find it [Oehler's words], in a French dissertation on Nietzsche and Stirner, which listed the books Nietzsche borrowed during his school years from the library at Pforta (1863-69) and subsequently from the University Library in Basel [1869–79].)

Ottman, Henning. *Philosophie und Politik bei Nietzsche.* Berlin: Walter de Gruyter, 1987.

Pasley, Malcolm, ed. *Nietzsche—Imagery and Thought: A Collection of Essays.* Berkeley and Los Angeles: University of California Press, 1978.

Perkins, Richard. "Analogistic Strategies in Zarathustra." In *The Great Year of Zarathustra,* 316–38. See Goicoechea 1983.

———. "MA 628 Preliminary Analysis of the Aphorism and its Precursors." *Nietzsche-Studien* 6: 203–39 (1977).

Pfeffer, Rose. *Nietzsche: Disciple of Dionysus.* Lewisburg: Bucknell University Press, 1972.

Pfotenhauer, Helmut. *Die Kunst als Physiologie: Nietzsches ästhetische Theorie von literarische Produktion.* Stuttgart: Metler, 1985.

Philonenko, Alexis. "Mélancholie et consolation chez Nietzsche." *Revue de la Métaphysique et de la Morale* 76/1: 77–98 (1977).

Pippin, Robert B. "Irony and Affirmation in Nietzsche's *Thus Spoke Zarathustra.*" In *Nietzsche's New Seas,* 45–71. See Gillespie and Strong, 1988.

———. "Nietzsche and the Origin of the Idea of Modernity." *Inquiry* 26: 154–80 (1988).

Pizer, John. "The Use and Abuse of 'Ursprung': On Foucault's Reading of Nietzsche." *Nietzsche-Studien* 19: 462–78 (1990).

Podach, E. F. *Nietzsches Zusammenbruch.* Heidelberg, 1930.

Pöltner, Gunther. *Zu Einer Phänomenologie Des Fragens: Ein Fragend-Fraglicher Versuche.* Freiburg: Verlag Karl Alber, 1972.

Pütz, Peter. "Nietzsche: Art and Intellectual Inquiry." In *Nietzsche: Imagery and Thought.* See Pasley 1978.

———. "Nietzsche im Licht de Kritische Theorie." *Nietzsche-Studien* 3: 175–91 (1974).

———. "The Problem of Force in Nietzsche and His Critics." In *Nietzsche: Literature and Values,* ed. Volter Dürr, Reinhold Grimm, and Katy Harms, 17–28. Madison: University of Wisconsin Press, 1988.

Quinot, Armand. *Friedrich Nietzsche: pages mystiques.* Paris: R. Laffont, 1945.

Reboul, Olivier. *Nietzsches critique de Kant.* Vendome: Presses Universitaires de France, 1974.

Rey, Jean-Michel. *L'Enjeu des signes: Lecture de Nietzsche.* Paris: Seuil, 1971.

Riehl, Alois. *Friedrich Nietzsche, der Künstler und Denker.* Stuttgart: Frommann, 1897.

Ries, Wiebrecht. "'Arbeit der Tiefe' und 'Unterirdischer Ernst': Ammerkungen zum Begriff des 'Subversiven' Im Denken Nietzsches." In *Friedrich Nietzsche: Perspektivität und Tiefe,* 149–62. See Gebhard 1982.

Roos, Richard, "Regles pour une lecture philologique de Nietzsche,"

283–324. In *Passion*. Proceedings of a colloquium; "Nietzsche aujourd'hui." Paris: 1973.

Rupp, Gerhard. *Rhetorische Strukturen und kommunikative Determinanz: Studien zur Textkonstitution des philosophischen Diskurses im Werk Friedrich Nietzsches*. Bern: Peter Lang, 1976.

———. "Text, Nicht Interpretation: Schleiermacher, Nietzsche. Das Ende des Auslegens und der Beginn des Schreibens." In *Friedrich Nietzsche: Perspektivität und Tiefe*, 225–64. See Gebhard 1982.

———. "Der 'ungeheure Consensus der Menschen über die Dinge' oder Das gesellschaftlich wirksame Rhetorische: Zum Nietzsche des Philosophenbuchs." *Literaturmagazin* 12: 179–203 (1980).

Salaquarda, Jörg, ed. *Nietzsche*. Darmstadt: Wissenschaftliche Buchgesellschaft, 1982.

Sallis, John. "Apollo's Mimesis." *Journal of the British Society for Phenomenology* 15/1: 16–21 (1984).

———. "Nietzsche and the Problem of Knowledge." *Tulane Studies in Philosophy* 18: 105–30 (1969).

———."Nietzsche's Underworld of Truth." *Philosophy Today* 16 (1972).

Savouret, Marie. *Nietzsche et Du Bos*. Paris: M. J. Minard, 1960.

Schacht, Richard. *Nietzsche*. London: Routledge & Kegan Paul, 1983.

———. "Nietzsche and Nihilism." *Journal of the History of Philosophy* 9/1: 65–144 (1973).

Scharff, Robert. "Nietzsche and the "Use" of History." *Man and World* 7: 67–77 (1974).

Schlüpmann, Heide. *Friedrich Nietzsches aesthetische Opposition*. Stuttgart: Metzler, 1977.

Schlechta, Karl. *Der Fall Nietzsche*. Munich: Carl Hanser, 1958.

———. *Nietzsche Chronik: Daten zu Leben und Werk*. Munich: Carl Hanser, 1975.

———. "Nietzsche über den Glauben an die Grammatik." *Nietzsche-Studien* 1: 353 (1972).

———, and Anni Anders. *Friedrich Nietzsche. Von den verborgenen Anfängen seines Philosophierens*. Stuttgart: Friedrich Frommann, 1962.

Schmid, Holger. *Nietzsches Gedanke der tragischen Erkenntnis*. Wurzburg: Königshausen & Neumann, 1984.

Schmidt, Alfred. "Zur Frage der Dialektik in Nietzsches Erkenntnistheorie." *Zeugnisse: Theodore W. Adorno zum sechszigsten Geburtstag*. ed. M. Horkheimer, 15–132. Frankfurt am Main: 1969.

Schmidt, Rüdiger. *"Ein Text ohne Ende für Denkenden."* *Studien zu Nietzsche*. Frankfurt am Main: Athenäum, 1989.

Schreiber, Jens. "Die Ordnung des Genießens. Nietzsche mit Lacan." *Literaturmagazin* 12: 204–35 (1980).

Schrift, Alan. "Between Perspectivism and Philology: Genealogy and Hermeneutics. *Nietzsche-Studien* 16: 91–111 (1987).

Schute, Ofelia. *Beyond Nihilism: Nietzsche Without Masks.* Chicago: University of Chicago Press, 1984.

Shapiro, Gary. "Festival, Parody, and Carnival in Zarathustra IV." In *The Great Year of Zarathustra*, 45–62. See Goicoechea 1983.

Siegfried, Hans. "Law, Regularity, and Sameness: A Nietzschean Account." *Man and World* 6: 372–89 (1973).

Silk, Michael S., and Joseph P. Stern. *Nietzsche on Tragedy.* Cambridge: Cambridge University Press, 1981.

Simmel, Georg. *The Problems of the Philosophy of History: An Epistemological Essay.* Trans. Guy Oakes. New York: Free Press, 1977.

———. *Schopenhauer und Nietzsche: Ein Vortragszyklus.* Leipzig: Duncker and Humblot, 1907.

Simon, Josef. "Grammatik und Wahrheit: Über das Verhältnis zur metaphysischen Tradition." *Nietzsche-Studien* 1: 1-26 (1972).

———. "Nietzsche und das Problem des Europäischen Nihilismus." In *Nietzsche kontrovers*, vol. 2, 9-38, ed. Rudolph Berlinger and Wiebke Schrader. Wurzburg: Königshausen & Neumann, 1984.

———. "Das Problem des Bewußtseins bei Nietzsche und der traditionelle Bewusbseinsbegriffe" In *Zur Aktualität Nietzsches*, vol. 2, ed. M. Djuric und J. Simon. Wurzburg: Königshausen & Neumann, 1984.

———. "Sprache und Sprachkritik bei Nietzsche." In *Über Friedrich Nietzsche*, 63–97. See Bachman 1985.

Sloterdijk, Peter. *Der Denker auf der Bühne. Nietzsches Materialismus.* Frankfurt am Main: Suhrkamp, 1986.

Small, Robin. "Three Interpretations of Eternal Recurrence." *Canadian Philosophical Review Dialogue* 22/1: 92–112 (1983).

Smith, Christopher P. "Heidegger's Break with Nietzsche and the Principle of Subjectivity." *The Modern Schoolman* 52: 227–48 (1975).

Sokel, Walter H. "The Political Uses and Abuses of Nietzsche in Walter Kaufmann's Image of Nietzsche." *Nietzsche-Studien* 12: 436–42 (1983).

Spiekermann, Klaus. *Naturwissenschaft als subjektlose Macht? Nietzsches Kritik physikalischer Grundkonzepte.* Berlin: Walter de Gruyter, 1992.

Stack, George J. *Lange and Nietzsche.* Berlin: Walter de Gruyter, 1983.

———. "Review of *Nietzsche's Theory of Knowledge*, by Rüdiger H. Grimm." *Man and World* 5/12: 254 (1979).

Stegmaier, Werner. "Nietzsches Neubestimmung der Wahrheit." *Nietzsche-Studien* 14: 69–95 (1985).

———. "Nietzsches Neubestimmung der Philosophie." In *Nietzsches Begriff der Philosophie*, ed. Mi Djuric, 21–36. Wurzburg: Königshausen & Neumann, 1990.

Stelzer, Steffen. *Der Zug der Zeit: Nietzsches Versuch der Philosophie.* Meisenheim am Glan: Anton Hain, 1979.

Stephens, Anthony. "Nietzsche: Die partielle Auferstehung. Zur Nietzsche-Renaissance in der BRD." In *Friedrich Nietzsche. Strukturen der Negativität*, 183–216. See Gebhard 1984.

———. "Nietzsche und die Poetische Metapher." In *Friedrich Nietzsche: Perspectivität und Tiefe*, 89–122. See Gebhard 1984.

Stern, Joseph. P. *A Study of Nietzsche.* Cambridge: Cambridge University Press, 1979.

Strong, Tracy B. *Friedrich Nietzsche and the Politics of Transfiguration.* Berkeley and Los Angeles: University of California Press, 1975.

Struve, Wolfgang. *Die neuzeitliche Philosophie als Metaphysik der Subjektivität: Interpretationen zu Kierkegaard und Nietzsche.* Freiburg: Kalber, 1948.

Taureck, Bernard H. F. *Nietzsche und der Faschismus.* Hamburg: Junius, 1989.

Taylor, Charles S. "Some Thoughts on Nietzsche, Kazantzakis and the Meaning of Art." *Nietzsche-Studien* 12: 379–86 (1983).

Thiele, Leslie Paul. *Friedrich Nietzsche and the Politics of the Soul: A Study of Heroic Individualism.* Princeton: Princeton University Press, 1990.

Thurner, Rainer. "Sprache und Welt bei Friedrich Nietzsche." *Nietzsche-Studien* 9: 38–60 (1980).

Ulmer, K. "Nietzsches Idee der Wahrheit und die Wahrheit in der Philosophie." *Philosophisches Jahrbuch* 70: 295–310 (1962–63).

Vaihinger, Hans. *Nietzsche als Philosoph.* Berlin. 1902.

———. *Die Philosophie des Als Ob. System der theoretischen, praktischen und religiösen Fiktionen der Menschheit auf Grund eine idealistische Positivismus: mit einem Anhang über Kant und Nietzsche.* Leipzig: F. Meiner, 1911.

Vattimo, Gianni. "Nietzsche and Contemporary Hermeneutics." In *Nietzsche as Affirmative Thinker*, 58–68. See Yovel 1986.

Warnock, Mary. "Nietzsche's Conception of Truth." In *Nietzsche: Imagery and Thought.* See Pasley 1978.

Wahl, Jean. "Le Nietzsche de Fink." *Revue de Metaphysique et de Morale* 68 (1962).

———. "Le problème du temps chez Nietzsche." *Revue de Metaphysique et de Morale* 67 (1961).

Warren, Mark. *Nietzsche and Political Thought.* Cambridge: MIT Press, 1988.

Westphal, Kenneth R. "Was Nietzsche a Cognitivist?" *Journal of the History of Philosophy* 23/1: 343–63 (1985).

White, Alan. *Within Nietzsche's Labyrinth.* New York: Routledge, 1990.

Wilcox, John T. "Nietzsche Scholarship and 'The Correspondence Theory of Truth': The Danto Case." *Nietzsche-Studien* 15: 337–57 (1986).

———. *Truth and Value in Nietzsche*. Ann Arbor: University of Michigan Press, 1975.

Wisser, Richard. "F. Nietzsche, Missverständnisse eines Denkerlebens." *Zeitschrift für Religion und Geistesgeschichte* 4 (1965).

Wolf, H. M. *Friedrich Nietzsche. Der Weg Zum Nichts*. Bern, 1956.

Wurzer, Wilhelm, S. "Nietzsche"s Hermeneutic of *Redlichkeit*." *Journal of the British Society for Phenomenology* 14/3: 258–70 (1983).

Yovel, Yirmiyahu, ed. *Nietzsche as Affirmative Thinker: Papers Presented at the Fifth Jerusalem Philosophical Encounter, April 1983*. The Hague: Martinus Nijhoff/Kluwer, 1986.

GENERAL BIBLIOGRAPHY

Bernstein, R. J. *Beyond Objectivism and Relativism: Science, Hermeneutics and Praxis*. Philadelphia: University of Pennsylvania Press, 1983.

Bleicher, Josef. *The Hermeneutic Imagination: Outline of a Positive Critique of Scientism and Sociology*. London: Routledge and Kegan Paul, 1982.

Brun, Jean. *Les masques du désir*. Paris: Buchet/Chastel, 1981

Bübner, Rüdiger. *Hermeneutik und Dialektik*. 2 vols. Tübingen. JCB. Mohr, 1970.

Chargaff, Erwin. *Unbegreifliches Geheimnis: Wissenschaft als Kampf für und gegen die Natur*. Klett: Cotta, 1980.

———. *The New Heraclitean Fire*. New York: Simon & Schuster, 1984.

Churchland, Paul M., and Clifford A. Hooker. *Images of Science: Essays on Realism and Empiricism*. Chicago: University of Chicago Press, 1985.

Connolly, William E. *Identity/Difference: Democratics Implications of Poltical Paradox*. Ithaca: Cornell University Press, 1991.

———. *Politics and Ambiguity*. Madison: University of Wisconsin Press, 1987.

Davidson, Donald. "On the Very Idea of a Conceptual Scheme." In *Reference Truth and Reality: Essays on the Philosophy of Langauge*, ed. D. Davidson, 183–98. London: Routledge Kegan Paul, 1980.

———. "What Metaphors Mean." In *Reference Truth and Reality: Essays on the Philosophy of Langauge*, ed. D. Davidson, 238-256. London: Routledge and Kegan Paul, 1980.

Feyerabend, Paul. *Against Method: Outline of an Anarchistic Theory of Knowledge*. Atlantic Highlands: Humanities Press; London: NLB, 1975.

———. "Explanation, Reduction and Empiricism." In *Minnesota Studies in the Philosophy of Science*, vol. 3: 28–97. Minneapolis: University of Minnesota Press, 1962.

———. *Farewell to Reason*. London: Verso, 1987.

———. "The Problems of Microphysics." In *Philosophy of Science Today*, 138–47, ed. S. Morgenbeser. New York: Basic Books, 1967.

———. *Three Dialogues on Knowledge*. London: Verso, 1990.

———. *Wider den Methodenzwang*, 3d ed. Frankfurt am Main: Suhrkamp, 1983.

van Fraassen, Bas C. *The Scientific Image*. Oxford: Clarendon, 1980.

———. "Theories and Counterfactuals." In *Action, Knowledge, and Reality*, 237–64, ed. H.-.N Castañeda. Indianapolis: Bobbs Merrill, 1975.

Fleck, Ludwik. *The Genesis of a Scientific Fact*. Trans. F. Bradley and T. Trenn. Chicago: University of Chicago Press, 1976.

———. *Enstehung und Entwicklung einer Wissenschaftlichen Tatsache: Einführung in die Lehre vom Denkstil und Denkkollektiv*. Frankfurt: Suhrkamp, 1980.

Fuller, Steve. *Philosophy of Science and Its Discontents*. Boulder: Westview Press, 1989.

———. *Social Epistemology*. Bloomington: Indiana University Press, 1988.

———. *Reason in Science*. Trans. Frederick F. Lawrence. Cambridge, Mass.: MIT Press, 1981.

Gadamer, Hans-Georg. "On the Circle of Understanding." In *Hermeneutics versus Science: Three German Views*, 68–78, ed. J. Connolly and T. Keutner. Notre Dame, Indiana: University of Notre Dame Press, 1988.

———. *Vernunft im Zeitalter der Wissenschaft*. Frankfurt am Main: Suhrkamp, 1976.

———. *Wahrheit und Methode. Grundzüge einer philosophischen Hermeneutik*. Tübingen. J.C.B. Mohr, 1975.

Gauquelin, Michel. *Birthtimes: A Scientific Investigation of the Secrets of Astrology*. Trans. Sarah Mathews. New York: Hill & Wang, 1983.

Gregory, Richard I. *Mind in Science: A History of Explanation in Psychology and Physics*. Cambridge: Cambridge University Press, 1981.

Gutting, Gary, ed. *Paradigms and Revolutions: Appraisals and Applications of Thomas Kuhn's Philosophy of Science*. Notre Dame: University of Notre Dame Press, 1980.

Hacking, Ian. *Representing and Intervening: Introductory Topics in the Philosophy of Natural Science*. Cambridge: Cambridge University Press, 1983.

Hanson, Norwood Russell. *Patterns of Discovery: An Inquiry into the Conceptual Foundations of Science*. Cambridge: Cambridge University Press, 1958.

Heelan, Patrick. "Hermeneutics of Experimental Science in the Context of the Life-World." In *Interdisciplinary Phenomenology*, ed. D. Ihde and R. Zaner, 7–50. The Hague: Martinus Nijhoff, 1975.

———. *Space-Perception and the Philosophy of Science*. Berkeley and Los Angeles: University of California Press, 1983.

———. *Quantum Mechanics and Objectivity*. The Hague: Martinus Nijhoff, 1965.

Heidegger, Martin. *Being and Time*. Trans. John Macquarrie and Edward Robinson. New York: Harper and Row, 1962.

———. "Die Frage nach der Technik." *Vorträge und Aufsätze*. Pfüllingen: Neske, 1954.

———. *The Question Concerning Technology and other Essays*. Trans. William Lovitt. New York: Harper & Row, 1977.

———. *Einführung in die Metaphysik*. Tübingen: Max Niemeyer, 1976.

———. *Sein und Zeit*. Tübingen. Neske, 1954. Niemayer. 1984.

———. *Die Technik und die Kehre*. Pfüllingen. Neske, 1962. (1982).

———. *Unterwegs zur Sprach*. Pfullingen: Neske, 1959.

Hesse, Mary. *Revolutions and Reconstructions in the Philosophy of Science*. Sussex: Harvester Press, 1980.

———. *The Structure of Scientific Inference*. London: Macmillan, 1974.

Hollis, M., and S. Lukes, eds.. *Rationality and Relativism*. Cambridge, Mass.: MIT Press, 1983.

Hübner, Kurt. *Critique of Scientific Reason*. Trans. Paul E. Dixon and Hollis M. Dixon. Chicago: University of Chicago Press, 1983.

Kahn, Charles. *Anaximander and the Origins of Greek Cosmology*. New York: Columbia University Press, 1960.

Kant, Immanuel. *The Critique of Judgment*. Trans. J. C. Meredith. Oxford: Clarendon Press, 1952.

———. *Kritik der Urteilskraft*. Stuttgart: Reclam, 1963.

Knorr-Cetina, Karen, and M. Mulkay. *The Manufacture of Knowledge*. Oxford: Pergamon, 1980.

———. *Science Observed: Perspectives on the Social Study of Science*. Beverly Hills: Sage Publications, 1983.

Kolb, Josef. "Hermeneutik in der Physik." In *Hermeneutik als Weg heutiger Wissenschaft*, 85–89. See Warnach 1971.

Kuhn, Thomas. *The Structure of Scientific Revolutions*. 2d ed. Chicago: University of Chicago Press, 1970.

Lacan, J. *Écrits: A Selection*. Trans. Alan Sheridan. New York: Norton, 1977.

———. *The Four Fundamental Concepts of Psychoanalysis*. Trans. Alan Sheridan. New York: Norton, 1981.

Lange, Friedrich Albert. *Geschichte des Materialismus und Kritik seiner Bedeutung in der Gegenwart*. 1875. Reprint. Frankfurt: A. Schmidt, 1974.

Lyotard, Jean-François. *The Post-Modern Condition: A Report on Knowledge*. Trans. Geoff Bennington and Brain Massumi. Minneapolis: University of Minnesota Press, 1984.

MacIntyre, Alasdair, "Epistemological Crises, Dramatic Narrative, and the Philosophy of Science." In *Paradigms and Revolutions*, 54–74. See Gutting 1980.

Newton-Smith, W. H. *The Rationality of Science*. Boston: Routledge and Kegan Paul, 1981.

Peirce, Charles Saunders. "The Fixation of Belief." In *Peirce: Collected Papers*, vol. 5, ed. P. Weiss and C. Hartshorne. Cambridge: Harvard University Press, 1960.

Polanyi, Michael. *Personal Knowledge*. Chicago: University of Chicago Press, 1958.

Popper, Karl R. *Conjectures and Refutations: The Growth of Scientific Knowledge*. London: Routledge & Kegan Paul, 1963.

———. *Logik der Forschung*. Vienna: Springer, 1959.

———. *The Logic of Scientific Discovery*. London: Hutchinson, 1968.

———. *Objective Knowledge*. London. Oxford University Press, 1972.

Prigogine, Ilya, and Isabelle Stengers. *La Nouvelle Alliance: Metamorphose de la Science*. Paris: Gallimard, 1979.

———. *Order Out of Chaos*. New York: Bantam Books, 1984.

Redner, Harry. *The Ends of Science*. Boulder: Westview Press, 1987.

Richardson, William J. *Heidegger: Through Phenomenology to Thought*. The Hague: Martinus Nijhoff, 1963.

———. "Heidegger's Critique of Science." *The New Scholasticism* 42/8: 511–36 (1968).

Rosteutscher, J. H. W. *Die Wiederkunft des Dionysos. Der Naturmystische Irrationalismus in Deutschland*. Bern: A. Francke, 1947.

Saussure, Ferdinand de. *Course in Formal Linguistics*. Ed. C. Bally, A. Sechehaye, and A. Riedlinger. Trans. Wade Baskin. New York: McGraw Hill, 1959.

Sellars, Wilfrid. "Counterfactuals, Dispositions and the Causal Modalities." In *Mind, Matter and Method: Essays in Philosophy and Science in Honor of Herbert Feigl*. Minneapolis: University of Minnesota Press, 1966.

———. *Philosophical Perspectives*. Springfield: Charles C. Thomas, 1967.

———. "The Refutation of Phenomenalism: Prolegomena to a Defense of Scientific Realism." In *Mind, Matter and Method: Essays in Philosophy and Science in Honor of Herbert Feigl*, Minneapolis. University of Minnesota Press, 1966.

———. *Science, Perception and Reality*. London: Routledge & Kegan Paul, 1963.

Serres, Michel. *Hermes: Literature, Science, Philosophy*. Josué V. Harari and David F. Bell. Baltimore: Johns Hopkins University Press, 1982.

Shapiro, Gary, and Alan Sica, eds. *Hermeneutics: Questions and Prospects*. Amherst: University of Massachusetts Press, 1984.

Sklar, Lawrence. "Perceived Worlds, Inferred Worlds, The World" (with P. Heelan reply: "Perceived Worlds are Inferred Worlds"). *The Journal of Philosophy* 81: 693–708 (1984).

Spariosu, Mihai I. *Dionysus Reborn: Play and the Aesthetic Dimension in Modern Philosophical and Scientific Discourse*. Ithaca: Cornell University Press, 1989.

Stegmüller, Wolfgang. "Walther von der Vogelweide's Lyric of Dream-Love and Quasar 3C 273." In *Hermeneutics versus Science: Three German Views*, 102–52, ed. J. Connolly and T. Keutner. Notre Dame: Unversity of Notre Dame Press, 1988.

Suppe, Frederick, *The Semantic Conception of Scientific Theories and Scientific Realism*. Urbana: University of Illinois Press, 1989.

———, ed. *The Structure of Scientific Theories*. Urbana: University of Illinois Press, 1974.

Warnach, Viktor, ed. *Hermeneutik als Weg heutiger Wissenschaft. Ein Forschungsgespräch*. Salzburg: Pustet, 1971.

Wiredu, J. E. "Kant's Synthetic A Priori in Geometry and the Rise of Non-Euclidean Geometries." *Kant-Studien* 61: 5–27 (1970).

NAME INDEX

SUBJECT INDEX